Critical Energy Issues in Asia and the Pacific

Also of Interest

*China Among the Nations of the Pacific, edited by Harrison Brown

China's Oil Future: A Case of Modest Expectations, Randall W. Hardy

*China's Economic Development: Growth and Structural Change, Chu-yuan Cheng

OPEC: Twenty Years and Beyond, edited by Ragaei El Mallakh

The Geopolitics of Energy, Melvin A. Conant and Fern Racine Gold

Energy from Biological Processes, Office of Technology Assessment, U.S. Congress

Indonesia's Oil, Sevinc Carlson

Nuclear Energy and Nuclear Proliferation: Japanese and American Views, Ryukichi Imai and Henry S. Rowen

*The Economics of Environmental and Natural Resources Policy, edited by J. A. Butlin

OPEC, the Gulf, and the World Petroleum Market, Fereidun Fesharaki and David T. Isaak

*Available in hardcover and paperback.

A Westview Special Study

Critical Energy Issues in Asia and the Pacific: The Next Twenty Years

Fereidun Fesharaki, Harrison Brown, Corazon M. Siddayao, Toufiq A. Siddiqi, Kirk R. Smith, and Kim Woodard

Confronting the critical energy issues that face both the developing and the industrial nations of Asia and the Pacific, the authors of this book discuss the changing political and economic situations in the region and the various implications of those changes. The topics addressed include the problems of access to crude oil, the price of crude, the historical dependence of the region on the major oil exporters, and the impact of oil prices on the foreign exchange earnings and foreign debt situation of the region's developing nations.

Nuclear power development and the environmental aspects of energy policies in the area are discussed in some detail, as are China's external energy relations and their regional importance. The book concludes with a chapter on the energy options open to the nations of Asia and the Pacific, suggesting that, given the economic, technical, political, and security considerations raised in the preceding chapters, cooperative programs may be the only way to ensure access to petroleum in the future.

Fereidun Fesharaki is a research associate at the East-West Resource Systems Institute in Honolulu and a consultant on energy. Previously he was energy advisor to the prime minister of Iran, 1977-1978, and research fellow of the Institute for International Political and Economic Studies in Tehran, 1975-1978. Harrison Brown, Corazon M. Siddayao, Toufiq A. Siddiqi, Kirk R. Smith, and Kim Woodard are also at the East-West Center.

Critical Energy Issues in Asia and the Pacific

The Next Twenty Years

Fereidun Fesharaki, Harrison Brown,
Corazon M. Siddayao, Toufiq A. Siddiqi,
Kirk R. Smith, and Kim Woodard

ENERGY PROJECT,
EAST-WEST RESOURCE SYSTEMS INSTITUTE

Routledge
Taylor & Francis Group
LONDON AND NEW YORK

First published 1982 by Westview Press

Published 2018 by Routledge
52 Vanderbilt Avenue, New York, NY 10017
2 Park Square, Milton Park, Abingdon, Oxon OX14 4RN

Routledge is an imprint of the Taylor & Francis Group, an informa business

Library of Congress Catalog Card Number: 82-70360

ISBN 13: 978-0-367-01897-9 (hbk)

Contents

Tables and Figures

Tables

Figures

Foreword

A book on energy in Asia deals with one of the most important subjects for the world today. Among resources, energy seems to be of the most critical importance, whereas among major regions, the global influence of Asia seems to be the most rapidly changing. To understand the evolution of the world's political and economic systems, it will be necessary to seek a more complete knowledge of how the energy systems of Asia are changing.

It is quite clear that the energy systems needed to create and sustain human welfare are in transition. In the last decade, much has been learned both about the supply alternatives that may eventually replace petroleum and the ways in which energy systems interact with human well-being through the economy and the environment. In the years ahead, we will see many more changes and gain new insights.

It also seems clear that the next two decades will see a dramatic change in Asia's status in the world. Because this region has most of the world's people, much of its untapped resources, and some of its fastest growing economies, a gradual shift of economic power towards Asia and the Pacific will occur. Great changes in energy systems will accompany this shift.

There is also the real possibility of counterproductive political upheavals, debilitating increases in international tensions, and even war in this region of complicated history, diverse cultures, and different levels of development. Conflicts over access to resources, such as energy, have the potential of leading the region towards such disruptions. Energy may well be used as a weapon or as a major tool of diplomatic leverage.

Perhaps the most critical problem facing the world today is that of raising the material standard of living of its poorest inhabitants to reasonable levels. In China, Southeast Asia, and South Asia live most of the world's poor. The viability of global development will be severely tested by what transpires in Asia during the rest of this century. Many observers hope that rationalization of global, national, and local energy systems will not jeopardize, and may even assist, this development process.

This book, which attempts to advance our understanding of the world's most rapidly changing resource issues as they are manifested in the world's most rapidly changing region, introduces Asia's energy problem within the global context and then explores a number of the most critical issues in depth. The final chapter draws together the entire range of energy issues that planners must consider in order to meet the challenge of the coming transition in energy--and in Asia--during the next few decades.

The book's authors are members of the Energy Project at the East-West Resource Systems Institute, a research organization devoted to the study of resource problems in the Asia-Pacific region. Born in 1977 under the directorship of Harrison Brown, the Resource Systems Institute is one of a few such organizations focusing on these critical issues and approaches them with a multidisciplinary and multinational team. In the few years of its existence, the Institute has also been able to build up an extensive regional network for exchange and consultation. Among its many participants, I have personally benefited a great deal from these regular contacts. Thus, it is for this reason that I have written this foreword with a sense of pleasure and gratitude.

Mohammad Sadli
Professor of Economics
University of Indonesia

Former Minister of Mines
and Petroleum, Republic
of Indonesia, 1973-1978

Jakarta

Preface

This book is the outgrowth of various research programs carried out within the Energy Project of the Resource Systems Institute (RSI) of the East-West Center.

The Resource Systems Institute, established in 1977, is devoted to the study of three broad-based resource areas: energy, food, and raw materials. In common with other RSI projects and other institutes within the East-West Center, the Energy Project focuses its research on policy-oriented issues in Asia and the Pacific. In early 1980, we decided to bring together portions of our ongoing work in book form. The present volume draws from the work of our researchers who each contributed a chapter, then helped to improve and revise the other chapters. One chapter on environmental issues was contributed by the coordinator of a related project at our sister institute at the Center, the Environmental and Policy Institute (EAPI). Our reference to the Asia-Pacific nations in the book corresponds to the United Nations' definition of the region--from Iran in the west to Japan in the east, from the Koreas in the north to Australia and New Zealand in the south, and including the United States on the other side of the Pacific. According to the dictates of the subject matter, minor additions to and deletions from this group occur in each chapter.

This book does not claim to deal with all the energy problems in the Asia-Pacific region. As the title implies, we have selected a number of issues that we feel are particularly important for the region. Neither did we try to include all energy issues under study at RSI--we did not, for instance, include the results of our work on energy and rural development problems; we felt that energy and rural development issues are of such significance that an entire book should be devoted to the study of this subject in the near future. At the same time, this book is not meant to be a fully integrated work, centering around a specific core issue. Rather, it is an attempt to provide an understanding of some of the major energy issues faced by the policy makers of the nations of Asia and the Pacific, offering options and policy choices where appropriate. In the last chapter, however, we try to present in capsule form the entire range of issues that might be addressed in a comprehensive look at energy in the region.

Because of the seriousness of the energy problems facing the developing nations of the region, and because there is no dearth of literature on the problems of the United States and other developed nations in this region, the emphasis of this book is on the developing countries in Asia.

On behalf of my coauthors, I wish to make the following acknowledgments. We are grateful to colleagues in Asia, the Pacific, and the United States who have either commented on parts of the manuscript or helped authors with information and discussions. In particular, I would like to thank Peter Hayes, David Isaak, Yoon Hyung Kim, Hoesung Lee, Don MacRae, Shuzo Nishioka, Guy Pauker, Sam Pintz, Qu Geping, Nochur Ramanathan, David Rose, Michael Santerre, Charles Schlegel, J. K. Sharma, and Richard Suttmeier. Special thanks are due to my research intern, Lena Low, who prepared many of the statistical tables and read through, revised, and improved many chapters of this book. Terence Friend, a former research intern, made valuable contributions to the statistical work in Chapter 3, while another intern, Thammanun Pongsrikul, helped with the preparation of tables in the same chapter. Thanks are also due to our secretarial staff, Dorothy Izumi, Jeni Miyasaki, and Jennifer Cramer, who had the overall responsibility of typing the different drafts of the manuscript; and Irene Crossman, Marietta Mendoza, Wanda McFall, and Ann Takayesu, who typed individual chapters and statistical tables. Finally, I wish to acknowledge the cooperation of EAPI and its project on the Environmental Dimensions of Energy Policies.

Fereidun Fesharaki
Honolulu, Hawaii

1 Introduction: Shifting Regional and Global Relationships

Harrison Brown

The large expansion in world oil trade has enormously complicated the relationships among nations and groups of nations. Indeed, it now seems clear that a resolution of the "energy crisis," either on a regional or a global level, will simply not be possible unless a variety of political and economic interactions, many of which are not directly related to oil, are also taken into consideration.

Before we discuss the Asia-Pacific region, the focus of this book, we need to understand the regional changes that have taken place around the world in the past two to three decades, changes that have made the study of discrete regions impossible without consideration of their global context.

THE NEW SET OF GLOBAL RELATIONSHIPS

Twenty-five years ago, the world was divided into three major interacting groups of nations: (1) the industrial democracies, now loosely banded together into the Organization for Economic Cooperation and Development (OECD), (2) the Soviet bloc nations, commonly referred to as the Council for Mutual Economic Aid (CMEA) or Warsaw Pact, all with centrally planned economies and dominated by the USSR, and (3) the less developed countries (LDCs), many of which, particularly in Asia and Africa, had been long-time colonies of Western European nations.

In 1955, the most important, single interaction was the military interchange between the grouping of industrial democracies of the North Atlantic Treaty Organization(NATO), led by the United States, and the nations of Eastern Europe (the Warsaw Pact), led by the USSR. The arms race that developed between these two groups still dominates the interactions between them, and the ideologies involved in their confrontations spilled over into the interactions between the LDCs and the industrial democracies on one hand, and the LDCs and the nations of Eastern Europe on the other. The two blocs of industrial nations vied with each other to curry the favors of specific developing countries, in part for military reasons and in part for economic or ideological ones.

In 1955, the gap between the rich and the poor countries was a matter of considerable discussion and some action. Programs of technical and economic assistance were established bilaterally by individual nations and multilaterally by international organizations. Economic development did indeed take place, but by 1980, the gap between the rich and the poor nations, in terms of per capita income, has actually widened.

By 1980, the major interacting groups of nations have increased from three to four. The members of the fourth group, the Organization of Petroleum Exporting Countries (OPEC),[1] control the greater part of crude oil exports in the world and in so doing wield a great deal of economic and political power. In 1955, when there were three major groups of nations, there were three major lines of interaction (OECD-CMEA, OECD-LDC, and CMEA-LDC). With the injection of a fourth major group into the system, the lines of interaction have been increased from three to six and the complexities of arriving at effective agreements have been correspondingly increased.

Further, there are interactions among nations within each grouping that give rise to tensions that must be resolved if the problems among the major groupings themselves are to be resolved. Among the OPEC countries, for example, to what extent will the anti-Western and xenophobic policies now characteristic of Iran spread throughout the Moslem world? To what extent can the OPEC nations, collectively and individually, exert self-restraint in the interest of the economic and political well-being of their customers? Can open warfare between members of the group (Iran and Iraq, for example) be contained?

Turning to the developing countries, how can they collectively develop both reasonable conditions for commodity pricing and satisfactory terms for gaining access to markets for their manufactured goods? Can regional economic groupings of LDCs such as the Association of South East Asian Nations (ASEAN) be made to function effectively? Can the LDCs really come to grips with their problems of stark poverty? Can they stop fighting among themselves as in Southeast Asia and the Horn of Africa?

In Eastern Europe, the satellites of the Soviet Union are showing visible signs of their desire to loosen their bonds with Moscow as well as to obtain more amenities and a greater degree of personal freedom than before. To what extent will this happen? Or will the USSR maintain its present tight grip? Can Soviet control be maintained while Russian workers endure a living standard lower than those of Czechoslovakia and Poland and much lower than that of East Germany?

As for the OECD nations, finally, although many have successfully banded together under the NATO military alliance, more often than not they tend to go their separate ways in economic matters. For example, they lack coordination of oil purchases from OPEC. Most suffer retarded rates of economic growth, most are experiencing unemployment and inflation, and many are substantial importers of food and raw materials. Problems of intra-OECD trade in manufactured goods are consequently becoming increasingly serious. All OECD nations are enormously complex technological systems, the workings of which are not fully understood. They are thus probably vulnerable to a wide variety of possible disruptions that can result from shocks imposed from within or without.

The populations and gross national products (GNPs) of these four groups of nations are shown in Table 1.1. Let us now examine briefly the interactions that occur among these groups.

OECD-OPEC

Not surprisingly, most of the interactions between OPEC and OECD nations are directly or indirectly oil related. As Table 1.2 shows, the industrial nations are by far the largest importers of oil, which now represents close to 20 percent of the dollar value of world trade. Never before in history has a single commodity loomed so large on the trade scene. Because substitutes cannot quickly be brought into use on a scale sufficient to lessen appreciably the need for oil, the OPEC nations are able to charge whatever they wish for their exports. The major effective operating restraint is fear of damaging seriously the economic and political well-being of their customers and eventually endangering their own security.

With inflation both widespread and unpredictable in magnitude, and with foreign exchange rates highly variable, many OPEC nations perceive oil left under the ground to be produced later as more valuable than the equivalent amount of money obtained by pumping and exporting today. Consequently, many OPEC nations are decreasing their rates of production--a development that is likely to have profound impact upon oil prices as well as upon the economies of the OECD nations.

In addition to oil prices, the OECD countries are also concerned about the reliability of their sources of supply. Remembering the Arab oil embargo of 1973-1974, OECD governments are concerned about the political terms that oil-producing countries could impose for their supplies. They are also concerned about the internal instability of most of the Middle East regimes and by the possible spread of religious fundamentalism and anti-Western policies from Iran to the rest of the Moslem world.

To help cope with problems of excessive dependence upon oil, the International Energy Agency (IEA) was set up in 1974 as an autonomous body within the OECD. Since its establishment, the IEA has set up information channels covering costs of crude oil and oil products, and it conducts periodic analyses of the oil market as well as of other energy markets. Perhaps most important, it is setting up a mechanism to deal with future supply emergencies.

A substantial proportion of the money paid to OPEC for its oil is used to purchase manufactured goods and agricultural products from OECD nations. The OPEC nations are becoming increasingly concerned about the effects of inflation upon the prices of imports. Further, OECD runs a balance of trade deficit with OPEC that in 1980 was about $50 billion, at the same time the balance of trade deficit of all oil importers was about $120 billion.

TABLE 1.1
Country Groupings, 1978

Group	Population Millions	Population Percent	Per Capita GNP ($US)	GNP $US Billions	GNP Percent
OECD	667.8	15.7	8,070	5,389	63.1
CMEA	369.2	8.7	3,750	1,365	16.0
OPEC					
a. Capital surplus[a]	60.1	1.4	3,340	201	2.4
b. Noncapital surplus	256.0	6.0	640	164	1.9
LDC (non-OPEC)	2,893.9	68.1	490	1,418	16.6
	4,247.0	100.0		8,537	100.0

Source: World Development Report 1980 (Washington, D.C.: World Bank, 1980).

[a]Iraq, Iran, Libya, Saudi Arabia, Kuwait.

TABLE 1.2
Shares of Net World Trade in Commercial Energy, 1977-1980

Country Group	Estimated 1977	1980
Share of exports		
Capital-surplus oil exporters	70	64
Oil-exporting developing countries	23	28
Centrally planned economies	7	8
Total	100	100
Share of imports		
Industrialized countries	79	78
Oil-importing developing countries	21	22
Total	100	100
Memo item		
Volume of total net trade (millions of barrels per day)	33.9	30.2
Of which bunkers and others[a]	4.6	4.6

Source: World Development Report 1980 (Washington, D.C.: World Bank, 1980).

[a]These imports are not allocated to country groups.

Over the years, OPEC nations have invested overseas oil profits that are not needed to pay for imports and development activities, primarily in OECD countries. A part of the investments are long term, consisting of purchases of real property and businesses. A substantial proportion of the investments, however, are short term, consisting of demand deposits in banks or short-term notes. The OECD banks in turn have placed much of this money on loan, a large proportion of it to oil-importing developing countries. Should world economic conditions lead to nonpayment of these debts on a substantial scale, the world financial system would be in deep trouble, as would the relationship between OPEC and OECD.

OECD nations have exported considerable quantities of highly sophisticated weaponry to OPEC nations. In 1980, some of this weaponry was used on a substantial scale in the war between Iran and Iraq. In 1976 alone, the capital-surplus oil exporters spent $20 billion on their military establishments. There is increasing military concern about the Persian Gulf, through which flows 30 percent of U.S., 60 percent of Western European, and 70 percent of Japanese oil imports. It is in this region that the possibility of an OECD-Eastern Europe confrontation looms large. Substantial Soviet and U.S. naval forces are already stationed there.

There is also concern about the instabilities of the regimes themselves, the Israeli-Arab confrontation over the Palestinians, the war between Iraq and Iran, tensions between Libya and Egypt and between the two Yemens. To this must be added the Soviet occupation of Afghanistan and Israel's destruction of an Iraqi nuclear reactor.

At this writing, a surplus of oil is available on the market, brought about in part by lessened OECD consumption that has resulted to some extent from conservation but also substantially from a slowing of OECD economic activity. For security reasons, some OECD nations want to create stockpiles of crude oil to tide them over in emergencies, but some OPEC nations strongly oppose such actions, because stockpiling would tend to lessen their economic and political leverage.

OECD-CMEA

The predominant interaction between the OECD and Eastern European groups of nations is military. Since the end of World War II, the United States and the Soviet Union have recognized that a major war between them is a possibility. As a result, both nations have attempted to put themselves in the position of winning that war if it should start.

A by-product of the "Cold War," which has now gone on for 35 years, has been the arms race, in which actions have led to reactions, which in turn have led to new actions. In the process, huge arsenals of thermonuclear weapons, together with the means of delivery, have been installed ready for use. Massive military research and development programs have resulted in a ceaseless flow of new weaponry that has rendered older weapons systems obsolete. As the technologies have become more sophisticated, costs have increased, with the result that both groups of nations are burdened by military budgets that absorb altogether several percent of their gross national products together with substantial manpower that, in principal, could be used in other productive ways.[2] It is estimated that in 1977 the USSR spent 13.3 percent and the United States, 5.4 percent of their GNPs on their military establishments.

Both NATO and Warsaw Pact groups now have the capability of eradicating all cities, industries, and military establishments in each other's territories. In the process, most people would be killed either directly as a result of thermonuclear explosions, or indirectly as a result of the aftereffects. Recognizing the inherent dangers in this situation, both sides have engaged in discussions aimed at stabilizing the available power of strategic weapons and eventually at bringing about some measure of disarmament. Thus far, the discussions (SALT I and II) have resulted in greater mutual understanding of the problems than before, but they have not yet been productive in actually lessening the intensity of the arms race.

Trade between the CMEA and OECD nations has thus been a relatively minor aspect of their interactions, although in recent years the Soviet Union has negotiated the purchase of considerable quantities of wheat from the major grain exporters, particularly the United States. During the period 1977-1979, the Eastern European nations collectively imported some 30 million tons of grain, primarily from the United States and Canada. The potential for even larger purchases in the future seems considerable. The flow of manufactured goods from OECD and CMEA is small but increasing, particularly from Western Europe. Sophisticated high technology manufactures appear to be in particular demand.

The Soviet Union is one of the world's richest nations in natural resources. It exports considerable quantities of gold and metallurgical metals, notably manganese and chromium, and has exported about 3 million barrels of oil per day to Western and Eastern European countries. Poland also exports considerable quantities of coal to Western Europe. It now appears, however, that the Soviet Union, the world's largest producer of crude oil, might possibly suffer serious oil shortages in the next few years. This turn of events would curtail Soviet oil shipments to other Eastern European countries and might even cause her to compete with OECD for OPEC oil.

The Soviet Union has extremely large potential reserves of oil in Siberia, but for geographical and technological reasons the exploitation of those resources has been slow. The Japanese have seriously explored, thus far unsuccessfully, the possibility of a joint Japan-USSR venture.

Finally, the Soviet Union is one of the world's largest and most efficient fishing nations. The new 200-mile limit has placed severe restrictions upon the fishing grounds available to Soviet fleets and has resulted in yet another major interaction between OECD and CMEA. Some of the richest fishing areas in both the North Atlantic and the North Pacific are now under U.S. jurisdiction. In most areas, the Soviets find themselves in direct competition with the Japanese over negotiated fishing rights.

OECD-LDC

When we compare OECD with developing nations, we are comparing rich nations with poor ones. The LDCs collectively (excluding the capital-surplus OPEC nations) have 70 percent of the world's population living on less than one fifth of the world's income. On a per capita basis, the large gap between the rich and the poor nations, which has been increasing since World War II, seems destined to continue to increase unless some dramatically new factor is introduced.

The LDCs are characterized for the most part by high birth rates and death rates, high levels of infant mortality, high rates of population growth, low literacy, and relatively low consumption of food calories. The greater part of the hunger and malnutrition in the world, brought on for the most part by abject poverty, occurs in the LDCs. Most of the hungry people in the world lack the resources either to grow food or to buy it.

Unless a developing country is well endowed with natural resources, it is virtually unable to carry out a viable development program without importing quantities of raw materials, machinery, energy--and often food as well. To pay for its imports, a country must either export commodities and manufactured goods of equivalent value, receive financial assistance from other countries, or go into debt.

Developing countries differ greatly from one another in their ability to pay for their imports. About 500 million persons in 20 developing countries (including OPEC nations) benefit from significant exports of raw materials. An additional billion or so persons live in developing countries that are reasonably self-sufficient in nonrenewable resources, including energy. Another 60 million persons live in developing countries that are closely integrated with the world economy almost entirely through the manufacture of goods. About 1,200 million persons, however, live in developing countries that are in serious straits over money and raw materials, particularly energy. These countries are the poorest nations on earth, and their prospects for future economic growth are dismal. Their annual bill for essential imports of petroleum, fertilizers, and food has been greatly increased in recent years, in no small measure because of increased costs of crude oil. This is money that they do not have, so they must borrow.

Beyond the basic inequality of economic strength between the rich and the poor nations, we are confronted by the fact that the rich countries dominate the international institutions of trade, money, and finance because they have the major part of the money and goods that flow through the system. Nevertheless, the dominant role of rich countries is a major source of irritation to the LDCs.

Related to the OECD-LDC economic disparities are the substantial migrations in recent years of persons from LDCs, both legal and illegal, in search of economic opportunities that are absent in the LDCs. Western Europe has made substantial use of labor from Turkey, Greece, Yugoslavia, and Algeria. It is estimated that between 6 and 12 million illegal aliens currently live in the United States, a large proportion from Mexico. During the period 1979-1980, substantial numbers of persons emigrated to the United States from Vietnam, Cuba, and Haiti. On the positive side, this migration provides the receiving country with relatively inexpensive labor and the sending country with foreign exchange, for the migrants often send money home. On the negative side, the migrations often give rise to serious social problems in the receiving countries and sometimes result in a drain of trained personnel from the sending countries.

Weapons of war, and war itself, have resulted in powerful interactions between rich and poor nations. In 1950 United Nations forces, composed primarily of U.S. military personnel, went to war against North Korea. Later that year, the People's Republic of China entered the same war on the side of North Korea. In 1956, England and France undertook military operations in the Sinai seeking control of the Suez Canal. In 1958, the United States sent troops to Lebanon, and Britain airlifted troops to Jordan. In 1961, the Bay of Pigs incident in Cuba occurred. In 1965, U.S. combat troops were sent to the Dominican Republic and had also become involved in a full-scale war against North Vietnam.

Since World War II, OECD nations have exported large quantities of sophisticated weaponry, about 75 percent of the total to the LDCs. Today the OECD nations, notably the United States, France, and the United Kingdom, appear to be shipping about $15 billion worth of arms each year to LDCs, of which a substantial proportion goes to the Middle East.

The rich-poor system as it now exists has enormous built-in instabilities. It is conceivable that the poor countries collectively could trigger the decline of the OECD countries, particularly within the context of the five other sets of interactions between groups of countries that are operating simultaneously. Rimmer de Vries, a respected forecaster of exchange rates and trade flows, has suggested that in the next five years the LDCs may need $70 billion more than international banks are willing to provide to help them meet their oil bills.[3] Inability to purchase adequate quantities of oil to meet their needs could lead to substantial political unrest among the LDCs.

One way to ameliorate this problem is to accelerate the economic and social development of the poor countries. As is well known, this is not a trivial task, for it involves transfers of capital and levels of technical assistance greatly exceeding those that have existed in recent years.

In part, transfers of capital can be made, as they have in the past, through mechanisms of international aid, either bilateral or multilateral. Several years ago, OECD proposed that total transfers for economic aid from OECD nations should be about 1 percent of their gross national products and that flows for official development assistance should be 0.7 percent of their GNPs. Some European nations met these goals; the United States has lagged far behind. In 1980, development assistance from OECD to LDCs is estimated to be 0.34 percent of GNP, ranging from a high of 0.45 percent (Sweden) to a low of 0.09 percent (Italy). The U.S. donation amounted to 0.18 percent of its GNP. The total of OECD official assistance came to $24.6 billion.

About two thirds of the exports of non-OPEC LDCs flow to the OECD nations. In 1978, this amounted to about $140 billion. In the same year, the OECD nations exported $193 billion worth of merchandise to the LDCs, resulting in an LDC balance of trade deficit of $53 billion, a figure more than twice as large as the flow of official development assistance. The greater part of this trade is accounted for by the richer LDCs that have high rates of economic growth--notably Brazil, South Korea, Taiwan, and Mexico, who also account for a dangerous amount of borrowing from OECD banks holding large deposits of OPEC surpluses.

Few disagree that the problem of the LDCs is a primary world concern, and that a large part of that problem involves the interactions between OECD and LDCs. Increased oil prices, increased imports of manufactured goods, and a mounting debt-service burden will produce increasingly large deficits in the LDCs' balance of payments.

In 1974, at a Special Session of the United Nations General Assembly, Algeria called for "a new international economic order." Since then, LDC spokesmen have made increasing demands upon the rich countries. The demands tend to concentrate on three themes: (1) obtaining needed foreign exchange, (2) obtaining the necessary technological capabilities for development, and (3) increasing the share of decision-making power of developing countries in the global economic system. To achieve this end, the LDCs are calling for: (1) greatly increased flows of investment capital to the LDCs, (2) higher and more stable prices for raw materials, (3) improved market access for their manufactured goods, and (4) significant representation on the governing bodies of such organizations as the World Bank and the International Monetary Fund.

Although many of the complaints the LDCs make about the OECD nations and their institutions seem valid, some of the greatest obstacles to development lie within the developing countries themselves. One factor is their substantial expenditures on military weaponry. Another lies with the privileged elites that exist in all countries, even the very poorest. More often than not, the privileged minority likes things the way they are. Many purchase imported luxuries, thus lessening the resources available for development. Many export money to private bank accounts in rich countries. Many accept substantial bribes from outsiders to influence their governments.

One of the major difficulties hindering the acceleration of the development process is the fact that developing countries have a low absorptive capacity for external funds. This circumstance suggests that access to capital is only one of the major problems of the LDC-OECD confrontation. The transfer of technology and management skills also appears to be a contentious and difficult process.

OPEC-LDC

Most LDCs are oil importers, so most of them have balance of trade deficits with OPEC; a substantial number of LDCs have had to borrow money to make up for this deficiency. Often these loans have been OPEC money cycled through OECD banks.

In relation to ability to pay, the OPEC countries have done well in providing economic assistance to LDCs. In 1979, OPEC contributions totaled $4.7 billion, corresponding to 1.3 percent of their GNP. Eliminating the non-Arab states, which were low contributors, the contributions amounted to 2.4 percent of the GNP of the Arab members of OPEC. The largest single contribution was Saudi Arabia's, which amounted to nearly $2.0 billion, corresponding to more than 5 percent of its GNP. All of these figures are very much larger than those of the OECD countries.

The Arab members of OPEC tend to channel the flow of OPEC economic assistance to Moslem LDCs. Middle East political considerations play an important role in determining the flow of funds, with the Israeli-Palestinian situation being the dominant factor. Egypt, for example, is no longer a major recipient of OPEC assistance.

Libya, which in 1979 contributed less than any other Arab OPEC state for LDC economic assistance, has apparently made sizeable contributions for the support of terrorist activities in a number of LDCs (as well as OECDs). Organizations fostering the liberation of Palestine have been substantial beneficiaries.

CMEA-LDC

The Eastern European countries, and particularly the USSR, are highly selective about their economic and technical assistance to developing countries. Virtually all assistance is given bilaterally and directed toward short- and long-term political, military, and ideological goals. In recent years, efforts have been concentrated in Cuba, Vietnam, Ethiopia, Mozambique, Congo, Yemen (People's Republic), and Syria. In earlier years, prior to the rupturing of political ties, the USSR had substantial programs in the People's Republic of China, Egypt, and Somalia. Ties between the USSR and India have varied in strength over the years; at present, they are becoming stronger. The evidence suggests, however, that the Soviet Union is unwilling to sustain programs of foreign economic assistance unless there is also a major military link.

The Soviet Union shares a 7,000-km border with the People's Republic of China, who clearly views its neighbor as the most serious potential threat to China today, in both political and military terms. The Chinese exploded their first nuclear device in 1964, and following the 1969 border clashes between Soviet and Chinese forces, the Soviets hinted that they were considering an attack on Chinese nuclear installations. The Chinese striking force now apparently consists of about 100 medium- and intermediate-range ballistic missiles, about 90 medium-range bombers, and possibly a few longer-range ICBMs. During the period 1969-1978, the Soviets reportedly increased their troop deployments along the Chinese border to about 700,000.

The most recent Soviet military involvement with an LDC has been in Afghanistan, which they invaded in late 1979. This action has been deeply disturbing to OECD nations as well as to Moslem LDCs, bringing back deep-seated memories of U.S. involvement in Indochina.

Finally, the USSR vigorously attempts to arrange for privileges from developing countries that will help it maintain a substantial fishing industry. These include fishing rights, permission to use dock facilities, and fueling privileges.

CMEA-OPEC

In 1979, the Soviet Union produced 11.8 million barrels of crude oil and natural gas liquids per day, maintaining its position as the largest producer in the world. About 3 million barrels per day were exported primarily to Eastern and Western Europe, and oil exports have been providing about 50 percent of the USSR's hard currency receipts.

The growth of Soviet oil production has slowed sharply since the mid-1970s and may well peak soon. Should this happen, the Soviet Union and other Eastern European countries will be faced with the decision of limiting domestic oil use, using substitute fuel supplies such as coal on a larger scale, or importing substantial quantities of oil from the Middle East.

The energy needs of Eastern Europe are a problem, for the Soviets have assured East Europeans of steady energy supplies. Indeed, the evidence suggests that the Soviet Union will become increasingly interested in OPEC oil in the 1980s. And if the other Eastern European countries are deprived of Soviet oil, they too will become increasingly interested in Middle Eastern oil.

The Soviet Union has vast reserves of coal, but full utilization of this resource will take much time and effort. In the meantime, the Soviet Union may well attempt to gain major access to Middle Eastern oil supplies and, being short of hard currency, may well attempt to do so by use of a combination of political maneuvering and force. Any such move would work against the interest of OECD nations, who would probably take countermeasures. But the Soviet Union, which shares a long common border with Iran and which has been a major supplier of arms to Iraq, would appear to have a clear geographic advantage.

THE ROOTS OF THE CRISIS

The crisis in energy has many roots, the most important being the extent to which nations permitted themselves to become dependent upon crude oil for their heat, light, and power. This dependence arose because oil, when compared with coal, the major fuel which it displaced, is cleaner, easier to handle, and (until recently) less expensive to use. Also, from the early days of large-scale consumption of crude oil, it appeared to be quite abundant on a global scale, although clearly inequitably distributed. For these reasons, decreasing consumption of coal over the years is directly linked to increasing consumption of crude oil, with a correspondingly vast amount of capital invested in a number of massive oil-based energy systems for power, heating, and transportation, all fed by the refineries that produce the specific fuels needed for each task.

Although energy demand rose steadily on both an absolute and a per capita basis following the onset of the Industrial Revolution, few persons anticipated the rate of increase of demand that would eventually be associated with the industrial development of Europe, North America, and Japan, especially after World

War II. The greater part of this burgeoning demand was satisfied by crude oil. But so great did the demand become by the late 1940s, that some fears were expressed by geologists that the resources of oil, both discovered and undiscovered, would not permit supply to keep pace with demand for very many decades. Indeed, production of crude oil in the conterminous United States (at that time the largest producer in the world) peaked in 1970 and has been decreasing steadily since then. By the mid-1970s, a number of respected specialists in the study of oil supply and demand were forecasting that global production would peak by the year 2000 and possibly as early as 1990.

The situation was further aggravated by the fact that most of the major consumers of oil had also become the major importers. By the mid-1970s, the United States was importing close to one half its oil, and Western Europe and Japan were importing nearly all of theirs. The exception was the Soviet Union, which had, like the United States in earlier years, become self-sufficient.

Thus, the stage was set for the major oil exporters, banded together as they were into the Organization of Petroleum Exporting Countries (OPEC), to take collective action. In late 1973, OPEC suddenly quadrupled the price of crude oil, a development that contributed to, but was by no means solely responsible for, the global recession of 1974-1975. Successive price increases brought the average price of crude oil to $35 per barrel in 1981 compared with $1.80 per barrel in 1970.

A combination of factors led a number of oil-exporting nations to reduce their production. In part, such policy decisions sprang from a realization that the amount of money that could effectively be absorbed in the development process was limited and that recycling petrodollars through Western banks could be both risky and costly. But the decisions to curtail production also resulted from a desire for oil prices not to fall in real terms. To put it simply, oil under the ground is worth more in the long run to many oil-exporting nations than the same oil above ground placed on the market.

In 1980, the situation was further aggravated by the aftereffects of the Iranian Revolution and by the war between Iraq and Iran, which resulted in a large decrease in exports from both countries.

Thus, 1979 may well turn out to be the year in which world petroleum production reached either a peak or a plateau. With a substantial fraction of production now being restrained more by policy than by resource availability, the life expectancy of petroleum for human use will, of course, be lengthened. But a substantial shortfall in petroleum availability relative to demand will hit the importing nations far sooner than if production were constrained solely by resource availability.

These factors are bound to have profound effects upon those nations which are, or are likely to be, oil importers. The industrial nations will be called upon to absorb constantly increasing crude oil prices and also to recycle, at not inconsiderable risk, money placed on deposit by oil-exporting countries. The oil-importing developing countries are finding themselves with rapidly increasing external debts and trade deficits that can substantially impede their development.

A potentially critical aspect of the oil situation lies in the increasing vulnerability of the oil-importing nations to disruptions in the flow of supplies. The curtailment of oil shipments was first used as a weapon of war in 1973-1974, when Arab oil exporters stopped shipment of oil to certain nations friendly to Israel. The effects of the embargo suggest that the oil-importing nations, and particularly the highly industrialized ones, are extremely vulnerable to substantial disruptions of oil shipments. Such disruptions can, of course, come about as the result of events other than the political decision to stop shipments. Perhaps the most important of these possibilities are military actions and internal upheavals. Following the Iranian Revolution in 1979, for example, oil production capacity dropped rather quickly from 7.0 to 4.5 million barrels per day because of the fall of gas pressure in the fields and lack of necessary maintenance. In 1980, as a result of military actions, production of oil in Iran and Iraq dropped significantly. Moreover, the sharp increase in direct sales by OPEC nations since 1970 to over 50 percent of the total means that embargoed nations will be hurt much more in the future than they were in 1973.

The oil-importing nations have long recognized that the main pathway out of their present and future energy difficulties lies in the diversification of their sources of energy supplies coupled with steps aimed at utilizing their available energy more efficiently. Unfortunately, the costs of utilizing virtually all forms of energy other than oil are high. Similarly, conservation itself often has high capital costs associated with it. In addition, there are environmental concerns about using coal and nuclear energy on a large scale. It is for such reasons that substantial programs in the industrial countries aimed at diversification and conservation have been more talk than action. Unfortunately, preoccupation with short-term political considerations has kept decision makers from developing adequate long-term policies.

In spite of these difficulties, the successful divorce of industrialized societies from overdependence upon crude oil will probably involve utilizing a multiplicity of energy sources in the years ahead, including solar energy for heat, coal for the generation of electrical power and the production of liquid fuels, uranium for the generation of electrical power, and biomass for the production of liquid, solid, and gaseous fuels. These approaches, combined with others, can give rise to a substantial measure of independence from petroleum.

The developing countries are confronted by a set of critical energy problems that differ from those faced by the industrial nations. The majority of the people in the developing countries live in rural areas. To increase agricultural production and develop small-scale industries in these areas, increasing quantities of energy must be injected into the system in the form of fertilizers, irrigation water, fuels for heating and cooking, and modest amounts of electrical power. The problem is how to provide such energy by utilizing indigenous resources to the fullest extent possible. Although this set of problems is critical for the economic and social development of the poor countries, it will not be treated in this book. Rather, the chapters that follow will consider the energy problems associated with large-scale energy systems of Asia and the Pacific, systems that are characteristic of the industrialized countries and the industrial-urban areas of the developing ones.

As for the energy policies that might be made by individual nations or by groups of nations, it is clear that none can be adopted in a vacuum--the complex interdependencies among nations preclude this easy solution. It is obvious that an energy policy made by one nation will affect, directly or indirectly, all others. And it is equally obvious that the problems and interests of certain large groups of nations differ from each other to such an extent that mutual policies are highly desirable. The OPEC nations have a broad range of legitimate problems and interests, as do the oil-importing industrialized countries. The resolution of those problems requires joint understanding and action, yet even this step would not be meaningful without consideration of the serious problems of the oil-importing developing countries. Thus, at a minimum, constructive discussions are needed among these three groups of nations if significant constructive policies are to be devised.

THE ASIA-PACIFIC REGION

Within this global context, the energy problems of the nations of the Asia-Pacific region loom particularly large. The region as a whole imports substantial quantities of crude oil, the greater part of it from the Middle East. Three nations--Indonesia, Malaysia, and the People's Republic of China--are at present oil exporters, but the quantities exported, measured on a global scale, are small. Domestic energy demands are increasing rapidly, so it is by no means clear what the export levels of these countries will be in the years ahead.

Indonesia is vigorously attempting to increase its oil production and does show possibilities. Some important new discoveries have already been made, and a large expansion of exploration and development activities is underway. The challenge of achieving a truly substantial increase in production is formidable, however, since up to 600 million barrels of new reserves must be found each year to maintain existing reserve-to-production ratios. In the meantime, domestic demands are increasing by about 12 percent each year, with the result that Indonesian oil available for export will continue to decrease.

The People's Republic of China, endowed with substantial reserves of crude oil, is at present exporting modest quantities, primarily to Japan. Using modern drilling technology, Chinese production could be increased considerably, particularly if the continental shelf areas are exploited. But considering China's own future substantial needs, it seems unlikely that exports will ever exceed about 500,000 barrels per day.

Thus, as long as the economies of the nations of Asia and the Pacific are primarily oil based, the region as a whole will continue to be a substantial net importer of crude oil, which for the most part will come from the Middle East. In view of the probable continued curtailment of production in that region, this could place substantial constraints upon the economic well-being of the oil importers of the region.

As Table 1.3 shows, the Asia-Pacific region, exclusive of the USSR, contains a broad spectrum of LDC and OECD nations. Although more than one half of the world's population lives in the region, it is unlike the rest of the world in many respects. The Asian part of the region, which is the most densely populated major area on earth, includes the world's two most populous nations, India and the People's Republic of China, which together have a population of 1,650 million persons (mid-1980). The World Bank projects that by the year 2000 this total may rise to 2,230 million persons.

In the years immediately following World War II, death rates in the LDCs of Asia dropped rapidly, but birth rates remained fairly constant. As a result, populations grew rapidly. Starting in the early 1970s, evidence suggested that birth rates were starting to drop. By 1978, the birth rate in India and Indonesia had fallen by about 20 percent, that in Malaysia by 25 percent, and those in Korea and the People's Republic of China by 50 percent. Although death rates continued to decrease, and the net rates of population growth in many countries rose, it became increasingly probable that a major demographic change was taking place which could lead to an overall slowing of population growth and eventually to stabilization. The trends suggest that the combined populations of India and China, once stabilized, might be upwards of 3,200 million persons (see Table 1.3).

The LDCs of Asia are becoming increasingly urban, but the process is taking place more slowly than in most of Latin America and parts of Africa. China and India are now about one quarter urbanized. Exceptions are those countries which have industrialized particularly rapidly, notably South Korea (55 percent urban) and Taiwan (77 percent urban). Singapore is, of course, a city-state.

The task of feeding such large populations is a difficult one, but the Asia-Pacific region houses the largest exporters of food in the world: the United States, Canada, Australia, and New Zealand--which also number among the world's most affluent nations. Japan, a leading OECD nation, is the largest food importer of the region, but many of the developing countries of Asia are also consistent importers--notably South Korea, Taiwan, the People's Republic of China, and Indonesia. Still others import sporadically as the need arises. In 1980, the People's Republic of China signed a major four-year wheat purchase agreement with the United States at a time when crop yields were poor in Australia, the Soviet Union, and the United States. The

TABLE 1.3
Selected Statistics for the Asia-Pacific Region

	Population, mid-1980 (millions)	Population Growth Rate (%)	Population Urban, 1980 (%)	Per Capita GNP, 1979(P)[a] (US$)	Per Capita GNP Growth Rate 1960-1978 (%)	PQLI, 1979[b]	Liquid Fuel Consumption, 1978 (10^6 mtce)	Liquid Fuel Production, 1978 (10^6 mtce)[c]
Developing								
Bangladesh	90.6	2.6	9	100	-0.4	32	1.82	0
Burma	34.4	2.4	24	160	1.0	50	1.52	2.15
China	975.0	1.2	26	230 (1978)	3.7	71	135.12	152.09
India	676.2	1.9	21	190	1.4	41	32.09	16.45
Indonesia	144.3	2.0	18	380	4.1	48	34.44	132.32
Malaysia	14.0	2.5	27	1,320	3.9	73	8.33	14.24
Pakistan	86.5	2.8	26	270	2.8	36	5.27	0.71
Papua New Guinea	3.2	2.5	13	650	3.6	42	0.84	0
Philippines	47.7	2.4	32	600	2.6	71	14.79	0
Singapore	2.4	1.2	100	3,820	7.4	86	5.73	0
South Korea	38.2	1.6	48	1,500	6.9	82	30.71	0
Sri Lanka	14.8	2.2	22	230	2.0	82	1.47	0
Taiwan	17.8	2.0	77	1,400 (1978)	6.6	87	17.12	0.32
Thailand	47.3	2.3	13	590	4.6	71	13.96	0.01
Industrial								
Australia	14.6	0.8	86	9,100	2.9	94	39.04	33.17
Canada	24.0	0.8	76	9,650	3.5	95	114.53	108.38
Japan	116.8	0.9	76	8,800	7.6	96	326.63	0.82
New Zealand	3.2	0.8	83	5,940	1.7	94	6.01	0.97
United States	222.5	0.7	74	10,820	2.4	95	1,177.76	709.89

Source: Population: Population Reference Bureau, Inc., *World Population Data Sheet* (Washington, D.C.: 1980); GNP: The World Bank, *Atlas: Population, Per Capita Product and Growth Rates* and *World Development Report* (Washington, D.C.: 1980); Liquid fuel consumption and production: United Nations, *World Energy Supplies 1973-1978*, Series J, no. 22 (New York: 1979).

[a](P) = preliminary.

[b]The Physical Quality of Life Index (PQLI), developed by the Overseas Development Council, Washington, D.C., combines three indicators--infant mortality, life expectancy at age one, and literacy--into a composite index. The index runs from 0 to 100, with 0 being the lowest level of well-being.

[c]mtce = metric tons coal equivalent.

effects of these developments both on food availability and prices could be profound.

In spite of difficulties with food and energy, parts of the region, notably the ASEAN countries,[4] Taiwan, and South Korea, have been experiencing rapid and sustained economic growth. In these countries, average growth rates of gross national product between 1970 and 1978 varied from 6.3 to 9.7 percent. Related to this rapid economic growth has been a rapid expansion of trade; during the period 1970-1978, exports in real terms grew at average rates of between 5.2 and 30 percent per year.

The OECD countries of the region have generally grown less rapidly than the LDCs and have felt greater shocks from the oil crisis. Even these countries, however, have done better than the OECD countries of Europe in sustaining real growth. In addition, the LDC countries of the Pacific region are becoming increasingly important trading partners of the OECD countries of the region. The trade of the United States, for example, is now more heavily directed across the Pacific than across the Atlantic.

An interesting feature of Asian developing countries is that some have attained relatively high levels of literacy and low death rates at surprisingly low levels of per capita GNP. An outstanding example in this regard is Sri Lanka, which has achieved a literacy rate of close to 80 percent and a life expectancy at birth of nearly 70 years with a per capita GNP of less than $200. The Philippines, with a per capita GNP of $500, has a literacy rate of nearly 90 percent and a life expectancy at birth of 60 years.

Since World War II, the countries of Asia have suffered numerous military confrontations, among them the wars in Korea and Indochina, the Chinese excursions into India and Vietnam, and the occupation of Cambodia by Vietnam. China's future military capabilities are viewed with concern by the Soviets to the north and by virtually all of the nations of Southeast Asia to the south. Soviet support of Vietnam coupled with the Chinese-Vietnamese confrontation over Cambodia could become serious destabilizing influences in the region. In this situation, it remains to be seen whether ASEAN, the United States, Japan, and Australia will be able to play a stabilizing role, as a counterbalance to the instabilities created by conflicts among communist countries. This involves, of course, both military and economic policies, including greater mutual cooperation in energy policy.

Clearly, if the nations of Asia and the Pacific are unable or unwilling to take major steps to decrease their dependence upon imported oil, they will remain extremely vulnerable to disruptions in supplies. Should a major military or political upheaval in the Middle East halt crude oil exports, the economic, social, and political effects upon the Asia-Pacific region could be disastrous. Even in the absence of such an upheaval, continuing increases in crude oil prices will have deleterious effects and will increase the likelihood that the volume of OPEC exports will remain policy limited, creating tensions and uncertainties. The effects of any of these eventualities upon the industrial giants of the region, notably Japan and the United States, would be particularly great in the event of a cutoff of Middle Eastern oil.

Even without a cutoff, continued dependence upon imported oil will have an increasingly harmful economic effect upon the oil-importing developing countries of the region as well as upon the industrial ones, but for the former, the cost of importing oil will probably continue to increase more rapidly than their ability to pay, even with borrowed money.

Because of their high degree of vulnerability to disruptions in oil supplies as well as to increasing oil prices, the nations of the Asia-Pacific region must diversify their energy supply as soon as possible, both in type of resource and in its geographic origin. In the next twenty years, a period too short to allow development of many alternative options, strenuous efforts should be made to expand known reserves of oil by accelerating exploration. Efforts to increase the efficiency of oil extraction from existing fields should also be stepped up. In the longer run, however--beyond the year 2000--strenuous efforts will be needed to replace oil as a primary energy source by natural gas, coal, nuclear power, and renewable energy resources such as water power, biomass, and direct utilization of solar energy in various ways. Recent experience suggests that such a transition will not be easy, in part because of high capital costs and in part because of technological complexities. Indeed, since the energy crisis of 1973-1974, efforts in this direction have made only a slight impact upon global oil demands.

Even if the energy diversification effort were to be a gigantic one, it is doubtful that much could be accomplished before the turn of the century at the earliest. Dependence upon imported petroleum in the region will remain substantial, although hopefully diminishing somewhat with time. During this period, the oil-importing nations of the region will have to take steps to use less oil by utilizing it more effectively. During the period 1979-1980, oil consumption in most industrial countries decreased appreciably. In part, this effect was a result of lessened economic activity, but a portion of the decrease seems to have resulted from more effective use, perhaps brought about by the greatly increased oil prices of the last two years.

Steps for lessening the impact of an oil cutoff from the Middle East should include stockpiling substantial reserves of crude oil in the region as well as developing both multinational plans for drastic cutbacks of liquid fuel consumption in the event of an emergency and some allocation agreements similar to those developed by the International Energy Agency.

Even if no supply disruptions occur, the oil-importing developing countries will find themselves in increasingly serious difficulties because of escalating oil prices. Their situation could be helped in part if major steps were taken to allow the oil-importing developing countries to expand their export markets substantially. Expanded programs of low-interest loans for fuel purchases could also be developed by OECD and OPEC governments and perhaps operated by intergovernmental institutions such as the World Bank. The borrowing countries should not be eligible for such loans, however, unless they demonstrate that they have developed well-conceived programs of diversifying their sources of energy supply, with special emphasis placed upon the effective utilization of indigenous resources so as to reduce their demand for imported oil.

The use of natural gas is likely to increase substantially in the Asia-Pacific region in the next few years. Australia, New Zealand, Indonesia, Malaysia, Thailand, Bangladesh, and Pakistan have substantial quantities available locally, which could be used in some cases for power generation, fertilizer production, and even automotive transport. Liquefied natural gas (LNG) is also becoming increasingly available from the major oil-producing regions of the Middle East, but the poor countries will not be able to afford the infrastructure costs required to utilize that resource.

Exports of coal for direct electric power generation seem destined to become a major activity in the region during the next two decades, with China, Australia, and the United States becoming the major exporters and Japan, South Korea, and Taiwan becoming the major importers. In addition, experimental development of conversion of coal to synthetic liquid fuels is beginning in the region. Coal conversion projects in the United States, China, Australia, and Indonesia both for domestic use and for export could begin to contribute to the solution of the energy crisis in the next decade.

In combination, these and other alternative energy developments could help to lessen the dependence of the nations of Asia and the Pacific upon imported crude oil. Nevertheless, a great deal of time and money will be required for adequate quantities of such oil substitutes to be produced. This will almost certainly also lead to increased pressures for the development of a nuclear power industry, although in the next period uranium will only substitute for oil in electricity generation.

Despite strong uneasiness caused by concerns about nuclear proliferation and waste disposal, the use of nuclear power seems destined to increase rapidly in Asia, particularly in Japan, South Korea, Taiwan, and India, but also in Pakistan, the Philippines, and some other countries.

Although not the focus of this book, the energy problems of the rural areas in developing countries are also serious. For large numbers of rural people, the only available energy supplies come from traditional biomass fuels--fuelwood, animal dung, and crop residues. It seems likely that development of rural areas will be accompanied both by greater uses of commercial fuels, such as petroleum products, coal, and electricity, and more sustainable and efficient utilization of traditional fuels. Planning for rural development is one of the most important tasks of the energy transition in the region.

Were the nations of the Asia-Pacific region to mount a major effort, they could reduce their energy vulnerabilities by the turn of the century, but there are valid historical reasons to assume that this will not happen. The steps that would have to be taken are complex, requiring the expenditure of a great deal of capital.

Above all, substantial collaboration across national boundaries would be required in adopting common policies on technology transfer, trade policy, loans, development finance, and monetary policy. In some way, all four interacting groups of nations (LDC, OECD, OPEC, and CMEA) would have to be involved. The complications arising from such an endeavor, it must be admitted, might well swamp the system's capacities.

NOTES

1. Although most OPEC nations have high per capita income because of oil exports, they are nevertheless LDCs when viewed in terms of such social indicators as literacy, life expectancy, and the prevalence of hunger and poverty.

2. It is estimated that some 25 percent of all the skilled scientific and engineering research workers in the United States are engaged by the military research and development effort.

3. *Time*, 13 October 1980.

4. Indonesia, Malaysia, the Philippines, Singapore, and Thailand.

2
Future Supplies of Petroleum: Implications for the Asia-Pacific Region

Fereidun Fesharaki

THE OIL MARKET OF THE 1970S

The decade of the 1970s was marked by major structural changes in the world petroleum market that affected every nation in the world. As the 1980s begin, new structural changes are emerging, most of which have their roots in the previous decade.

Events in the 1970s centered upon the fundamental shift in the control of supply from the operating oil companies to the members of the Organization of Petroleum Exporting Countries (OPEC). In 1973, an Arab oil embargo and a quadrupling of oil prices led to an increase in the price of oil from $1.80 per barrel in 1970 to $11.65 per barrel in 1974. The following period, 1974-1979, began with many pledges and commitments by many oil consumers to develop alternative sources of energy, increase the efficiency of their energy consumption, and devise contingency plans to protect their economies from another price shock. The industrial world created the International Energy Agency (IEA), while the developing nations continued to deal individually with their energy problems. Despite some progress, most oil consumers had not achieved their stated objectives by the end of this period.

The sense of urgency surrounding the initial price increase began to fade away as a world recession and changes in energy/gross domestic product relationships reduced the demand for oil and as the real price of oil declined. In 1975, OPEC production fell by 11 percent and prices increased thereafter at nominal rates below the level of world inflation. As a result, the real price of oil fell for four straight years. A study examining the impact of dollar devaluation and inflation on the cost of crude oil shows that the real cost of crude for Japan and West Germany had declined by around 20 percent in the four-year period. Only by the third quarter of 1979 had oil price increases fully compensated for the declines in the preceding years. At $7 to $11 per barrel (in 1974 dollars) in the third quarter of 1979, the price of oil reached the 1974 level in real terms.[1] Also, the large OPEC trade surplus of around $60 billion in 1974 had, by the end of 1978, been reduced to only $1 billion.[2] The international banking system proved to be far more flexible than many had expected, successfully channeling the surplus capital to both developed and developing nations.

In retrospect, both the industrial and developing nations managed to successfully escape major economic problems in the face of rapid changes in oil prices. The industrial economies grew at an average annual rate of 3.4 percent during 1970-1978, whereas the rate of economic growth for the developing world was 5.7 percent for 1970-1976 and 4.8 percent for 1977.[3]

In general, the 1974-1979 period began with fears of further price rises, which did not materialize, and ended with an atmosphere of complacency typical of the previous decades of cheap and abundant oil. The February 1979 revolution in Iran reversed this trend, bringing home again the message that the problem of oil supplies and prices would not go away by itself.[4] In just one year, January 1979 to January 1980, oil prices increased by 120 percent to an average of $30 per barrel. The OPEC surplus of $1 billion in 1978 rose to around $60 billion in 1979, and exceeded $100 billion in 1980.[5] The second oil price shock added a new dimension to the fears of oil importers: access to oil. Before 1979, the oil importers were concerned more about the impact of oil prices on their economies and less about the problem of availability of crude. In the aftermath of the Iranian Revolution, fears about security of supply became the major preoccupation of the oil-importing nations.

In the 1980s, the dependence of the world economy on petroleum trade from OPEC oil will remain strong. In 1980, over 80 percent of the world petroleum trade originated from OPEC, of which around 60 percent came from the politically volatile Persian Gulf. Not only is the Persian Gulf the most important oil-exporting region of the world, it is also the area where the largest potential for expansion of capacity exists. Non-OPEC oil production is unlikely to have a major influence on the world trade for petroleum in the 1980s. Some developing nations may well become self-sufficient and some new oil-bearing areas will no doubt be discovered. It would take more than a new North Sea or a new Alaska, however, to make a dent in the supply pattern.

For the Asia-Pacific nations, the dependence on Middle East oil will continue. The purpose of this chapter is to examine OPEC policy plans and options covering oil production, exports, and prices in the 1980s, indicating the available supplies and the problems which the importing nations are likely to face in the future. In light of the historical pattern of Asia-Pacific dependency and the emerging OPEC supply outlook, we will examine first, the specific circumstances of Asia-Pacific nations and second, a set of policy options that minimize the difficulties of securing access to petroleum supplies.

FUTURE WORLD OIL BALANCE

There are basically two approaches to projecting world oil balance: global energy modeling and "bean-counting."[6] These two approaches have led to different conclusions, mainly because of the way OPEC oil is viewed by forecasters.

Traditionally, many Western forecasters have chosen to consider OPEC production as a residual of world energy supply and demand. When world energy demand is projected for the future, all nonoil alternative sources of energy are first deducted from the total, then non-OPEC oil production possibilities are subtracted to obtain the required level of production from OPEC. For instance, eight forecasts from respectable sources in 1977/1978 projected that required OPEC production in 1985 would be between 35 to 45 million barrels per day (mb/d).[7] Such a forecasting method is certain to be misleading. First, it does not take into consideration OPEC's desired production, which can be significantly different from what the world "requires" OPEC to produce, or from OPEC's own production capacity. Second, it does not allow for the possibility that OPEC policy itself dictates the development of other energy sources, not vice versa.

It is the author's belief that the bean-counting approach provides a more useful guide to the supply of petroleum from OPEC. This approach consists of a country by country assessment of production and exports from OPEC. It assumes that a large number of OPEC nations, if not all, are guided by economic, political, and technical domestic considerations as well as by interactions among themselves in determining their production levels. World demand for oil is an important, but not decisive factor in their decisions. In other words, sovereign nations, unlike the private oil companies, base their decisions on a host of economic and noneconomic factors that do not necessarily correspond to world demand for oil.

The bean-counting approach can be far more accurate than global energy modeling in making projections because it takes account of each nation's specific perceptions, plans, and options.[8] Sophisticated modeling is not required. What is needed is a thorough understanding of the technical, economic, and political situation of each OPEC nation as well as its development plans and institutional factors. This approach, initially used by the CIA in 1977, has been refined each year by the Agency; the last version of the CIA studies in 1979/1980 remains one of the best on record.[9] More recently, the U.S. Department of Energy has published projections based on this approach. But, surprisingly, this new approach has been virtually ignored by many oil companies, private consultants, and international oganizations. The bean-counting approach remains to a great extent the domain of governments and some academicians. In 1980, there were about a half dozen projections of crude oil supplies, of which the CIA and DOE projections seemed to be the most useful. As oil market behavior becomes more and more difficult to explain by the traditional approach, however, the bean-counting approach is likely to be used more frequently in policy decision making.

Table 2.1 draws on five studies published in 1980 showing demand and supply for oil in the next 20 years. Exxon[10] and IEA[11] studies use the conventional approach, whereas the CBO,[12] DOE,[13] and OTA[14] studies use the bean-counting approach in their

TABLE 2.1
Noncommunist World's Oil Balance (mb/d)

	1979	1985				1990			2000	
	Actual	OTA	DOE	IEA	CBO[a]	Exxon[b]	IEA	CBO[a]	OTA	Exxon[b]
OECD	14.9	13.0-15.5	14.4-16.9	17.1[c]	15.1 [43.9]	13.0	17.1[c]	14.6 [46.8]	7.5-13.0	13.0
OPEC	31.3	28.5-35.0	24.6-31.4[d]	30.8	30.1 [4.3]	30.0[e]	31.6	31.3 [6.3]	27.0-37.0	29.0[e]
Non-OPEC LDC	5.1	7.5-9.0	7.2-9.4	11.0	10.3 [10.7]	10.0	11.0	11.7 [13.3]	7.5-11.5	13.0[f]
Total	51.3	49.0-59.5	46.2-56.7	58.9	55.5	53.0	59.7	57.6	42.0-61.5	55.0
Net supplies from CPEs[g]	1.0	[-1.9]-0	0	0.4	-1.0	0	-1.1	-1.8	[-2.0]-0	0
Free world total supply/demand	51.3	47.0-60.0	46.2-57.7	59.5	54.5	53.0	58.6	55.8	40.0-62.0	55.0
(M = Median)		M = 53.5	M = 52.0		[58.9]			[66.4]	M = 51.0	

Source: *World Energy Outlook* (New York: Exxon Corp., December 1980); International Energy Agency, *Energy Policies and Programmes of IEA Countries* (Paris: OECD, 1980); Congressional Budget Office, *The World Oil Market in the 1980s* (Washington, D.C.: U.S. Government Printing Office, May 1980); Office of Technology Assessment, *World Petroleum Availability 1985-2000* (Washington, D.C.: U.S. Government Printing Office, October 1980); J. Sawhill, "U.S. Assessment of the Free World's Dependence on Persian Oil in the 1980s,"(U.S. Senate Foreign Relations Committee, Washington, D.C., March 1980).

[a] CBO figures represent supply of oil. Figures in brackets represent projected demand.

[b] Exxon also estimates production of synthetic oil will rise to 2 mb/d in 1985 and 6 mb/d by the year 2000.

[c] IEA countries, not OECD. Except for the exclusion of minor oil production in France (not an IEA member), the totals are identical.

[d] Excludes 1.5-1.8 mb/d of natural gas liquids (NGL).

[e] Excludes NGL production of 3-5 mb/d.

[f] Includes higher rate of Mexican production than assumed elsewhere (perhaps 6-9 mb/d).

[g] Centrally planned economies.

projections. Except for the CBO study, which distinguishes between different nations' supply and demand, other projections deal with "oil balance" or supply and demand put together. The latter group's method avoids the difficulty of projecting shortages and anticipating the necessary price adjustments. The CBO study, on the other hand, projects a 4.5 mb/d shortfall in 1985 and a nearly 11 mb/d shortfall in 1990. It is important to again emphasize that any shortfall is, in effect, theoretical--what is not available cannot be consumed. Table 2.2 projections are summarized below:

1. Oil supply and demand for the noncommunist world is likely to be in the range of 52-55 mb/d throughout the next 20 years.
2. OECD supplies will remain relatively stable during 1980-1990 but will fall by a few mb/d by the year 2000.
3. The supply from non-OPEC oil producers will rise substantially, as will the domestic demand of these countries.
4. OPEC supplies are expected to be in the range of 24.6-31.6 mb/d throughout the next 20 years, with a high estimate of 27-37 mb/d by the OTA study.
5. Centrally planned economies (CPEs) are expected to be a net drain on world supplies, except for the Exxon study, which sees a balance for these countries throughout the year 2000. All of the studies examined forecast that the Soviet block will either be self-sufficient or a net importer after 1985. All studies forecast that China will remain in the export market throughout the year 2000.

OPEC Production and Exports

OPEC nations were never able to agree on a unified production policy. Prorationing (production programming) proposals first brought up in 1967 and then frequently during the 1970s were rejected by the member nations. Indeed, by the late 1970s, it had become clear that OPEC could not enforce a unified approach to production control. This is not to say that the individual member nations did not or will not respond to market changes by individually adjusting their production levels. Indeed, the oil price increase of 1979/1980, which has put most OPEC members in financial surplus, gives them the flexibility to adjust their production to correspond with their own domestic, international, political, and economic requirements.

The structural changes in the world oil market during the 1970s shifted the burden of decision making to OPEC governments, who became responsible to the world and their own people for the consequences of their actions. No longer could the purely profit-motivated courses of action followed by the oil companies dictate production and export policies within OPEC. Instead, the member governments had to take into account noneconomic factors in deciding on production levels. In facing up to their new responsibilities, they have to consider the following factors:

TABLE 2.2
Projections of OPEC Production and Exports (mb/d)

	1979		1985		1990		2000	
	Production	Exports[a]	Production	Exports[a]	Production	Exports[a]	Production	Exports[a]
Actual	31.3	29.0	-	-	-	-	-	-
IEA	-	-	30.8	26.9	31.6	25.3	-	-
OTA	-	-	28.5-35.0	24.6-31.1	-	-	27-37	9.4-19.4
DOE	-	-	24.6-31.4	20.7-27.5	-	-	-	-
Exxon	-	-	-	-	30.0	26.1	29	22.7
CBO	-	-	30.1	26.2	31.3	25.0	-	-
CIA	-	-	-	-	-	17.5	-	-
Fesharaki	-	-	26.1	22.2	23.9	17.6	-	-

Source: World Energy Outlook (New York: Exxon Corp., December 1980); International Energy Agency, Energy Policies and Programmes of IEA Countries (Paris: OECD, 1980); Congressional Budget Office, The World Oil Market in the 1980s (Washington, D.C.: U.S. Government Printing Office, May 1980); Office of Technology Assessment, World Petroleum Availability 1985-2000 (Washington, D.C.: U.S. Government Printing Office, October 1980); J. Sawhill, "U.S. Assessment of the Free World's Dependence on Persian Gulf Oil in the 1980s," (U.S. Senate Foreign Relations Committee, Washington, D.C., March 1980); J. Eckland, "The CIA's View of World Oil Outlook," in Symposium Proceedings: California and World Oil (Sacramento: California Energy Commission, 1981); F. Fesharaki, "The Future of OPEC Policy," in Symposium Proceedings: California and World Oil (Sacramento: California Energy Commission, 1981).

[a] Export figures are obtained by subtracting estimated OPEC demand from output.

1. The original hopes of modernization and industrialization through the use of oil income in a relatively short period have now faded away. It has become clear that the injection of oil money will not by itself bring about rapid development. Self-sustained growth is an evolutionary process and for that reason the life span of oil revenues must be extended to ensure a continual flow of foreign exchange.
2. Countries with surplus funds invested abroad face inflation and declining value of the dollar while making themselves hostage to the holders of their assets. The freeze of Iranian assets by the United States had a major negative psychological impact on the oil exporters. Storing their wealth under the ground has an attraction far greater than many alternatives.
3. Political pressures in many of these countries are mounting in favor of conservation (lower production). Partly, this is because of the fear that excessive domestic investments, as in the case of Iran, may disrupt the very fabric of social order in their countries. Even without revolutions or change of governments, many of the conservative regimes are likely to cut production to appease the opposition.
4. Rising domestic demand for oil products at home is also a major consideration in reducing oil exports. OPEC domestic demand rose from 770,000 b/d in 1970 to around 2.2 mb/d by the end of the 1970s. It is expected to reach 3.9 mb/d in 1985, 6.3 mb/d in 1990, and 16.7 mb/d by the turn of the century.[15]
5. The oil exporters realize that there are no major substitutes for their oil in the next 10 to 20 years. They also know that supply reduction is the key to price increases. At the same time, reduced supplies give them enormous political power, which they can exercise to their own benefit as well as to obtain some concessions for other LDCs.

Out of thirteen OPEC nations, six are located in the Persian Gulf. The Persian Gulf oil producers are the key to the future of the world oil market and, perhaps, the future of world economic and political security. In 1979, these nations produced one third of the world's oil, supplied around 62 percent of the world petroleum trade, and controlled more than 50 percent of the world's proven reserves. The Persian Gulf nations' role within OPEC is predominant. Not only do they produce more than two thirds of OPEC oil, but they also have the largest potential for expanding production through both primary and secondary enhanced recovery techniques.[16] The Persian Gulf oil exporters have divergent political and economic ideologies. On foreign policy issues, Saudi Arabia, Kuwait, the United Arab Emirates (UAE), and Qatar are pro-Western; Iraq is nonaligned, leaning towards the Soviet Union and belonging to the radical Arab Rejectionist Front; Iran has no foreign policy posture as yet. Three of the six Persian

Gulf oil exporters are tiny sheikdoms or city states. All the Persian Gulf oil exporters suffer from severe internal political opposition and are unstable. The possibilities of revolution, civil war, or major civil disobedience loom large. Even without major upheavals, however, some countries may be forced to reduce their oil production and exports to appease opposition forces.

Almost all OPEC nations have either or both resource or policy constraints on their oil production. Within the Persian Gulf, for example, all of the oil exporters have self-imposed production ceilings. Other OPEC nations face mainly technical and resource difficulties, either at present or in the near future.[17]

For the near term, there is likely to be a sufficient volume of oil produced by the OPEC countries to meet world demand. The oil glut of 1980/1981 created by declining petroleum consumption in the industrial countries, world recession, conservation, and overproduction by Saudi Arabia (above its self imposed ceiling of 8.5 mb/d) is unlikely to persist beyond 1981/1982. The oil glut has served to divert attention from the fundamental problem of petroleum supply, much in the same way as did the oil glut of 1975. This time, however, the life span of the glut promises to be short lived. With Saudi Arabia's expected reduction in output by 2 mb/d by 1982, and with the world economy moving out of the recession, a balanced market is expected by 1982.

For the longer term, over the next 10 to 20 years, the outlook seems grim. Just how grim depends mainly on the stability and oil production policy of Saudi Arabia and, to a lesser extent, on world oil demand. At the time of writing, the industrial world is in a jubilant and optimistic mood resulting from an oil market glut and the downward pressure on real prices. Industry analysts and governments are hoping that the decline in demand will continue and that demand for OPEC oil will be reduced to somewhat below the 1980 level of 26.8 mb/d over the next decade or two.

Declining demand in the industrial world may indeed take place, but most analysts believe that increasing demand in the developing world could well compensate for conservation in the industrial world.[18] In any case, a return by Saudi Arabia to its former production ceiling or to a lower production level could change the picture overnight, could lead to panic which has so often characterized the oil market, and could push up prices. At the same time, OPEC's increase in domestic consumption could lead to large reductions in exportable crude.[19] Table 2.2 provides a comparison of various projections of future OPEC production.

There are basically three schools of thought on future OPEC production:

1. The first school believes that OPEC production will rise to its technical limit of 35-38 mb/d. This assumption has now lost its credibility and is rarely used by forecasters. Most people are convinced that purely economic and technical considerations will not determine OPEC's behavior.

2. A second school of thought assumes generally steady levels of production compared to 1978/1979. This is the majority view reflected in the Exxon, CBO, IEA, and OTA studies.
3. A third school of thought assumes continual cutbacks in production to 1990. This view is supported by the U.S. Department of Energy (to some extent), the CIA,[20] and the author.[21] Although the CIA's detailed country-by-country analysis of technical and political factors has not been made available, there is a striking similarity between the aggregate CIA projection for 1990 and the author's projection for the same year. This view, however, is still the minority view among petroleum analysts.

With steady production, the rise in OPEC's own consumption would by itself take away an additional 4 mb/d from the market by 1990. In other words, under the most optimistic scenario (steady production), there will be around 15 percent less OPEC oil available in international trade in 1990 compared to 1979. By the year 2000, 16.7 mb/d of OPEC consumption will result in a cut of more than 50 percent in OPEC exports. Under the declining production scenario, by 1985 OPEC supplies will shrink by 17 to 25 percent; by 1990, around 40 percent of OPEC supplies will be lost.

Market developments during 1981 indicate that OPEC production is likely to decline from its 1978/1979 levels because of structural downward shifts in demand for oil by the industrial world. The extent of the decline in demand for OPEC oil is as yet unclear, depending on demand for oil from LDCs, the development of alternative energy sources, and the growth of the world economy. Whether this downward shift is more than compensated by OPEC production cutbacks is a matter that will determine the course of events in the next decade or two. In this situation, the politics and economics of the Persian Gulf nations, particularly Saudi Arabia, remains the key factor--not the physical availability of oil. It is the author's belief that political and economic pressures will lead to a drastic production cutback (see Table 2.2) that is likely to be far greater than any decline in demand. Indeed, production cutbacks by individual oil producers, and not necessarily through a unified policy, will be imposed in such a way as to create a persistently tight oil market.

Non-OPEC Supply Potential

Much has been said about the prospects of non-OPEC supplies during the next 10 to 20 years. This is an important issue because of the general confusion in the literature concerning how much the high rate of growth of non-OPEC production could reduce dependence on OPEC oil. The reason for the confusion is that, in considering increased non-OPEC production, due attention is not paid to the fast rates of growth of consumption in the emerging non-OPEC producing areas. Clearly, as oil becomes available, the producing nations, particularly those in the developing world, will increase their consumption immediately to give themselves a competitive edge in import-substitution and export markets.

Increased reliance on oil will be easier and less politically troublesome than using other sources of energy. Past experience shows that a 7 to 8 percent annual rate of growth of oil consumption in the non-OPEC oil-producing developing world can be expected, very much in line with OPEC's own rate of growth of oil consumption.

A general view of non-OPEC production by the developing world was presented in Table 2.1. Except for the CBO study, none of the studies accounts for increased domestic oil demand. All agree that the production of non-OPEC developing countries will double from the current 5.1 mb/d to 10-11 mb/d by 1990 and possibly reach 13 mb/d by the year 2000. The CBO study, however, projects domestic demand to exceed supply by 0.4 mb/d by 1985 and 1.6 mb/d by 1990. In short, as a whole, the developing world will remain an oil importer.

Table 2.3 provides a complete picture of major non-OPEC producers in the industrial world, centrally planned economies, and the developing world using the bean-counting approach. Despite a large increase in non-OPEC production, almost all of the increase will be consumed by this group of countries. Clearly, this does not imply that distribution of exports and consumption in these nations will be based on current patterns. For instance, although Mexico's output will initially lead to a large volume of oil that can be exported to a number of countries, domestic consumption will eventually rise to limit exports. On the whole, the major non-OPEC producers exported around 2.4 mb/d in 1979, which corresponded to only 7 percent of world oil trade. By 1990, the same countries are expected to export 2.1-3.1 mb/d or about the same as 1979. The world economy cannot, in consequence, expect much relief from non-OPEC suppliers.

The Asia-Pacific Supply Picture

Petroleum supply and demand in the Asia-Pacific region covers a wide spectrum of countries. The region includes two OPEC members, Iran and Indonesia; the United States (the third largest oil producer and the largest oil consumer in the world); China (a large oil producer and the largest oil consumer in the developing world); and Japan (an industrialized but oil-poor country).

The oil-producing nations of Asia include six net exporters of oil and four net importers of oil. Table 2.4 shows the demand and supply picture of the developing Asian nations that have a current oil production level of at least 10,000 b/d. In 1979, a total of 7.7 mb/d of oil was produced in these countries, 4.4 mb/d were consumed, and there was a net export of 3.3 mb/d. By 1990, production is expected to rise to 10-11 mb/d, consumption will double to 9 mb/d, and net exports will be less than one half the 1979 level. To be more realistic, Indonesia and Iran should be excluded, as these two OPEC nations are expected to act more in harmony with the other OPEC nations. Excluding Iran and Indonesia, the Asia-Pacific nations had an oil deficit of 410,000 b/d in 1979. Despite a doubling of output, their own domestic consumption is expected to offset the growth in output and

TABLE 2.3
Present and Future Export Potential of Major Non-OPEC Oil Producers (mb/d)

	1979			1990		
	Production	Consumption	Net Exports	Production[a]	Consumption[b]	Net Exports
Industrial						
Canada	1.5	1.8	-0.3	1.7	1.8[c]	-0.1
Norway[d]	0.4	0.2	0.2	0.7	0.3[c]	0.4
United Kingdom[e]	1.6	1.9	-0.3	2.0-2.5	1.9[c]	0.1-0.6
Total	3.5	3.9	-0.4	4.4-4.9	4.0	0.4-0.9
Centrally planned economies						
China	2.1	1.8	0.3	4.0-5.0	3.5	0.5-1.5
Soviet bloc[f]	11.9	11.5	0.4	8.5-10.5	12.8	-2.3 to -4.3
USSR	11.5	8.9	2.6	8.0-10.0	10.0	0.0 to -2.0
East Europe	0.4	2.4	-2.0	0.5	2.5	-2.0
Other	0.0	0.2	-0.2	0.0[c]	0.3	-0.3
Total	14.0	13.3	0.7	12.5-15.5	16.3	-1.8 to -2.8
LDC net exporters						
Angola	0.1	0.0	0.1	0.3	0.0[c]	0.3
Brunei	0.3	0.0	0.3	0.3	0.0	0.3
Egypt[g]	0.5	0.3	0.3	1.0-1.5	0.6	0.4-0.9
Malaysia	0.3	0.2	0.1	0.5	0.4	0.1
Mexico[h]	1.6	1.1	0.5	3.5-4.5	2.0	1.5-2.5
Oman	0.3	0.0	0.3	0.4	0.0[c]	0.4
Peru	0.2	0.1	0.1	0.2[c]	0.2	0.0
Syria	0.2	0.1	0.1	0.2	0.1[c]	0.1

Table 2.3 (cont'd)

	1979			1990		
	Production	Consumption	Net Exports	Production[a]	Consumption[b]	Net Exports
LDC net exporters (cont.)						
Trinidad and Tobago	0.2	0.0	0.2	0.2[c]	0.0[c]	0.2
Tunisia	0.1	0.0	0.1	0.2	0.0[c]	0.2
Total	3.8	1.8	2.1	6.8-8.3	3.3	3.5-5.0
TOTAL	21.3	19.0	2.4	23.7-28.7	23.6	2.1-3.1
Percent of World	32	30	7			

Source: Major non-OPEC oil producers are defined as those countries that produce more than 100,000 barrels a day. Figures may not add up due to rounding. U.S. Central Intelligence Agency, World Oil Market in the Years Ahead (ER 79-10327U), August 1979; U.S. Central Intelligence Agency, Some Perspectives on Oil Availability for Non-OPEC LDCs (ER 80-10493), September 1980; Congressional Budget Office, The World Oil Market in the 1980s (Washington, D.C.: U.S. Government Printing Office, May 1980); F. Fesharaki, "The Future of OPEC Policy," in Symposium Proceedings: California and World Oil (Sacramento: California Energy Commission, 1981); D. Ronfeldt, R. Nehring, and A. Gandara, Mexico's Petroleum and U.S. Policy (Santa Monica, Ca.: Rand Corp., 1980); British Petroleum Co., Ltd., B.P. Statistical Review of the World Oil Industry (London: B.P., 1979); U.S. Central Intelligence Agency, International Energy Statistical Review (Washington, D.C.: CIA, bimonthly); Philippine Government, Ministry of Energy, Ten-Year Energy Development Program (Manila: 1978); Petroleum and Gas in Non-OPEC Developing Countries (Washington, D.C.: World Bank, 1978).

[a] Projected production levels.

[b] Projected consumption levels.

[c] Projections are based on the assumption that current levels are maintained or do not increase significantly.

[d] Norway has imposed a production ceiling for reasons of "social policy." The ceiling is expected to be maintained during the 1980s.

[e] The United Kingdom's output of oil is expected to peak by 1982 at 2.8 mb/d before it begins to decline, unless unexpected new fields are found. There are strong indications that the government will impose a production ceiling on output in

Table 2.3 (cont'd)

the early 1980s in order to increase the life span of its exports and avoid becoming a net importer of oil in less than a decade. Another consideration is flaring of the associated gas, which the government wishes to avoid. Already, the government has ordered reduction of output from 185,000 to 100,000 b/d for Esso/Shell's Brent oilfield to avoid flaring. Also, a now-radicalized Labour Party victory in the next election is sure to result in some kind of production ceiling. It is assumed here that production will be steadily maintained at 2-2.5 mb/d during the 1980s.

[f]The Soviet Union is expected to be able to produce 10 mb/d, at most, between 1985 and 1990. If its domestic demand grows at over 5%, as in 1973-1978, domestic demand will reach nearly 14 mb/d, making the USSR a major oil importer. It is assumed optimistically here that the Soviet Union will restrict the growth of demand to 2% annually. The same goes for Eastern Europe: a traditional demand growth of 5.5% would imply imports of nearly 4 mb/d. It is assumed here that demand growth is held down to 2% a year.

[g]Egypt's production is expected to reach 1 mb/d by 1982. It may wish to follow other producers by imposing a ceiling or stretch itself to the limit by producing 1.5 mb/d. No official or unofficial hints of future policy can be discerned.

[h]Mexico is expected to impose a production ceiling in the early 1980s. The limit is likely to be in a 3.5-4.5 mb/d range. The government has often hinted that its production policy will be based on "domestic economic needs" with a view to spreading the life span of reserves--a borrowed OPEC catch phrase.

---

maintain their net oil imports. Table 2.4 does not give a complete picture of oil dependency since countries that have little or no oil production are excluded. It is useful, however, in showing that there is little hope that oil production in the already producing areas of the region is likely to be sufficient even for their own needs, let alone for exports to other importing nations of the region.

As for the giant oil importers, the United States and Japan, the future is unclear because of the possibilities of declining demand and declining energy/GDP ratios. Although there is some agreement that Japan's demand for oil in 1990 is not likely to fall below the 1980 level of around 5 mb/d and may indeed rise above this level, the U.S. demand picture remains a puzzle. In 1979, the United States demand totaled 18.5 mb/d (including natural gas liquids, or NGL), but in 1980, demand fell by 7 percent to 17.1 mb/d.[22] There is little or no consensus about whether or not the structural downward shift is so pronounced as to allow a continual decline in demand. In the first quarter of 1981, demand continued to decline by a further 4 percent. If one assumes that the 1981 first quarter decline in demand of 4 percent

TABLE 2.4
Present and Future Petroleum Potential of Developing Asia-Pacific Countries (thousand b/d)

	1979			1990		
	Production	Consumption	Exports	Production[a]	Consumption[b]	Exports
Asia-Pacific (net exporters)						
Brunei	260	c	260	300	c	300
Burma	30	30	0	60	40	20
Malaysia	280	160	120	500	400	100
China	2,100	1,800	300	4,000-5,000	3,500	500-1,500
Iran	3,130	640	2,490	2,700	1,700	1,000
Indonesia	1,590	360	1,230	1,800	920	880
Asia-Pacific (net importers)						
India	250	660	-410	400	1,100	-700
Pakistan	10	110	-110	50	140	-90
Philippines[d]	30	250	-220	100	400	-300
Taiwan	10	370	-360	10	850	-840
Total Asia-Pacific	7,690	4,380	3,310	9,920-10,920	9,050	870 to 1,870
Total non-OPEC Asia-Pacific	2,970	3,380	-410	5,420-6,420	6,430	-10 to -1,010

Source: F. Fesharaki, "Global Petroleum Supplies in the 1980s," OPEC Review 4 (1980): 27-49; U.S. Central Intelligence Agency, World Oil Market in the Years Ahead (ER 79-10327U), August 1979; U.S. Central Intelligence Agency, Some Perspectives on Oil Availability for Non-OPEC LDCs (ER 80-10493), September 1980; Congressional Budget Office, The World Oil Market in the 1980s (Washington, D.C.: U.S. Government Printing Office, May 1980).

[a] Projected production levels for countries with production of more than 100,000 barrels a day are derived from Congressional Budget Office, The World Oil Market in the 1980s (Washington, D.C.: U.S. Government Printing Office,

Table 2.4 (cont'd)

May 1980). Countries with production of less than 100,000 barrels a day are assumed to increase production at historical rates.

[b]Projections of consumption levels are based on projected growth rates from U.S. Central Intelligence Agency, World Oil Market in the Years Ahead (ER 79-10327U), August 1979 and Some Perspectives on Oil Availability for Non-OPEC LDCs (ER 80-10493), September 1980.

[c]Negligible.

[d]Projections are derived from F. Fesharaki, "World Oil Availability," Annual Review of Energy 6 (1981): 267-308, and Philippine Government, Ministry of Energy, Ten-Year Energy Development Program (Manila: 1978).

---

will be carried through the end of 1981 and then stabilize, we can get an estimate of U.S. demand in 1990: oil production (including NGL) in the United States would be expected to fall from nearly 10 mb/d in 1980 to 9.2 mb/d in 1982 and to 8.5-9.0 mb/d in 1990.[23] This would mean net imports of around 6.5 to 7 mb/d in 1990.

The other two major LDC importers, South Korea and Taiwan, will continue to import large quantities of oil. Korea's oil demand rose by 14.7 percent a year between 1978-1980, but this growth rate is expected to decline to 8.4 percent between 1980-1986, and 7.1 percent a year between 1986-1990. This will lead to the increase of oil imports from around 600,000 b/d in 1980 to nearly 1.5 mb/d in 1990.[24] Although Taiwan is a smaller importer than Korea, it is, nevertheless, the second largest oil importer in the developing Asia-Pacific region. If we assume that Taiwan's demand for oil will rise at the same rate as Korea's, Taiwan's oil imports of around 400,000 b/d in 1980 will more than double to 850,000 b/d in 1990.

Table 2.5 shows the likely position of the large oil importers in the Asia-Pacific region. Between 1979 and 1980, there was a drastic decline in oil demand of 2.3 mb/d. As the United States' demand continues to decline, however, South Korea's and Taiwan's demand will increase. The 1990 aggregate demand for these four large importers is likely to be 1.6 mb/d above the 1980 level. This illustrates how critical the oil imports of these four Asia-Pacific countries are, not only for the region, but also for the world. In 1980, the imports of these countries amounted to 45 percent of OPEC output and 40 percent of the total oil imports of the world. Tables 2.4 and 2.5 show that there is no possibility that the nations of the region can become self-sufficient in oil output. Their dependence on imported oil is likely to continue, unless drastic actions are undertaken to curtail demand and to switch to other energy sources.

TABLE 2.5
Large Oil Importers of the Asia-Pacific Region (mb/d)

Oil Importers	1979	1980	1990
United States	8.0	6.2	6.5
Japan	5.6	5.0	5.0
South Korea	0.54	0.6	1.5
Taiwan	0.37	0.4	0.84
Total	14.51	12.2	13.84

Source: U.S. Central Intelligence Agency, World Oil Market in the Years Ahead (ER 79-10327U), August 1979; Congressional Budget Office, The World Oil Market in the 1980s (Washington, D.C.: U.S. Government Printing Office, May 1980); Office of Technology Assessment, World Petroleum Availability 1985-2000 (Washington, D.C.: U.S. Government Printing Office, October 1980); U.S. Central Intelligence Agency, International Energy Statistical Review (Washington, D.C.: CIA, bi-monthly); United Nations and World Bank, Considerations Affecting Korea's Energy Objectives, Policies and Strategy (New York: 1981).

MARKET STRUCTURE AND EXPORT STRATEGY

As previously shown, OPEC will remain the dominant force in the world oil trade during the foreseeable future. It is, therefore, useful to discuss the emerging OPEC export strategy in the light of the structural changes in the world oil market so that we may understand the problems that lie ahead.

There were two major structural changes in the oil market that will have significant importance in the 1980s: (1) the declining supply of oil to the major oil companies and the rising direct sales by OPEC nations, and (2) the emergence of large-scale product exports from OPEC. These two developments will have significant implications for all oil importers.

In 1973, 93 percent of OPEC production, or 27.9 mb/d, were made available to the major oil companies on a long-term or preferential basis (e.g. equity crude or buy-back crude). This amounted to 90 percent of the world oil trade. By 1979/1980, this had dropped to around 50 percent of OPEC exports and 42 percent of the world oil trade. At the same time, direct state-to-state sales by OPEC rose from 1.5 mb/d in 1973 to 5.0 mb/d in 1979 and possibly 7.8 mb/d in 1980. Also, increasing volumes of oil were sold directly by OPEC national oil companies to the oil companies under short-term or spot sales. In total, OPEC nations' ownership/entitlements to their own crude rose from 2 percent in 1970 to 80 percent in 1980. They sold directly around 50 percent of this entitlement crude in 1980, compared to near zero volume in 1970.

One important casualty of this structural change is the so-called "third-party" sales, that is, the surplus oil sold by the major oil companies (the "majors") to other smaller oil companies or state-owned oil companies of the developing and developed world. Third-party sales are extremely important for the oil-deficit companies or countries that do not possess tanker transport or distribution facilities. Between 1973 and 1979, third-party sales were slashed from 6.8 mb/d to 3.4 mb/d. For 1980, the estimates are below 1 mb/d. Table 2.6 provides a rough estimate of these changes. It also shows that in 1980 the majors had to go into the spot market to buy 1.2 mb/d to meet their integrated network product commitments. This trend is likely to continue in the future.[25]

Since 80 percent of OPEC oil is already owned by the member countries, they could easily expand their direct sales to that level. By 1990, OPEC may well be handling over three quarters of its exports directly. State-to-state deals are encouraged, particularly when the second party is from the developing world. Third-party sales are likely to be eliminated altogether and the majors will themselves become crude short. Third-party purchasers can no longer depend on the majors and will have to make their own arrangements with OPEC nations.

In addition to these changes in sales patterns, the oil market will be faced with the large-scale entry of OPEC refined products. In 1980, OPEC refining capacity was 6 million barrels per stream day (mb/sd). By 1990, the refining capacity is expected to reach 11.4-13.4 mb/sd. This expansion of capacity comes at a time when world refining capacity utilization is only 68 percent, and there is little likelihood of a surge in demand in the industrial world. Since most of the refining capacity is concentrated in the industrial world, the economic impact of such activity is going to be mainly felt in the OECD. The emergence of OPEC product exports from 1.6 mb/d in 1980 to 3.5 mb/d in 1985, and possibly up to 5.7 mb/d in 1990, however, will drastically affect current export patterns.[26] Since every barrel of refined product export replaces a barrel of crude export, the volume of crude supplies will decline drastically, and products will be made available to companies and countries that have their own processing facilities. Indeed, by 1990, between 25 to 30 percent of OPEC exports are expected to be in product form.[27]

This development also has major implications for the world tanker industry. With rising state-to-state sales, demand for smaller carriers is rising while demand for the larger carriers is declining. At the same time, the emergence of OPEC product exports means that 400 to 1,000 new product carriers will be needed. OPEC nations are themselves involved in a massive ordering of product and crude carriers with the intention of using their own means of transport for part of their exports.[28]

TABLE 2.6
Structural Changes in the World Petroleum Market

	1970	1973	1978	1979	1980[a]
OPEC sales of crude (mb/d)					
To majors:					
Affiliates	-	21.1	14.5	14.1	-
Third party	-	6.8	4.8	3.4	0.9
Total	-	27.9	19.3	17.5	-
Direct sales:					
State-to-state	-	1.5	4.6	5.0	6.8-7.8
Commercial	-	0.9	5.1	7.8	-
Total	-	2.4	9.7	12.8	-
TOTAL	-	30.3	29.0	30.3	-
Other indicators (%)					
OPEC ownership/entitlements to own crude	2	20	75	80	80
Direct OPEC exports	1	7	33	42	50+
Majors' share in OPEC oil	99	93	67	58	50
Majors' share in world oil trade	92	90	50	42	-

Source: *Petroleum Intelligence Weekly* (February 25, 1980); J. H. Mohnfeld, "Changing Pattern of Trade," *Petroleum Economist* 67 (August 1980); OPEC Secretariat, "A Statistical Approach to Analyze the Evolution of Major Oil Companies' Control of the World Market," unpublished (August 1980); *Annual Statistical Bulletin* (Vienna: OPEC Secretariat, 1980).

[a] Estimates.

The structural changes in the oil market provide some indication of OPEC's emerging export strategy (though it is by no means expected to be applied uniformly). This export strategy has the following characteristics:

1. Oil supplies will be made available on a short-term basis, say, six to twelve months. Long-term contracts are unlikely to be awarded.
2. More and more, oil sales will be destination controlled. The loss of flexibility of the international oil companies, the emergence of product exports, and partial control of transportation mean that destination controls will be easier to impose and monitor.
3. Oil sales will be in package deals that may include the following:[29]
   a. Oil sales linked to investment in exploration, establishment of petrochemical, refining, or other industries
   b. Oil sales linked to sales of refined products and petrochemicals, even though there is surplus capacity available in the consuming countries
   c. Oil sales linked to transport in OPEC-owned tankers
   d. Oil sales linked to natural gas pricing and sales policies
   e. Oil sales linked to arms sales, technology transfer, and the like
   f. Oil sales linked to major concessions from the industrial world to the poor nations in a North-South dialogue
   g. Oil sales linked to indexation of OPEC investments in the industrial world

Package deals are likely to become the dominant characteristic of oil sales. Since 1979, we have seen Saudi Arabia's linkage of some oil sales to investment in refining and petrochemicals through the so-called "incentive crude program," Libya's and Algeria's linkage of oil sales to oil exploration, and Algeria's partial linkage of some oil sales to purchases of higher-priced liquefied natural gas.

OIL PRICES

The Iranian Revolution of February 1979 led to significant changes in the oil market, with a price increase of 120 percent from January 1979 to January 1980. The price increase itself, however, may in fact be less significant than the structural changes following the price increase. It is important to recognize that the Iranian Revolution was not the cause of these changes. On the contrary, these changes were expected by many experts within OPEC and the oil industry because of the continuous decline in the real price of oil. The Iran revolt simply ignited the potentially explosive situation in the oil market. The most important feature of the 1979 price increases was that they were motivated by the fear of supply interruptions and spot market movements. In the previous period, supply problems and the spot market had not played a major role. Rather, it was OPEC's decision to increase prices at a time when no lasting shortages were in sight. The 1979 changes signaled a period when market prices rose so fast that OPEC nations, some caught by surprise, had to raise their prices to keep up with the market. In 1979, spot prices reached $45, more than double the $18/b price of marker crude. Though spot sales represent a small portion of the world oil trade, they are significant in signaling the market's ability to absorb high prices.

In 1979, there was more oil produced and exported than in 1978, despite the Iranian supply cutback. So why did prices rise so sharply? The factors behind the sharp increases were as follows:

1. The breakdown of the conventional oil distribution mechanisms. Traditionally, the international oil companies received the bulk of oil exports and distributed them to those who needed supplies, their own affiliates, and third-party sales. The Iranian situation denied the large oil companies their usual access to crude. Similar patterns were seen in other OPEC nations, who preferred to sell their oil directly to state oil companies and smaller independents. For instance, Japan was denied 1 mb/d of oil by the majors. Since there was no mechanism to fill the vacuum left by the large oil companies, the Japanese went into the open market and bought oil at spot prices, further pushing up the spot prices. In short, the number of actors in the market increased substantially. While there were sufficient volumes of oil around, those who needed it could not get access to it without using the volatile spot market.

2. The Iranian crisis. This event signaled a period of supply instability from the most important oil region of the world, the Persian Gulf. The fear of more interruptions from this region, whether imaginary or real, has pushed up stockpiling to record volumes of 5 billion barrels. Unlike the past, such stocks will probably not be drawn down substantially, and there is evidence that the feeling of supply insecurity will push the private oil companies as well as governments to maintain high stocks throughout the 1980s.

3. The disappearance of Iran as the second largest exporter of crude and the natural, if unwilling, ally of Saudi Arabia (since 1975) in pushing for moderate policies on production and prices. This has substantially reduced Saudi flexibility. Now Saudi Arabia is left alone to carry the burden of moderation, with occasional help from the United Arab Emirates. In 1979, the Saudis kept their prices at $18/b, $4/b below similar-quality crudes, in order to prevent prices from rising sharply. They failed. Later, they had to raise their prices to $24/b and then to $26/b, $28/b, $30/b and, finally, to $32/b in December 1980, $3 to $4/b below similar-quality crudes. Though they export more than one third of OPEC oil, their role as the "swing producer" has been seriously curtailed. By stretching their output to capacity,

they have left themselves little room for maneuvering. And though they may succeed in temporarily stalling the rise in prices, in the medium and long term they are not likely to succeed. They may soon find this battle not worth fighting and will bow to domestic and OPEC pressure to reduce supplies, eliminating any surplus oil that may be available in the market.

The result of events since 1979 has been the disappearance of a unified price policy. Indeed, the December 1980 and May 1981 OPEC Ministerial Conferences set up a three-tiered price system with $9/b differentials. This was adopted to give the appearance of price agreement. There was, in fact, no agreement with such a large range. And there is no need to have a uniform policy. Indeed, it would be economically irrational for OPEC nations to agree on a unified approach if they are able to sell at higher prices.

As the oil market tightens, OPEC pricing and production policy will be formed in such a manner as to eliminate any need for collective decision making. The system will operate as follows: As spot prices rise as a result of real or anticipated supply cutbacks, official price increases will follow (not necessarily to the full extent). Once official prices are raised, they will be administratively maintained. If demand declines, lifting from various exporters will decline. This year's lower production will become the guideline for next year's ceiling on output, ensuring persistent shortages in the market. Since the oil price increases have left almost all OPEC nations in a surplus situation, and since the rate of imports by OPEC from the industrial nations is not likely to follow the wild spending spree of the post-1973 era (in fact, OPEC imports in 1979 were below 1978 levels in current prices), such operations will not exert undue economic pressure on OPEC nations to expand exports. The price-output signals can work effectively in this manner, without unified policies. Indeed, lower production and exports will, in turn, reduce domestic political pressures on oil exporters.

In predicting the extent of the price increases, one can only say with certainty that real prices will not be allowed to decline again (beyond the short term). Oil price modeling based on demand and supply considerations or on OPEC's need for foreign exchange (based on a questionable set of assumptions about the absorptive capacity in OPEC) are misleading and invariably lead to false expectations. The decision-making process within OPEC is, in fact, far less complicated. Decisions are made by consensus or on an individual basis. The upper ceiling on oil prices is determined by international security considerations of one or all member nations. Oil will not be priced so as to invite wars or catastrophic changes that may endanger the nation's security. Of course, close ties with consumer nations may induce some exporters to be more moderate than others, although, as we have seen, the market can accommodate even a $10/b difference between similar quality crudes. As long as there is a likelihood of a shortage, there is no need for unity in prices.

There is a consensus within the Western world that oil prices will rise by 2 to 3 percent annually in real terms during the

1980s. It is, however, quite likely that oil prices will rise by 5 to 10 percent a year in real terms. Table 2.7 shows three price scenarios for the 1980s: 3 percent, 5 percent, and 10 percent annual price increases. These price increases do not take into account major political upheavals or revolutions. They are, however, the natural evolution of current trends. The table shows that real prices are likely to be in the $42 to $70/b range and current prices, $75 to $123/b by 1990 (assuming a 7 percent annual inflation rate). These prices will not, of course, increase in a regular manner, but they represent the order of magnitude that may be expected over the decade.

IMPLICATIONS FOR THE ASIA-PACIFIC COUNTRIES

As previously discussed, oil exports from the OPEC nations, and particularly the Middle East, will remain the dominant force in the oil trade for the 1980s. The implication for the Asia-Pacific nations is extremely serious, especially because of their historical dependence on the politically volatile Middle East supplies.

Historical Dependency Patterns: The Developing Asia-Pacific Nations

Tables 2.8 and 2.9 provide historical data on the source of crude oil imports for various Asia-Pacific nations (see also Chapter 3, Tables 3.7 and 3.11). The data are by no means complete; they cover only imports of crude oil and not imports of products. In view of the characteristics of the petroleum product trade, no reliable recording system is available and all such data are best left aside. For crude trade, various sources within the OPEC Secretariat and the industrial nations as well as international organizations were consulted. It was found that only the data prepared by the United Nations are in a form which, despite its shortcomings, can provide a useful picture.

South Asia. In India, the historical pattern of dependence on OPEC/Middle East sources has changed only slightly in the 1970-1977 period. In 1977, 88 percent of India's crude oil imports came from the Middle East. Within the Middle East, India has managed to diversify its sources of crude. In 1970, India's imports were dominated by Iran (69 percent) and Saudi Arabia (31 percent). By 1977, less than one half of India's Middle East imports came from Iran, and oil was also imported from Iraq and the United Arab Emirates. Because of India's diversified economy and its own energy resources, it depended on petroleum for only 28 percent of its commercial energy consumption. Around 56 percent of India's oil consumption in 1978 (equal to 28 percent of its total commercial energy consumption) was imported. Oil imports constituted 26 percent of India's total import bill in 1978. The data should not be interpreted to minimize the importance of oil imports for India. Despite a major shift to coal,

TABLE 2.7
OPEC Oil Price Scenarios in the 1980s ($US/b)

	1981[a] (Actual)	1982	1985[b] Low	1985[b] Medium	1985[b] High	1990[b] Low	1990[b] Medium	1990[b] High
Average OPEC price (Constant 1981 prices)	35	33[c]	36	38	43	42	48	70
Average OPEC price (Current prices)[d]	35	35[c]	47	49	56	75	87	123

[a] Average OPEC price in January 1981 was $35/b. This includes some premium charges.

[b] Low: 3% real price increase. Medium: 5% real price increase. High: 10% real price increase.

[c] Assumes a decline in the real price of oil for 1982.

[d] Assumes 7% annual inflation for 1981-1990.

TABLE 2.8
Developing Asia-Pacific Countries' Imports of Crude Oil, by Source
(10^3 metric tons)

To \ From	Year	Middle East		Far East		CPEs[a]		Western Europe		Other		Total Imports (100%)	OPEC	
		Volume	Percent of Total Imports	Volume	Percent of Total Imports	Volume	Percent of Total Imports	Volume	Percent of Total Imports	Volume	Percent of Total Imports		Volume	Percent of Total Imports
South Asia[b]														
India	1977	12,560	(88)	0		1,650	(12)	0		0		14,210	12,560	(88)
	1973	13,020	(98)	0		280	(2)	0		0		13,300	13,020	(98)
	1970	11,700	(100)	0		0		0		0		11,700	11,700	(100)
Pakistan	1977	3,270	(100)	0		0		0		0		3,270	3,270	(100)
	1973	3,160	(100)	0		0		0		0		3,160	3,160	(100)
	1970	4,010	(100)	0		0		0		0		4,010	4,010	(100)
Sri Lanka	1977	1,460	(95)	0		0		70	(5)	0		1,530	1,460	(95)
	1973	1,800	(100)	0		0		0		0		1,800	1,800	(100)
	1970	1,820	(100)	0		0		0		0		1,820	1,820	(100)
South East Asia[c]														
Malaysia	1977	3,730	(93)	280	(7)	0		0		0		4,010	3,730	(93)
	1973	2,600	(67)	1,260	(33)	0		0		0		3,860	2,600	(67)
	1970	2,740	(29)	6,620	(71)	0		0		0		9,360	2,740	(29)
Philippines	1977	7,100	(75)	1,680	(18)	750	(8)	0		0		9,530	8,290	(87)
	1973	7,620	(82)	1,640	(18)	0		0		0		9,260	8,850	(96)
	1970	5,470	(60)	3,590	(40)	0		0		0		9,060	7,540	(83)
Singapore	1977	18,180	(77)	5,450	(23)	0		0		0		23,630	23,290	(99)
	1973	19,100	(88)	2,570	(12)	0		0		20	(d)	21,690	19,240	(89)
	1970	10,100	(96)	400	(4)	0		0		0		10,500	10,100	(96)
Thailand	1977	6,860	(84)	1,110	(13)	230	(3)	0		0		8,200	6,860	(84)
	1973	6,400	(85)	1,110	(15)	0		0		0		7,510	6,400	(85)
	1970	3,630	(92)	310	(8)	0		0		0		3,940	3,630	(92)
East Asia[e]														
South Korea	1977	20,740	(100)	0		0		0		0		20,740	20,740	(100)
	1973	14,040	(100)	30	(c)	0		0		0		14,070	14,040	(100)
	1970	8,780	(99)	100	(1)	0		0		0		8,880	8,780	(99)
Oceania[f]														

Table 2.8 (cont'd)

Source: United Nations, World Energy Supplies, Series J, nos. 19, 22 (New York: 1976, 1979).

[a]Centrally planned economies.

[b]Time series data are not available for Bangladesh. Afghanistan and Nepal import no crude oil but only refined products.

[c]Time series data are not available for Burma and Indonesia. Brunei imports no crude oil but only refined products.

[d]Less than 1 percent.

[e]Time series data are not available for North Korea, Hong Kong, and China.

[f]Time series data are not available for Fiji, Papua New Guinea, and Tonga.

---

reducing India's dependence on petroleum is critical for the country's economic development program.

Pakistan imports 100 percent of its petroleum, all originating from the Middle East. This dependency has remained unchanged over the 1970s. Until the mid-1970s, the bulk of Pakistan's imports came from Iran. By 1977, Saudi Arabia had become Pakistan's main supplier of crude (69 percent), followed by the UAE (31 percent). In Pakistan, 40 percent of commercial energy consumption came from oil and 86 percent of oil consumption from imports. Discovery of gas resources has allowed Pakistan to begin a shift in consumption to natural gas.

Sri Lanka's dependence on Middle East oil has declined slightly, from 100 percent in 1970 to 95 percent in 1977. Within the Middle East, Sri Lanka has depended on Saudi Arabia (53 percent) and Iraq (42 percent) for its imports. Ninety percent of this nation's commercial energy consumption came from oil, all of which was imported. Sri Lanka is one of the most petroleum dependent nations in the world.

For Bangladesh and Nepal, time series data are not readily available. In 1978, 51 percent of Bangladesh's and 81 percent of Nepal's commercial energy consumption came from oil. In both countries, all of the oil consumed was imported. In Nepal, no refining facilities exist and thus all imports are refined products.

Southeast Asia. During the 1970-1977 period, Malaysia's dependence on the Middle East increased from 29 percent to 93 percent, with the balance coming from the Far East. Within the

TABLE 2.9
Developing Asia-Pacific Countries' Imports of Crude Oil from the Middle East
(10^3 metric tons)

From / To	Year	Total Middle East		Iran		Iraq		Kuwait		Qatar		Saudi Arabia		United Arab Emirates	
		Volume	Percent of Total Imports	Volume	Percent of Total Imports	Volume	Percent of Total Imports	Volume	Percent of Total Imports	Volume	Percent of Total Imports	Volume	Percent of Total Imports	Volume	Percent of Total Imports
South Asia[a]															
India	1977	12,560	(88)	6,870	(48)	1,860	(13)	0		0		2,870	(20)	960	(7)
	1973	13,020	(98)	9,620	(72)	0		0		0		3,400	(26)	0	
	1970	11,700	(100)	8,100	(69)	0		0		0		3,600	(31)	0	
Pakistan	1977	3,270	(100)	0		0		0		0		2,270	(69)	1,000	(31)
	1973	3,160	(100)	2,470	(78)	0		260	(8)	0		370	(12)	60	(2)
	1970	4,010	(100)	n.a.		n.a.		n.a.		n.a.		n.a.		n.a.	
Sri Lanka	1977	1,460	(95)	0		650	(42)	0		0		810	(53)	0	
	1973	1,800	(100)	1,800	(100)	0		0		0		0		0	
	1970	1,820	(100)	n.a.		n.a.		n.a.		n.a.		n.a.		n.a.	
South East Asia[b]															
Malaysia	1977	3,730	(93)	810	(20)	0		810	(20)	0		2,110	(53)	0	
	1973	2,600	(67)	0		0		1,440	(37)	0		1,160	(30)	0	
	1970	2,740	(29)	0		0		1,300	(14)	0		1,440	(15)	0	
Philippines	1977	7,100	(75)	820	(9)	1,100	(12)	1,880	(20)	0		3,300	(35)	0	
	1973	7,620	(82)	890	(10)	30	(c)	1,240	(13)	190	(2)	5,060	(55)	210	(2)
	1970	5,470	(60)	2,390	(26)	30	(c)	2,280	(25)	770	(8)	0		0	
Singapore	1977	18,180	(77)	3,080	(13)	2,250	(10)	1,180	(5)	150	(1)	11,330	(48)	190	(1)
	1973	19,100	(88)	4,270	(20)	670	(3)	6,550	(30)	0		5,800	(27)	1,810	(8)
	1970	10,000	(96)	1,170	(11)	60	(1)	7,820	(74)	0		1,050	(10)	0	
Thailand	1977	6,860	(84)	0		0		550	(7)	2,350	(29)	3,910	(48)	50	(1)
	1973	6,400	(85)	40	(1)	340	(5)	2,230	(30)	1,700	(23)	2,090	(28)	0	
	1970	3,630	(92)	100	(3)	160	(4)	0		1,550	(39)	1,820	(46)	0	
East Asia[d]															
South Korea	1977	20,740	(100)	2,140	(10)	450	(2)	6,810	(33)	0		11,340	(55)	0	
	1973	14,040	(100)	540	(4)	1,270	(9)	4,210	(30)	0		8,020	(57)	0	
	1970	8,780	(99)	3,180	(36)	0		2,500	(28)	0		3,100	(35)	0	
Oceania[e]															

Table 2.9 (cont'd)

Source: United Nations, World Energy Supplies, Series J, nos. 19, 22 (New York: 1976, 1979).

[a]Time series data are not available for Bangladesh. Afghanistan and Nepal import no crude oil but only refined products.

[b]Time series data are not available for Burma and Indonesia. Brunei imports no crude oil but only refined products.

[c]Less than 1 percent.

[d]Time series data are not available for the Democratic Republic of Korea (North Korea), Hong Kong, and China.

[e]Time series data are not available for Fiji, Papua New Guinea, and Tonga.

n.a. = data not available.

Middle East, over one half of the imports came from Saudi Arabia and the remainder from Iran and Kuwait. In Malaysia, 88 percent of commercial energy consumption has been based on petroleum. Since Malaysia is becoming a net oil exporter, and has a diversified export base, there is no immediate need for concern.

The Philippines' dependence on OPEC oil has risen from 83 percent to 87 percent, and on Middle East oil from 60 percent to 75 percent of total oil imports in the 1970-1977 period. In 1970, a quarter of the Philippines' imports came from Iran and the balance from Kuwait and Qatar. By 1977, 35 percent of imports came from Saudi Arabia, 20 percent from Kuwait, 12 percent from Iraq, and only 9 percent from Iran (Iran's oil exports to the Philippines were embargoed after the 1979 Revolution). The Philippines' dependence on the Far East declined from 40 percent in 1970 to 18 percent in 1977. In the Philippines, 94 percent of all energy consumption came from oil. In 1977, all of the Philippines' oil needs were imported. Recent development in domestic production, however, will reduce dependence on imported oil slightly in the short run. Alternative energy production may further reduce oil import dependency in the 1980s.

Singapore's dependence on OPEC oil has slightly increased, from 96 percent in 1970 to 99 percent in 1977. Dependence on Middle East oil has declined from 96 percent in 1970 to 77 percent in 1977, the balance being supplied by the Far East. About one half of Singapore's imports came from Saudi Arabia, while the remainder was supplied by other Middle East oil exporters. Singapore's commercial energy consumption was totally based on oil, all of which was imported. Singapore is a particularly important nation since it is the main refining center in the region.

Thailand's dependence on OPEC and Middle East oil has slightly declined, from 92 percent in 1970 to 84 percent in 1977, the balance coming from the Far East. Just under one half of the imports came from Saudi Arabia, 29 percent from Qatar, and 7 percent from Kuwait. Around 95 percent of all commercial energy consumption came from oil, all of which was imported.

East Asia. South Korea's dependence on Middle East oil is total and has not changed over the 1970s. In 1970, Iran and Saudi Arabia were its major sources of crude, followed by Iraq. By the late 1970s, imports from Iran dropped to 10 percent, but imports from Saudi Arabia rose to 55 percent, the balance coming from Iraq. Sixty-one percent of South Korea's commercial energy use came from oil in 1978, all of which was imported.

North Korea's main sources of crude are China and the Soviet Union. According to U.N. statistics, North Korea's dependence on oil as a percentage of its total commercial energy consumption was 4 percent - a figure that is highly suspect. All oil consumed in North Korea was imported. Similarly, Hong Kong depends on imported oil for all of its commercial energy.

Historical Dependency Patterns: The Developed Nations of Asia and the Pacific

Like the developing Asia-Pacific nations, the developed countries of this region are also greatly dependent on imported oil. Tables 2.10 and 2.11 summarize their dependency patterns.

Australia is blessed with domestic oil production, which provides about 80 percent of its consumption. In 1979, Australia produced 440,000 b/d of oil and imported just over 100,000 b/d. Between 1970 and 1979, over 90 percent of Australia's oil imports came from OPEC, and about 80 percent originated from the Middle East. During 1973-1977, its import dependence on Middle East oil rose to 99 percent, but in 1979, its Middle East imports declined to 80 percent and its imports of Indonesian oil reached 18 percent. Within the Middle East, Australian imports were well diversified. Since 1970, Kuwait has remained its largest supplier of oil, with Iraq, UAE, and Saudi Arabia supplying the rest. Australia's current level of imports and its dependency pattern is expected to continue unless shale oil production plans are significantly accelerated. Because of its large exports of meat and wheat to the Middle East, Australia is in a good position to ensure continued access to crude in the 1980s.

Japan imported 5.6 mb/d of oil in 1979, 4.8 mb/d of which was imported as crude. Its dependence on OPEC declined from 96 percent in 1970 to 87 percent in 1979. Its dependence on the Middle East was reduced by 9 percent since 1970 to 76 percent in 1979. Japan has tried hard to diversify its imports by increasing imports from China and Indonesia. The disappointing performance of China's oil production and the probable drop in Indonesia's export levels, however, will mean its dependence on the Middle

TABLE 2.10
Developed Asia-Pacific Countries' Imports of Crude Oil by Source
(10^3 metric tons)

From / To	Year	Middle East		Far East		CPEs		Western Europe		Others		Total Imports	OPEC	
		Volume	Percent of Total Imports	Volume	Percent of Total Imports	Volume	Percent of Total Imports	Volume	Percent of Total Imports	Volume	Percent of Total Imports		Volume	Percent of Total Imports
Australia	1979	4,208	(80)	936	(18)	0		0		125	(2)	5,269	4,791	(91)
	1978	4,791	(87)	677	(12)	0		0		55	(1)	5,523	5,134	(93)
	1977	8,460	(98)	80	(1)	50	(1)	0		0		8,590	8,540	(99)
	1973	8,420	(99)	60	(1)	0		0		0		8,480	7,800	(92)
	1970	11,540	(79)	3,150	(21)	0		0		0		14,690	13,840	(94)
Japan	1979	178,085	(76)	42,449	(18)	7,321	(3)	0		6,638	(3)	234,493	203,632	(87)
	1978	174,868	(78)	37,046	(16)	7,520	(3)	0		5,149	(2)	224,583	196,252	(87)
	1977	188,880	(79)	43,020	(18)	7,080	(3)	0		1,500	(a)	240,480	213,050	(89)
	1973	195,310	(78)	44,760	(18)	2,180	(1)	0		7,800	(3)	250,050	232,400	(93)
	1970	145,680	(85)	21,330	(13)	500	(a)	0		3,080	(2)	170,590	163,490	(96)
New Zealand	1979	n.a.		n.a.		n.a.		n.a.		n.a.		n.a.	n.a.	
	1978	n.a.		n.a.		n.a.		n.a.		n.a.		n.a.	n.a.	
	1977	2,350	(87)	170	(6)	0		0		180	(7)	2,700	2,350	(87)
	1973	3,100	(98)	20	(a)	0		0		50	(2)	3,170	3,120	(98)
	1970	2,130	(91)	200	(9)	0		0		0		2,330	2,130	(91)
United States	1979	107,702	(32)	18,999	(6)	563	(a)	14,014	(4)	195,360	(58)	336,638	264,831	(79)
	1978	113,937	(35)	27,355	(8)	40	(a)	13,680	(4)	171,392	(53)	326,404	266,579	(82)
	1977	125,590	(38)	27,720	(9)	90	(a)	3,390	(1)	169,180	(52)	325,970	281,220	(86)
	1973	32,620	(20)	10,080	(6)	0		0		117,800	(73)	160,500	102,690	(64)
	1970	9,580	(14)	3,580	(5)	0		60	(a)	53,870	(80)	67,090	31,840	(48)

Source: 1970-1977 figures from United Nations, World Energy Supplies 1973-1978, Series J, no. 22 (New York: 1979); 1978-1979 figures from World Oil Trade (June 1980). Figures may not add up due to rounding.

[a] Less than 1 percent.

n.a. = data not available.

TABLE 2.11
Developed Asia-Pacific Countries' Oil Imports from the Middle East
(10^3 metric tons)

To \ From	Year	Iran Volume	Iran Percent of Total Imports	Iraq Volume	Iraq Percent of Total Imports	Kuwait Volume	Kuwait Percent of Total Imports	Qatar Volume	Qatar Percent of Total Imports	Saudi Arabia Volume	Saudi Arabia Percent of Total Imports	United Arab Emirates Volume	United Arab Emirates Percent of Total Imports	Other Volume	Other Percent of Total Imports	Total Middle East Volume	Total Middle East Percent of Total Imports	Total Imports Volume (100%)
Australia	1979	95	2	871	16	1,479	28	-	-	627	12	782	15	354	7	4,208	80	5,269
	1978	503	9	1,474	27	1,305	24	50	1	662	12	463	8	334	6	4,791	87	5,523
	1977	960	11	1,150	13	1,560	18	480	6	4,030	47	280	3	-	-	8,460	98	8,590
	1973	500	6	1,680	20	3,060	36	540	6	1,880	22	120	1	640	8	8,420	99	8,480
	1970	450	3	1,300	9	4,420	30	1,330	9	3,470	24	570	4	-	-	11,540	79	14,690
Japan	1979	23,311	10	13,073	6	21,957	9	6,952	3	78,704	34	24,586	10	9,502	4	178,085	76	234,493
	1978	40,169	18	7,629	3	18,575	8	5,398	2	70,885	32	23,496	10	8,715	4	174,868	78	224.583
	1977	40,670	17	7,450	3	18,320	7	1,750	1	85,240	35	26,700	11	8,750	4	188,880	79	240,480
	1973	83,730	33	1,400	(a)	23,020	9	110	(a)	58,470	23	23,380	9	5,200	2	195,310	78	250,050
	1970	74,850	44	-	-	33,020	19	170	(a)	24,900	15	8,620	5	4,120	2	145,680	85	170,590
New Zealand	1979	n.a.		n.a.		n.a.		n.a.		n.a.		n.a.		n.a.		n.a.		n.a.
	1978	n.a.		n.a.		n.a.		n.a.		n.a.		n.a.		n.a.		n.a.		n.a.
	1977	1,010	37	-		210	8	-		1,130	42	-		-		2,350	87	2,700
	1973	600	19	-		2,100	66	-		300	9	100	3	-		3,100	98	3,170
	1970	520	22	-		1,380	59	-		230	10	-		-		2,130	91	2,330
United States	1979	16,364	5	4,606	1	339	(a)	1,788	1	66,901	20	15,005	4	2,699	1	107,702	32	336,638
	1978	28,859	9	2,609	1	473	(a)	3,337	1	56,120	17	18,964	6	3,576	1	113,937	35	326,404
	1977	29,040	9	3,620	1	2,030	(a)	3,200	1	67,870	21	16,070	5	3,710	1	125,590	38	325,970
	1973	10,860	7	270	(a)	2,050	1	350	(a)	15,110	9	3,690	2	290	(a)	32,620	20	160,500
	1970	1,670	2	-		2,830	4	-		1,980	3	3,100	5	-		9,580	14	67,090

Source: 1970-1977 figures from United Nations, World Energy Supplies 1973-1978, Series J, no. 22 (New York: 1979); 1978-1979 figures from World Oil Trade (June 1980). Figures may not add up due to rounding.

[a] Less than 1 percent.

n.a. = data not available.

East will not decline further. Currently, Japan is importing some oil from Mexico and is likely to seek increases in Mexican oil exports. Mexican oil, however, is likely to mainly replace imports from China and Indonesia. Within the Middle East, Japan's largest source of imports shifted from Iran to Saudi Arabia during 1970-1979. In 1979, over one third of Japan's imports came from Saudi Arabia.

New Zealand's oil consumption is one of the smallest in the region. New Zealand imports around 55,000 b/d and has a small level of domestic oil production. Its dependence on the Middle East has declined from 91 percent in 1970 to 87 percent in 1977. Within the Middle East, Saudi Arabia and Iran were its major oil suppliers in 1979. Its current imports and dependence pattern is likely to continue in the future. Like Australia, New Zealand is an important exporter of meat and dairy products to the Middle East. Because of these exports and its relatively small import needs, New Zealand is not likely to face difficulties in securing access to Middle East crude.

Because the United States is the third largest producer of oil in the world and the largest consumer and importer of oil in the world, its oil import levels and dependency pattern affect the whole world. In 1979, the United States imported just under 50 percent of its consumption. Crude oil imports of 6.5 mb/d amounted to 77 percent of its imports, and the balance was imported as refined products. The United States dependence on OPEC imports has risen significantly during 1970-1979, from 48 percent to 79 percent. At the same time, dependence on Middle East oil rose 2.5 times, from 14 percent in 1970 to 32 percent in 1979. After Saudi Arabia, the largest exporters of oil to the United States are Nigeria, Venezuela, and Libya. As stated before, the United States is experiencing a radical downward shift in its demand for oil. If the decline stabilizes, however, the current dependency pattern can be expected to continue. Because of the United States' need for light African crudes, Nigeria and Libya will continue to be important sources of supply during the foreseeable future.

THE FUTURE OUTLOOK

The pattern of dependency for the Asia-Pacific nations clearly shows that, despite slight changes, the dependence on OPEC oil, and particularly Middle East oil, has remained extremely important during the 1970s. Indeed, the rising price of oil and the political instability of the Middle East, has led to little or no change in the sources of crude during the last decade.

As the Asia-Pacific nations enter the 1980s, the implications of rising prices and difficulties of access to crude are extremely serious for oil importers. The impact of rising prices on the economies of the Asia-Pacific nations is discussed in detail in Chapter 3. As for the problem of securing access to crude, it is misleading to mix the developed and developing nations of the region together. Australia and New Zealand are small importers of oil and both are well placed to secure access to oil. The United States and Japan, the two largest importers of oil in the world, will compete for access to oil using their economic, diplomatic, and--in the case of the United States--military influence to satisfy all or part of their needs. At the same time, they can fall back on the IEA or their petroleum stockpiles to withstand short-term disruptions. It is the developing nations of the regions, however, who will have to consider various options to minimize their difficulties in securing access to oil.

Future Sources of Supply

That the developing Asia-Pacific nations continued their unabated dependence on the Middle East during the 1970s was due less to their own free choice than to the fact that Middle East oil was the only oil they could obtain. At the same time, most of the importing nations had little to say regarding the source of their crude. The international oil companies that controlled the bulk of Middle East oil trade provided the importing nations with the volume demanded. Thus, the changing structure of supply sources within the Middle East in the 1970s had less to do with the deliberate choice of the importing countries than with the oil companies' decisions on oil distribution.

The picture is radically different in the 1980s. First, the declining access of the international oil companies to Persian Gulf oil implies that most developing nations will have to use their national oil companies, or direct government-to-government trade deals, to obtain crude to supplement supplies from the oil companies.

Second, given the pattern of future production and export scenarios presented earlier, declines in available exports from the Persian Gulf will seriously hurt the oil importers. To minimize the impact, <u>relatively</u> safer sources should be concentrated upon. It is the author's opinion that Iraq, Kuwait, and the UAE (in that order) are safer bets than Saudi Arabia and Iran in the medium term. It must be kept in mind that political and economic leverage must be used to obtain long-term contracts for crude supplies (nothing longer than one- to three-year supply contracts should be expected).

Third, the petroleum products market could become highly complicated for the countries that are dependent on refined products. As a result of declining crude availability, the international oil companies are beginning to pull back from their refining operations by offering their refineries for sale, lease, or partnership. Thus, the product-importing nations will begin to face difficulties in getting access to refined products. At the same time, as discussed earlier, OPEC nations' massive expansion of refining facilities at home is likely to lead to crude oil/refined products packages, both for the international oil companies and for state oil companies. This will result in difficult policy choices. Should new domestic refineries be constructed? If so, what happens if importers are forced to buy refined products with crude? Should importers continue to rely on the international oil companies to supply them with refined products from Singapore? How long is this situation going to last?

Fourth, along with the changes in crude and product supply patterns, the shifts in tanker transport patterns could cause potentially serious problems. With the independent oil tanker owners likely to be squeezed out in the 1980s, the major oil companies and OPEC will control most of the tanker fleet. With growing crude and product transport on state-to-state deals, average tanker size is likely to fall drastically to about 80,000 to 100,000 dead weight tons (dwt) for crude carriers, and 35,000 to 40,000 dwt for product carriers. Should new tankers be purchased or leased by oil importers? Should oil importers seek to obtain OPEC tanker transport before it is forced upon them? How long can they rely on traditional sources of tanker transport?

Possible Options

The preceding analysis emphasizes one point: the developing Asia-Pacific nations are going to face major difficulties both in securing access to crude oil and products and in raising funds to pay for these imports. To a considerable extent, these developments are beyond their control. The task facing them is to prepare themselves to minimize the damage to their economies. The options suggested here are to be treated as exploratory or even hypothetical. There are a number of political, economic, and sociocultural problems that such proposals create. The cost of doing nothing, however, may far outweigh the cost of taking positive action, even when other difficulties arise.

Regional Crude Purchasing Authorities. The declining availability of crude to the international oil companies, which have been the traditional suppliers of crude to the region, and the rise of state-to-state crude sales, means that individual nations are likely to become involved in direct crude purchases. Most nations' crude requirements, however, are small and individual purchases of small volumes lead to diseconomies in tanker transport. At the same time, an individual nation with small crude purchasing needs has little leverage in bargaining with Middle East oil exporters, both in securing supply contracts and in obtaining a better price. Regional groupings of a few nations in a crude purchasing authority could significantly increase the bargaining position of the member nations in obtaining a supply contract for a larger volume of crude. At the same time, such an organization could minimize tanker transport costs.

A regional tanker transport authority under the umbrella of a purchasing authority could also prove useful. Importers could lease the excess capacity of independent tanker owners at reasonable rates. Since OPEC sales of crude may be linked to transport in the OPEC tanker fleet by the late 1980s, purchase of tankers is not advisable except for small-to-medium-sized tankers that may be deemed necessary in specific regions. If regional groupings are large enough to warrant use of large tankers, advantage should be taken of the difficult situation facing very large crude carriers (VLCCs). Many VLCCs are likely to be sent to scrap yards before 1985, and oil importers may be able to buy them cheaply. It is, however, important that this regional authority allow itself a large degree of flexibility with regard to tanker transport.

Regional Refining Authority. The excess refining capacity of the international oil companies that promises to increase in the 1980s, together with the increasing need for refined products by the importing nations, provides a strong case for a regional refining authority. This authority ideally could coordinate its activities with crude purchasing and transport authorities. For the nations dependent on imports of petroleum products, there are two sets of problems to be considered: first, how to get the refined products and second, how to get the right mix of petroleum products. Clearly, the demand pattern for products is different in each of the countries concerned. The growing demand, however, for the middle distillates, particularly kerosene has meant that domestic refinery output mix is not in harmony with the demand pattern. A study of the demand and import mix of seven developing nations of the region was carried out in 1980 based on United Nations statistics.[30] Since the U.N. data do not separate diesel and gas oil from fuel oil, it is not possible to say with precision what the state of middle distillates supply and demand is. Figures for kerosene and jet fuel, however, are available from the United Nations. They show that, except for Sri Lanka, the countries have succeeded in maintaining the same proportion of kerosene and jet fuel use in 1978 as existed in 1973. The growth in demand meant, however, that domestic refining facilities could no longer provide the required kerosene and jet fuel and large imports were necessary. For instance, in Indonesia, consumption of kerosene and jet fuel doubled in five years, maintaining its 25 percent share in petroleum consumption. Imports of these products, however, rose from 8,000 tons in 1973 to 812,000 tons in 1978 to keep up with demand. Similarly, in Sri Lanka, imports of these products doubled; in Thailand, they tripled (see Tables 2.12 and 2.13).

The symptoms of the middle distillates problem can be seen clearly in these countries. First, these countries have been successful in not allowing the share of kerosene and jet fuel in total consumption to rise. This may be regarded as less the result of deliberate policy actions than of the low standard of living and the slow effects of urbanization on the consumption of these products in each period. The middle distillate problem is sure to worsen in the 1980s. Second, the imbalance between domestic demand and domestic supply of kerosene and jet fuel, which has come to the fore in three of the countries discussed, has resulted in large increases in the imports of these products. This situation will worsen in those countries that are already facing the problem and is likely to occur in others.

Kerosene demand in the developing world is a politically sensitive issue. Kerosene is the most popular commercial fuel for residential purposes, being used for heating, lighting, and

TABLE 2.12
Consumption of Petroleum Products (10^3 metric tons)

Country	Year	LPG	Percent of Total	Gasolines[a]	Percent of Total	Kerosene and Jet Fuel	Percent of Total	Residual and Distillate Fuel Oils	Percent of Total	Total All Refined Products	100%	Petroleum Consumption as of Total Commercial Energy Use
Bangladesh	1978	0		78	5	370	31	753	63	1,186	100	50.5
	1973	0		51	9	130	22	415	70	596	100	38.8
Average annual growth rate, 1973-1978										(12.2%)		
India	1978	400	2	1,526	7	3,935	19	15,114	72	20,994	100	28.2
	1973	259	1	1,643	9	3,958	22	12,262	68	18,142	100	30.9
Average annual growth rate, 1973-1978										(2.5%)		
Indonesia	1978	44	0.2	2,578	11	5,555	25	14,181	63	22,434	100	84.4
	1973	6	0.1	1,366	13	2,583	25	6,366	61	10,516	100	92.0
Average annual growth rate, 1973-1978										(13.5%)		
Nepal	1978	0		19	25	23	31	33	44	75	100	80.7
	1973	0		20	25	25	31	35	44	80	100	83.8
Average annual growth rate, 1973-1978										(-1.1%)		
Philippines	1978	175	2	1,743	18	605	6	7,121	74	9,675	100	94.1
	1973	152	2	1,927	24	591	8	5,173	66	7,875	100	96.1
Average annual growth rate, 1973-1978										(3.5%)		
Sri Lanka	1978	5	0.5	105	11	310	33	531	56	951	100	90.3
	1973	8	0.7	122	11	272	25	671	63	1,073	100	94.3
Average annual growth rate, 1973-1978										(-2.0%)		
Thailand	1978	127	1	1,436	16	900	10	6,670	73	9,133	100	94.7
	1973	75	1	1,147	16	913	13	5,120	71	7,255	100	96.6
Average annual growth rate, 1973-1978										(3.9%)		

Source: United Nations, *World Energy Supplies 1973-1978*, Series J, no. 22 (New York: 1979).

[a]Consumption figures are not disaggregated.

TABLE 2.13
Net Imports of Petroleum Products (10^3 metric tons)

Country	Year	LPG	As Percent of Consumption	Gasolines[a]	As Percent of Consumption	Kerosene and Jet Fuel	As Percent of Consumption	Residual and Distillate Fuel Oils[a]	As Percent of Consumption	All Refined Products	As Percent of Consumption
Bangladesh	1978	0		+8	13	+25	7	+80	11	+113	10
	1973	0		+20	39	+51	39	+31	7	+102	17
India	1978	0		+20	1	+885	22	+1,415	9	+2,320	11
	1973	0		+16	1	+826	21	+2,454	20	+3,296	18
Indonesia	1978	-1	2	+125	5	+812	15	+620	4	+1,556	7
	1973	-2	33	-146	11	+8	0.3	-3,967	62	-4,107	39
Nepal	1978	0		+19	100	+23	100	+33	100	+75	100
	1973	0		+20	100	+25	100	+35	100	+80	100
Philippines	1978	+25	14	-57	3	-15	2	+751	11	+704	7
	1973	-20	13	-12	1	-59	10	-293	6	-384	5
Sri Lanka	1978	0		+3	3	+45	15	+50	9	+98	10
	1973	0		0		+20	7	-10	1	+10	1
Thailand	1978	+2	2	+56	4	+25	3	+1,510	23	+1,593	17
	1973	-31	41	-40	3	+8	1	+450	9	+387	5

Source: United Nations, *World Energy Supplies 1973-1978*, Series J, no. 22 (New York: 1979).

[a] Trade figures are not disaggregated.

cooking. In the urban areas where electric lighting is available, kerosene is used for heating and cooking. In the rural sector, the use of kerosene increases as traditional fuels are abandoned. Also, increases in the level of income per head does not significantly decrease kerosene use. For instance, in Iran a rise in income per capita from $200 in 1960 to $2,000 a year in 1978 led in fact to a higher percentage of kerosene consumption in the total energy demand.[31] As incomes increase, more of the rural and urban population abandon traditional fuels and shift to kerosene. Given this fuel's widespread use, kerosene prices are politically sensitive. There have been many instances of civil disobedience and strikes in Asia after a kerosene price increase. Many governments, therefore, have been reluctant to raise kerosene prices.

Traditionally, the growth in kerosene demand has been handled by: (1) modification of existing refineries by installing high-cost isomerization units to increase middle distillate yield, (2) construction of new complex refineries that maximize middle distillate output, and (3) importation of the middle distillates. The problem is common to most LDCs, even OPEC members (Iran and Indonesia). In the past, the international oil companies used their refining facilities in places such as Singapore to match supply with demand. If there was a surplus of a product, they would move it to where it was needed. In this way, the oil companies enjoyed a flexibility that is now absent from the system because of their loss of crude supplies.

A regional refining authority could attempt to replace this flexibility. To do so, it is probably not advisable to construct costly refining facilities at home. This would not eliminate the problem but simply postpone it. It is also not advisable to construct new refineries on a regional basis, since refining centers in Asia, such as Singapore, are now available. The rational course of action seems to be to lease the excess capacity of Singapore's refining facilities through the so-called processing deals. If necessary, the authority might go into partnership with some of the refinery owners. The idea is to return flexibility to the system. In general, heavy investments for construction or purchase of refineries may prove to be a mistake, in view of the expectation that, by 1990, one third of the world petroleum trade may be in products refined outside of Asia.[32] Thus, the smaller the financial commitment, the better it is for the countries concerned. The issues are control and coordination, not ownership.

Emergency Sharing Programs. It is not practical to assume that the regional authorities could exercise total monopoly over oil flows into the importing countries. Clearly, other factors, including oil companies and state-to-state deals with oil exporters, may well play an important role. In such situations, the authority may consider oil import sharing programs along the lines prepared by the International Energy Agency for times of crisis, that is, when one nation's imports fall below a certain percentage point. Generally, when emergencies arise, the international oil companies declare *force majeure* and cut off supplies from nations which they consider unimportant. Regional authorities could prove to be a useful buffer in such circumstances.

Bargaining with OPEC. The developing nations can exert tremendous political pressure on the OPEC nations to obtain concessions. So far, such political muscle has not been fully exercised, as evidenced by OPEC's weak economic assistance programs. Regional authorities representing the poor nations could bargain effectively with OPEC nations, especially those in the Persian Gulf. They could seek price discounts, OPEC investments in the development of their oil and gas resources through the World Bank or individually, and long-term supply contracts. Even without regional authorities, many developing Asian nations are in a good position to bargain with OPEC nations with their skilled labor and technical know-how. In 1979, over $1 billion was remitted to South Korea through export of labor and technology in the Middle East. As the large Western labor force in the Persian Gulf continues to make an impact on the lifestyle and traditional values of the Islamic nations, many of the Persian Gulf nations find it preferable to replace them with foreign labor from the developing world. Indeed, preference is now given to Koreans, Filipinos, Indians, and Pakistanis over Americans and Europeans. This flow is likely to continue to grow and may prove to be a useful and long-term link for state-to-state arrangements.

Regional authorities need not necessarily involve political or economic communities. The authorities could be based on a loose association of the organizations within each country that are responsible for petroleum supplies. There is also no particular need to fit such authorities within the already established regional economic groupings, some of which have been bogged down for a variety of other reasons. Indeed, a recent ESCAP report recommends new regional arrangements for oil purchases, refining, and transportation.[33]

CONCLUSION

The oil market of the 1980s will continue to undergo structural changes that stem from the price shocks of 1973 and 1979. There is likely to be a continuing reduction in available oil from OPEC members, while non-OPEC exports are not likely to add significantly to the volume of traded oil. This decline in the future availability of crude oil has to be assessed within the context of declining demand for oil in the industrial world. It is not yet clear whether the decline in demand will continue in the 1980s, or whether the demand is going to stabilize. In any case, demand for oil by the developing world is expected to continue to grow in the next two decades.

The nations of the Asia-Pacific region will continue to be major oil importers in the 1980s and beyond. They will not become self-sufficient, but in fact will continue to be very dependent on the Middle East for their crude. This geographical dependency is not likely to change significantly without major measures to curtail demand and to develop alternative sources of energy. Moreover, the development of major refinery capacity in the Middle East poses an additional problem. By 1990, up to 30 percent of OPEC exports may be in refined form. This development has major implications for domestic refineries in the oil-importing nations of the region. The developing nations of the region face the added problem of imbalance between their refinery output mix and their demand pattern for petroleum products, a problem that is further complicated by OPEC's exports of refined products.

In considering the possible options to minimize the expected difficulties, it is important not to lose sight of the root of the problem--oil import dependency. Measures such as regional groupings only postpone the day of reckoning; they do not avoid it. It is thus critical that all such steps be accompanied by major efforts in production of alternative sources of energy, energy conservation, attempts at domestic oil and gas production, and well-formulated policies on domestic energy pricing.

The price of oil is expected to rise again from 1982/1983 onwards. The extent of the increase, however, will depend greatly on demand management in the oil-importing nations, and on the economic and political situation in the Middle East. The impact of these price increases on the economic growth, balance of payments position, and energy demand in the Asia-Pacific region is discussed in detail in Chapter 3.

NOTES

1. Petroleum Intelligence Weekly, 19 November 1979, pp. 7-9.
2. Morgan Guaranty Trust Company of New York, World Financial Markets, November 1979, pp. 1-19.
3. World Bank, World Development Report (Washington, D.C.: World Bank, 1979).
4. Fereidun Fesharaki, Revolution and Energy Policy in Iran (London: Economist Intelligence Unit, 1980).
5. See World Financial Markets (monthly).
6. Fereidun Fesharaki, "World Oil Availability: The Role of OPEC Policies," Annual Review of Energy 6 (1981): 267-308.
7. Fereidun Fesharaki, "Global Petroleum Supplies in the 1980s: Prospects and Problems," OPEC Review 4, no. 2 (1980): 27-49.
8. Fereidun Fesharaki, Analysis of Petroleum Sector Development Plans and Options of Major Oil Exporters (New York: United Nations Energy Division, 1980).
9. See U.S. Central Intelligence Agency, Office of Economic Research, The World Oil Market in the Years Ahead (ER 79-10327U), August 1979; and U.S. Central Intelligence Agency, Some Perspectives on Oil Availability for Non-OPEC LDCs (ER 80-10493), September 1980.
10. World Energy Outlook, Exxon Background Series (New York, December 1980).

11. International Energy Agency, Energy Policies and Programmes of IEA Countries: 1979 Review (Paris: Organization for Economic Cooperation and Development, 1980).

12. Congressional Budget Office, The World Oil Market in the 1980s: Implications for the United States (Washington, D.C.: U.S. Government Printing Office, May 1980).

13. Office of Technology Assessment, World Petroleum Availability 1985-2000 (Washington, D.C.: U.S. Government Printing Office, October 1980).

14. John Sawhill, "U.S. Assessment of the Free World's Dependence on Persian Gulf Oil in the 1980s," testimony to the U.S. Senate Foreign Relations Committee, Washington, D.C., March 1980.

15. S. A. R. Kadim and A. Al-Janabi, "Domestic Energy Requirements of OPEC Member Countries," in OPEC Papers (Vienna: OPEC Secretariat, 1981).

16. See Fesharaki, "Global Petroleum Supplies," p. 36.

17. Ibid.

18. See notes 10 to 14.

19. See note 15.

20. James Eckland, "The CIA's View of World Oil Outlook," in Symposium Proceedings, California and World Oil: The Strategic Horizon (Sacramento, Ca.: California Energy Commission and the University of Southern California Institute of Politics, 1981).

21. Fereidun Fesharaki, "The Future of OPEC Policy," in Symposium Proceedings, California and World Oil: The Strategic Horizon (Sacramento, Ca.: California Energy Commission and the University of Southern California Institute of Politics, 1981).

22. U.S. Central Intelligence Agency, National Foreign Assessment Center, International Energy Statistical Review (bimonthly).

23. See U.S. Central Intelligence Agency, The World Oil Market in the Years Ahead; see also notes 12 and 13.

24. United Nations and World Bank, Considerations Affecting Korea's Energy Objectives, Policies and Strategy, vol. 1 (New York: United Nations and World Bank, 1981).

25. See Petroleum Intelligence Weekly, 25 February 1980, pp. 3-4; OPEC Secretariat, Annual Statistical Bulletin (Vienna: OPEC Secretariat, 1980); OPEC Secretariat, "A Statistical Approach To Analyze the Evolution of Major Oil Companies' Control of the World Market," unpublished (August 1980); J. H. Mohnfeld, "Changing Pattern of Trade," Petroleum Economist 67, no. 8 (August 1980): 329-332.

26. Fereidun Fesharaki and David T. Isaak, "The Emerging Product Market," Petroleum Intelligence Weekly, supplement, 22 June 1981, pp. 1-7.

27. Ibid.

28. Ibid. See also note 8.

29. See Fereidun Fesharaki, "Evolution of Petroleum Contracts" (Paper presented at the Workshop on Mineral Policies to Achieve Development Objectives, Resource Systems Institute, East-West Center, Honolulu, June 1980); see also note 8.

30. Richard Morse and Fereidun Fesharaki, *Assessing Alternative Resources, Technologies and Organizational Means for Meeting Rural Energy Needs: Program Report* (Honolulu: East-West Resource Systems Institute, 1980).

31. See note 4.

32. See note 26.

33. United Nations, Economic and Social Commission for Asia and the Pacific, *Short Term Economic Policy Aspects of the Energy Situation in the ESCAP Region* (Bangkok: United Nations, 1981).

3
Energy Demand and Rising Oil Prices: Economic Implications

Corazon M. Siddayao

INTRODUCTION

After the price increases of 1973-1974, oil-importing developing countries were pushed into massive current account deficits, their terms of trade worsening with the sharp increases in the costs of energy imports and the rise in the prices of imported manufactured products from the industrialized countries. Adjustment measures had to be taken. Whereas some countries were able to weather these crises without suffering slowdowns in growth, others suffered a loss in momentum in their development programs and developmental investments had to be curtailed. The new set of oil price increases in 1979-1980 reemphasizes the plight of the net-oil-importing countries (henceforth to be referred to as NOILDCs)[1] in their efforts to achieve sustained progress in economic development.

This chapter covers four major topics: (1) the value[2] of energy and the implications of rising oil prices, especially for the oil-importing developing countries of the region; (2) dependency on oil and indigenous supply in the region; (3) the constraints faced by the NOILDCs in meeting their oil requirements and the implications of such constraints; and (4) a concluding section on the prospects for adjustment by the NOILDCs. The adjustment options adopted by these developing countries in the 1970s, we will suggest, may not be feasible solutions in the 1980s and beyond. The search for and development of indigenous sources, both fossil and nonfossil, appear to be the best option over the longer term. In addition, other structural changes may be called for involving the determined efforts of industrialized and developing countries alike.

DEMAND FOR ENERGY AND THE ECONOMY

There would be no need for concern about high oil prices or even about the availability of petroleum resources if a demand for oil did not exist. Since the demand for oil is really merely a more specific aspect of the demand for energy, the issue requires an understanding of the demand for energy sources generally and of the value placed on this commodity.

Factors Influencing Demand

The demand for energy is a derived demand. Its value lies basically in the contribution energy makes to the achievement of some level of satisfaction--in this case, in the utilization of goods produced with or used with some form of energy to improve the "quality of life." When mechanical and electrical energy are substituted for human labor in fields, factories, service establishments, and even in homes and offices, the driving motivation is the desire to increase efficiency in the use of human labor and time. Underlying this motivation is the desire to derive increased satisfaction from the consumption of goods resulting from such improved productivity.

Oil is thus not wanted for itself but, as a form of energy, simply for use in the production of goods and services. Until the 1970s, however, oil was so cheaply and easily available to almost any country that it was not explicitly considered as an input, either in measuring factor productivity or in planning economic development. The price increases in the last decade have significantly altered the weight assigned in such planning to energy in general, and to oil in particular. The role of energy in the attainment of some level of aggregate and individual satisfaction is the basis for concern about the effects of rising oil prices.

The rate of growth of commercial[3] energy consumption in developing countries is usually projected to be higher than that of already developed economies. The potential for such growth is basically related to the potential for increasing the employment of electrical and mechanical energy in industrial and other economic applications. Productivity is usually linked with the use of capital equipment, and energy use is likewise linked closely to the stock of capital equipment.

Table 3.1 shows the energy end-use structure in Asia. The first part shows that whereas in South Asia the share of industry in total energy use rose from 40 to 52 percent, in East Asia the share remained at close to 60 percent. The lower part of the table shows the demand for commercial fuels in Bangladesh and the Philippines for 1978-1979, as well as sector shares of petroleum usage in these and other countries. Because transport energy use involves activity related to production--namely, the distribution of products--and because personal transport for other than nonproductive activity will be low in these countries, one can expect the proportion of energy use devoted to productive activity to be in the range of 60 to 80 percent.

As economic development proceeds, the primary sector grows less slowly than the secondary and tertiary sectors.[4] Because the secondary and tertiary sectors will tend to use more commercial rather than traditional energy, and because shifts from traditional to commercial energy occur in all sectors, the energy/gross domestic product ratio would tend to follow an upward trend in the

TABLE 3.1
Asia: Energy End-Use Structure (percent)

By Region, 1960 and 1975	Industry		Transport		Domestic[a]	
	1960	1975	1960	1975	1960	1975
All Energy						
South Asia	40	52	36	30	24	18
East Asia	58	57	31	28	11	15

By Country, 1978-1979	Industry	Transportation	Power	Agriculture	Household/Commercial
1. Petroleum					
South Asia					
Bangladesh (1978/79)	20.1	30.9	13.4	5.4	30.0
natural gas	(60.0)	(-)	(29.7)	(-)	(1.3)
India (1978)	28.7	39.5	8.1	4.6	19.1
Pakistan (1978/79)	5.2	57.2	(-)	6.0	31.2
Southeast Asia					
Indonesia (1978)	24.5	29.5	7.7	[→ 38.3 ←]	
Philippines (1979)	35.8	34.3	23.0	n.a.[b]	6.8
Thailand (1978)	17.0	44.4	23.3	8.9	6.4
2. All Energy					
Korea	37.0	10.5	(c)	3.2	49.3
3. Commercial Energy					
Bangladesh (1978/79)	61.1	17.5	(c)	3.0	18.5
Philippines (1979)[d]	44.0	19.4	-	n.a.[b]	36.6

Table 3.1 (cont'd)

Source: L. Hoffman, "Energy Demand and Its Determinants in Brazil," preliminary draft (Regensburg, Germany: University of Regensburg, 1979, cited with permission); and C. M. Siddayao, Fossil Fuel Pricing Policies in the Asia-Pacific Region, Resource Systems Institute Working Paper WP-81-3 (Honolulu: East-West Center, 1981).

[a]Includes residential, agriculture, commerce, utilities, government, and other services.

[b]Possibly included in industry data.

[c]Distributed to sectors.

[d]Estimated from electricity and petroleum sectoral consumption data.

---

early stages of economic development (see Appendix Table A.2 and compare the industrialized and developing countries).

Thus, the rate at which energy consumption will grow cannot be understood separately from an understanding of the rate at which an economy develops, including the accompanying structural changes. This does not mean that economic growth is the sole factor that determines energy consumption growth. There are other developments that could influence changes in energy consumption. One can, however, start by identifying the most important factors that determine and influence economic growth, and then pinpoint those that determine and influence energy consumption growth.

In summary, given any supply situation, the rate at which consumption of energy will change for any given period will be associated with several variables: (1) the price of energy relative to other goods and services; (2) the geographic and economic structure of the country, including the level and distribution of income, and the product and output mix of industry; (3) technical change (which affects the energy intensiveness of consumption); (4) changes in the stock of energy-consuming goods, and utilization rate; (5) population growth and age distribution; (6) consumer price and income expectations; (7) tastes and preferences (sometimes referred to as "lifestyles"); (8) public policy changes that would affect spending and savings patterns; and (9) other institutional and organizational arrangements that might affect energy use or supply. These variables are not mutually exclusive; in fact, several are highly correlated. Detailed discussion of each of these variables will not be presented here. This discussion and the relevant issues concerning proper identification and meaningful measurement of the various factors affecting energy demand are taken up in a parallel work.[5]

Analysis of the Energy-Development Link

The energy consumption/GDP ratio is a useful first approach to reviewing the dependence of a country on energy in its economic development. Table 3.2 shows how the ratios between commercial energy and gross domestic product changed over the years 1965 to 1978 in the Asia-Pacific countries. For all middle-income and certain lower-income developing countries, the trend has been upward; the same is generally true for the industrialized countries, although by 1978 both the United States and Canada show a decline relative to 1972 (see Appendix Tables A.1 and A.2 for the time series data). An historical growth of the energy/GDP ratio implies that the income elasticities of commercial energy demand are greater than unity; a declining trend--especially after 1972--implies the reverse. A relatively stable ratio, on the other hand, suggests one of close to unity. (Caution is necessary in drawing such conclusions, as will be shown shortly.)

Table 3.2 provides an indication of which countries have responded significantly to the dramatic increases in energy prices since 1973. Although there were decreases in the ratio, they were not significant, and in some cases there were increases. The relative stability and even growth in the ratios, especially in the lower-income countries, may be explained by several factors. First, most developing countries cannot substantially alter their demand structure for final goods in a short period. Second, most of their energy consumption is in the industrial sector. Third, conservation implies substitution of more energy-efficient capital goods to reduce energy consumption, and such substitution requires time to allow for new investments to be made.

While many studies project future demand for energy by analyzing the historical relationship between energy consumption and changes in the real GDP,[6] such analysis is only a starting point. To go deeper, as shown earlier, one would have to assess future changes in the determinants of energy use. A few reasons for caution are listed herein; a fuller discussion is available elsewhere.[7]

First, although energy consumption is necessary to output growth, some of its use may not necessarily be connected with productive activity. For example, the sharp rise in the ratios for South Vietnam may be considered an aberration because of the abnormal conditions prevailing in that region during the period. Second, incomplete data distort the ratio series, both by itself or in comparison to others; China is a good case in point. Obviously, poor information on GDP and/or energy data create an unrealistic picture of China's energy/GDP ratio. Third, the ratio is highly sensitive to the monetary base used, and cross-country comparisons can be misleading; conversion factors also distort the time series picture.[8]

Fourth, other very important variables need to be taken into consideration; two important variables are energy price responsiveness and structural change. Thus, a correlation between energy consumption and GDP may not be taken as causation. Too many factors are involved, and the differences among countries are

TABLE 3.2
Energy/GDP Ratios for the Asia-Pacific Region

Country	Energy Consumption per Capita (kce)[a]			Energy/GDP (per capita in kce per US$)		
	1965	1972	1978	1965	1972	1978
South and West Asia						
Afghanistan	30	37	n.a.	0.264	0.309	n.a.
Bangladesh	n.a.	32[b]	43	n.a.	0.260[b]	0.284
India	174	193	178[e]	1.415	1.403	1.180
Nepal	9	15	11[e]	0.075	0.127	0.089
Pakistan	82	177	172	n.a.	1.153	1.024
Sri Lanka	118	138	109	0.623	0.614	0.364
Iran	395	993	1,542	0.528	0.714	0.926
Southeast Asia						
Burma	49	61	60[e]	0.412	0.504	0.461
Indonesia	112	128	278	0.758	0.565	1.076
Malaysia	357	686	738	0.676	0.390	0.792
Philippines	218	287	339	0.763	0.874	0.830
Singapore	749	1,466	2,461	0.723	0.690	0.807
Thailand	130	264	327	0.588	0.873	0.802
Cambodia	46	24	n.a.	0.053	0.035	n.a.
Laos	35	82	n.a.	0.521	0.990	n.a.
South Vietnam	149	425	n.a.	0.997	3.050	n.a.

East Asia (LDCs)						
China	456	572	837	1.856	1.734	1.889
South Korea	435	796	1,359	1.605	1.764	1.754
Pacific Islands (LDCs)						
Fiji	365[c]	491	466	0.567[c]	0.578	0.373
Papua New Guinea	85[d]	253	293[e]	0.231	0.565	0.623
Industrialized Countries						
Australia	4,659	5,640	6,622	0.971	0.946	1.007
Canada	7,074	9,612	9,930	1.459	1.513	1.272
Japan	1,817	3,557	3,825	0.738	0.784	0.780
New Zealand	2,437	3,178	3,555	0.631	0.740	0.777
United States	9,176	11,617	11,374	1.529	1.619	1.430

Source: Basic data from United Nations, World Energy Supplies, Statistical Series J, nos. 19, 21, 22 (New York: 1976, 1978, 1979).

[a] kce = kilograms of coal equivalent.

[b] 1973 data.

[c] 1968 data.

[d] 1966 data.

[e] 1977 data.

n.a. = data not available.

wide. In fact, any attempt to use constant income and price elasticities of energy demand over a period of years is unrealistic and misleading.[9] Rather, elasticities should be distinguished by country, per capita income groupings, energy price levels, and over time.

Table 3.3 presents some early results of the author's work in progress. The results shown are for two periods, 1965-1978 and 1973-1978.[10] The year 1973 has been taken as a turning point.

For the period 1965-1978, the income elasticity coefficients for all but five developing countries were greater than unity.[11] Those of the industrialized countries were split, but those below unity were close to 1.0.

Estimates for the shorter period 1973-1978 showed changes both upwards and downwards. For most countries, the shift was upward. In four out of five of the industrialized countries, the income elasticity coefficients dropped; the exception was Australia. In five out of twelve developing countries where these comparisons were possible, the coefficients also were higher. Among these countries were India, Indonesia, and China. In the remainder, the value of the coefficients either dropped or were negative. Those of the non-OPEC East and Southeast Asian countries (Korea, Burma, Malaysia, the Philippines, Singapore, and Thailand) dropped, although for the net oil importers these coefficients remained above 1.0. For Burma and Malaysia, these coefficients dropped to around 0.50.

The increase in the coefficient values for Indonesia, Iran, and China reflect the capability of these oil-producing countries to satisfy the shift to commercial fuels despite the rising prices after 1973. The declines in the coefficient values for the other countries, whose growth rates remained high, may be interpreted as reflecting energy conservation efforts in production and consumption.

Estimation of the price elasticity coefficients produced less clear results. The estimated coefficients for several countries, industrialized and developing, were statistically insignificant for either the long period or the short period, or both. There were also several unexpected results, namely, positive values. In the category of statistically insignificant results were those of all net oil exporters. The statistically insignificant results during the period 1973-1978 for most of the countries affected may reflect the policies adopted by these countries during the period that depressed domestic energy prices below their true economic values. The generally low values of the coefficients that were statistically significant reflect the inability of the countries to switch to other fuels during the short period, given the oil-based technology in place.

As in the case of the energy/GDP ratio, the statistical analysis cannot provide the reason for the relations observed or the interpretation of cause and effect. It can only establish the facts of the relation, given the data. The explanation, or the meaning of the facts, of such relation must be sought elsewhere; this circumstance illustrates the need, among other things, for a technological basis for the interpretation of the results and for

TABLE 3.3
Estimates of Income and Price Elasticities
for the Asia-Pacific Region

Country/Year	Elasticity Coefficients[a] GDP (t-ratio)	Price (t-ratio)	R^2 (adjusted)	Standard Error of the Estimate
South and West Asia				
Afghanistan				
1966-1971[b]	*	-4.64	0.8577	0.10
		(-4.22)	(0.7628)	
Bangladesh				
1973-1978	3.71	-0.40	0.9997	0.03
	(18.48)	(-8.65)	(0.9995)	
India				
1965-1977	0.48	-0.08	0.9564	0.05
	(1.86)	(-2.83)	(0.9476)	
1973-1977	1.57	0.01?	1.0000	0.0008
	(126.7)	(8.18)	(1.0000)	
Nepal				
1965-1977	*	*	-	-
1973-1977	0.11	-0.09	1.0000	0.0006
	(5.54)	(-72.70)	(1.0000)	
Pakistan				
1972-1978	-1.83?	0.08?	0.9998	0.02
	(-3.18)	(3.68)	(0.9996)	
Sri Lanka				
1965-1978	0.50	-0.24	0.9391	0.07
	(1.45)	(-2.85)	(0.9280)	
1973-1978	-0.65?	-0.10	0.9988	0.02
	(-2.56)	(-2.66)	(0.9981)	
Iran				
1965-1977	1.71	*	0.9284	0.12
	(7.08)		(0.9141)	
1973-1977	2.31	0.06	1.000	0.02
	(9.90)	(1.57)	(1.000)	
Southeast Asia				
Cambodia				
1965-1972	2.60	-1.96	0.8646	0.15
	(4.12)	(-1.90)	(0.8104)	
Laos				
1965-1973	3.61	*	0.5606	0.27
	(2.59)		(0.4141)	

Table 3.3 (cont'd)

Country/Year	Elasticity Coefficients[a] GDP (t-ratio)	Price (t-ratio)	R^2 (adjusted)	Standard Error of the Estimate
Burma				
1966-1977[b]	1.27 (2.77)	-0.07 (-1.65)	0.4652 (0.3464)	0.06
1973-1977	0.49 (1.26)	*	0.9995 (0.9990)	0.03
Indonesia				
1965-1978	1.36 (3.42)	*	0.9566 (0.9487)	0.10
1973-1978	4.09 (9.69)	*	0.9986 (0.9976)	0.08
Malaysia				
1965-1978	1.38 (5.70)	*	0.9492 (0.9340)	0.08
1973-1978	0.51 (5.07)	*	0.9998 (0.9997)	0.02
Philippines				
1965-1978	1.45 (5.08)	-0.09 (-1.94)	0.9943 (0.9933)	0.04
1973-1978	1.03 (17.69)	-0.06 (-4.76)	1.0000 (1.000)	0.01
Singapore				
1965-1978	1.10 (13.28)	*	0.9942 (0.9931)	0.07
1973-1978	1.20 (4.54)	-0.07 (-1.20)	0.9260 (0.8766)	0.05
Thailand				
1965-1978	2.04 (6.03)	-0.14 (-1.46)	0.8427 (0.8141)	0.13
1973-1978	1.05 (14.82)	*	1.000 (0.9999)	0.02
East Asia (LDCs)				
China				
1965-1978	1.14 (5.96)	*	0.9647 (0.9583)	0.08
1973-1978	1.63 (9.82)	0.04? (1.[illegible])	0.9964 (0.[illegible])	0.03

Table 3.3 (cont'd)

Country/Year	Elasticity Coefficients[a] GDP (t-ratio)	Price (t-ratio)	R^2 (adjusted)	Standard Error of the Estimate
Korea				
1965-1978	1.15	-0.08	0.9718	0.05
	(12.76)	(-1.70)	(0.9667)	
1973-1978	0.83	*	0.9999	0.02
	(16.17)		(0.9998)	
Pacific Islands				
Fiji				
1968-1978	0.59	0.14	0.9825	0.10
	(2.0)	(-1.26)	(10.9781)	
1973-1978	0.87?	0.24?	0.9907	0.09
	(-1.05)	(1.24)	(0.9846)	
Papua New Guinea				
1966-1977	3.50	0.17?	0.9851	0.10
	(10.13)	(2.90)	(0.9818)	
Industrialized Countries				
Australia				
1965-1978	0.89	0.04?	0.9997	0.01
	(11.27)	(3.46)	(0.9996)	
1973-1978	2.22	0.01	1.000	0.01
	(13.46)	(1.13)	(1.000)	
Canada				
1965-1978	0.08	-0.04	0.9995	0.02
	(7.24)	(-1.69)	(0.9994)	
1973-1978	0.25	*	0.9999	0.03
	(6.25)		(0.9999)	
Japan				
1965-1978	1.17	*	0.9973	0.02
	(34.13)		(0.9968)	
1973-1978	0.52	-0.03	1.000	0.01
	(3.66)	(-1.96)	(1.000)	
New Zealand				
1965-1976	1.91	-0.02	0.9996	0.02
	(12.79)	(-1.24)	(0.9995)	
United States				
1965-1978	0.79	*	0.9994	0.02
	(4.98)		(0.9993)	

Table 3.3 (cont'd)

Country/Year	Elasticity Coefficients[a] GDP (t-ratio)	Price (t-ratio)	R^2 (adjuted)	Standard Error of the Estimate
United States (cont'd)				
1965-1973	1.22 (6.18)	-0.15 (-1.76)	0.9998 (0.9997)	0.02
1973-1978	0.35 (3.00)	-0.04 (-2.90)	1.000 (1.000)	0.01

Source: Data from United Nations and International Monetary Fund publications. Detailed approach description appears in C. M. Siddayao, "The Demand for Energy and Economic Growth: Some Measurement Issues," unpublished working paper.

[a]Autoregressive, Cochrane-Orcutt technique unless otherwise indicated, estimating the following equation:

$E = aY^bP^ce$ and $e = pe(t-1) + u$ where E = energy per capita consumption

Y = GDP per capita

P = the price of Saudi Arabian "light" crude

e = a stochastic variable

p = the autoregressive parameter

u = a new independent disturbance

[b]Ordinary least squares with a lagged price variable.

? = unexpected sign.

* = statistically insignificant result.

more detailed studies of the structural composition of various economies.

A major shortcoming of the analysis is also that substitution elasticities are not captured in the equations. Furthermore, lags in the adjustments of technology, as well as of responses to changes in the prices of fuel, are extremely long. Such adjustments could take as much as ten years, as evidenced in the transportation industry.

For purposes of understanding the importance of energy to economic growth, however, three things are clear: (1) price plays a role in improving the efficiencies of energy use in economic activities, although consumers have been shielded--at a cost--from the full impacts of world price increases by government intervention in domestic pricing; (2) the income elasticities of the countries are in general close to or greater than unity, implying a high degree of responsiveness in the relation between economic growth and commercial energy consumption; and (3) the industrializing developing countries have higher GDP elasticities than the industrialized countries during recent years--their years of high growth also--implying that, all other major factors remaining unchanged, a significant drop in energy consumption can only be accommodated with a lower target in economic growth.

OIL DEPENDENCY AND POTENTIAL AVAILABILITY

Regional Dependency

Table 3.4 shows the patterns of energy consumption for 1970 and 1978 globally for the three main groups--industrialized, developing, and nonmarket economies--and individually for countries in the Asia-Pacific region.

Globally, the developing countries are, on average, over two thirds dependent on oil for their energy needs, whereas the industrialized countries are only about 50 percent dependent. Both sets of countries, however, are about 75 to 80 percent dependent on oil and gas.

In the Asia-Pacific region, all but five of the countries depend on oil and gas for more than 50 percent of their primary energy sources; all but two in this group are developing countries, some producers included. Table 3.4 shows that two thirds of the 21 developing countries listed had oil dependence levels higher than the global average of 45 percent in 1978--a range of 68 to 100 percent, with 10 dependent for 80 or more percent of their energy requirements on oil.

In 1979, the patterns remained essentially unchanged, with the exception of marked increases in the share of natural gas in Afghanistan and Indonesia, and a significant decline in natural gas use in Iran. Table 3.5 shows that consumption of oil in the 14 countries with only 60 percent dependence grew at rates ranging from 1.8 percent (Burma) to 15.8 percent (Indonesia), with some countries showing declining rates (Nepal and Sri Lanka). China's consumption of oil grew at 19.3 percent during the period. Except for Japan, which was three-quarters dependent on oil, the developed countries of the region were, on the other hand, relatively less dependent on oil than the developing countries. Consumption grew between 1970 and 1978 at 2 to 4 percent.

Further comparisons of the patterns of oil and energy consumption in individual countries for the years 1970 and 1978 yield very mixed results. In South Asia, there was no generally significant change in the proportions but a significant decline in

TABLE 3.4
Asia-Pacific Patterns of Commercial Energy Consumption, 1970, 1978, 1979 (percent of total)[a]

Country	1970				1978				1979			
	Coal	Oil	Natural Gas	Hydro/ Nuclear	Coal	Oil	Natural Gas	Hydro/ Nuclear	Coal	Oil	Natural Gas	Hydro/ Nuclear
Afghanistan	28	64	...	1	20	68	4	8	21	43	30	7
Bangladesh	9	53	36	2	10	51	37	2	7	53	37	2
India	74	22	1	3	66	28	1	5	63	30	2	5
Nepal	6	90	-	4	6	81	-	13	6	81	0	14
Pakistan	12	48	37	3	6	40	49	5	6	38	48	7
Sri Lanka	1	94	-	5	1	90	-	9	0	88	0	12
Iran	2	46	51	1	2	53	44	1	2	72	25	1
Brunei	25	-	75	-	-	8	92	-	0	16	84	0
Burma	15	81	1	3	10	74	11	5	11	66	18	5
Cambodia	3	96	-	1	-	100	-	-	n.a.	n.a.	n.a.	n.a.
Indonesia	1	69	29	1	1	84	14	1	1	78	20	1
Laos	-	98	-	2	-	95	-	5	n.a.	n.a.	n.a.	n.a.
Malaysia	1	95	1	3	1	88	10	1	1	87	11	2
Philippines	<1	97	-	2	2	94	-	4	3	95	0	2
Singapore	...	100	-	-	-	100	-	-	-	100	-	-
Thailand	2	96	-	2	1	95	-	4	3	95	0	2
Vietnam	21	78	-	1	84	15	-	1	78	21	0	1
Fiji	-	100	-	-	4	96	-	-	7	93	-	-
Papua New Guinea	-	95	-	5	-	96	-	4	0	96	0	4
China	92	7	...	1	80	18	1	1	79	17	3	1
Japan	26	70	1	3	17	74	5	3	16	73	7	4
South Korea	48	52	-	...	38	61	-	1	36	62	0	1
Canada	13	52	25	10	11	49	27	13	13	47	28	12
United States	21	42	36	1	21	47	30	2	22	45	30	3
Australia	48	47	3	2	47	41	10	2	46	41	11	2
New Zealand	21	58	2	19	14	51	17	17	15	47	18	20
Developed countries	26	49	22	3	22	52	22	4	23	49	23	4
Developing countries	21	62	14	3	15	67	14	4	15	67	13	4
Centrally planned economies	59	24	16	1	54	27	18	1	52	27	20	2
World	35	43	20	2	32	45	20	3	31	44	21	3

Source: Based on United Nations, World Energy Supplies, Statistical Series J, nos. 19, 22 (New York: 1976, 1979) and United Nations, Yearbook of World Energy Statistics (New York: 1981).

[a]Details may not add up to 100 because of rounding.

.. = less than or equal to 0.5 percent.

TABLE 3.5
Asia-Pacific Commercial Energy Consumption Growth Rates, 1970-1978 (10^6 metric tons of coal equivalent)[a]

Country	Total	Coal	Oil	Natural Gas
Afghanistan	+4.6	+0.8	+5.4	-
Bangladesh	+6.5	+8.7	+5.96	+6.9
India	+1.9	+0.4	+5.0	+12.5
Nepal	-1.7	0	-2.8	-
Pakistan	+0.5	-7.4	-2.0	+4.1
Sri Lanka	-2.1	-23.7	-2.5	-
Iran	+4.7	+7.9	+11.6	+10.6
Brunei	+23.8	-	+8.7	+26.4
Burma	+3.0	-1.7	+1.8	+38.6
Cambodia	-31.7	-	-31.1	-
Indonesia	+13.2	+5.3	+15.8	+4.5
Laos	-4.0	-	-4.2	-
Malaysia	+7.9	-2.8	+6.8	+34.5
Philippines	+4.4	+19.7	+4.0	-
Singapore	+8.5	+10.1	+8.5	-
Thailand	+6.2	+5.1	+6.1	-
Vietnam	-9.4	+7.8	-30.3	-
Fiji	+1.42	-	+1.0	-
Papua New Guinea	+11.9	-	+11.7	-
China	+7.6	+6.0	+19.3	+13.3
Japan	+3.0	-1.9	+3.8	+18.6
South Korea	+8.5	+5.6	+10.6	-
Canada	+2.9	+1.4	+2.1	+3.9
United States	+1.3	+1.2	+2.7	-1.0
Australia	+4.1	+3.8	+2.4	+19.2
New Zealand	+4.6	-0.5	+3.0	+33.6
Developed countries	+1.7	-0.05	+2.4	+1.6
Developing countries	+6.0	+2.0	+7.0	+5.8
Centrally planned economies	+4.6	+3.4	+6.3	+5.8
World	+3.0	+1.8	+3.7	+3.0

Source: Based on data in United Nations, *World Energy Supplies*, Statistical Series J, nos. 19, 22 (New York: 1976, 1979).

[a] Computed using the formula: $Y_t = YO^{e^{rt}}$ or $r = \left[\dfrac{\ln y_1}{\ln y_0}\right] \div t$.

the rate of growth of oil consumption. These declines did not necessarily match the growth rates for total energy consumed, so that in some instances a redistribution away from oil was taking place. In 1979, the most significant changes occurred in Afghanistan's use of natural gas for 30 percent of its commercial energy requirements from a share of only 4 percent in 1978.

In Southeast Asia, the market economies showed much higher growth rates for both energy and oil consumption than South Asia and the nonmarket economies of Southeast Asia. By 1978, there was no significant change in the relative importance of oil, compared to 1970, except for Indonesia and Malaysia, where marked shifts took place. Indonesia showed a shift towards greater dependence on oil consumption. Malaysia showed the reverse with natural gas share rising. In 1979, Indonesia's increased usage of natural gas produced a large change in proportions, but oil remained the source of 78 percent of the country's commercial energy.

The two rapidly industrializing NOILDCs in Asia--Singapore and South Korea--increased energy and oil consumption between 1970 and 1979 at an annual rate of 9 percent and 11 percent, respectively. Korea relied on coal for about 38 percent of its energy in 1978, and on oil for only 61 percent; Singapore was almost 100 percent dependent on oil. For Korea, however, the 1978 distribution showed a severe shift to oil from coal. In 1979, Singapore's energy use rose by 41 percent over 1978 consumption, whereas Korea's rose by 21 percent.

In summary, over the 1970-1978 period the following features can be identified: (1) an average growth rate of energy consumption of 3 percent for all developing countries in the region; (2) an average of 1.6 percent in South Asia, but a negative growth rate of 1.5 percent for oil; (3) an average energy growth rate of 9.5 percent in the developing market economies of East Asia and the Pacific and an average of 8.1 percent growth in oil consumption, with some of the difference in growth rates being accounted for by shifts to natural gas; (4) an average energy growth rate of 3.9 percent in the industrialized countries of the region outside the American continent, and a 3.1 percent oil growth rate for the same countries; and (5) growth rates in the United States of 1.3 for energy and less than 3 percent for oil, with Canada's growth rates at 2.9 percent for energy and 2.1 percent for oil. In general, the oil dependency levels either increased or remained stable, with only five countries showing slight declines.[12] These were Nepal, Malaysia, the Philippines, Australia, and New Zealand.

Patterns of Consumption in Large-Scale Systems

Now, how was the petroleum consumed by these countries used in large-scale energy systems? By "large-scale systems" is meant energy usage in the following sectors: transportation, industry, and electric utilities. In certain industrialized countries, highly mechanized agriculture and mining activities may justify inclusion of these latter sectors in the category under discussion. The specification of the sectors is not really of major significance for our purposes. Besides, comparative data by sectors are generally not available in the countries under review. In the absence of such data, gasoline and fuel oil consumption are taken as indicators of the level of petroleum use in large-scale systems.

Table 3.6 shows the consumption of fuel oils[13] and gasoline[14] during the years 1970 and 1978 in South Asia, some countries in East Asia, and the industrialized countries in the region. This table shows the relation of each product type to total energy consumed and the growth rates for both products over the period 1970-1976. The bulk of oil consumption in LDCs is in large-scale uses.

In general, fuel oil consumption grew faster in the developing countries of the region than in the industrialized countries (see particularly Japan, Canada, and Australia). In three South Asian countries, negative growth rates for fuel oils coincided with positive, though small, GDP per capita growth. In two out of three cases, these negative growth rates occurred at rates slower than those of the decline in total oil consumption, and as dependence on nonfossil fuels increased. In the rest of the developing countries, positive growth rates in fuel oil consumption occurred as GDP per capita grew, even though no identifiable direct relationship could be observed. With the exception of the South Asian countries and Australia, fuel oil use grew at rates ranging from 1 to 20 percent. In several countries, however, such growth rates were slower than those for either coal or natural gas, suggesting a shift away from fuel oil.

The products under review here are those related to economic activity, whether in the production or transportation/distribution of goods, hence their importance in the total energy picture. We can, therefore, expect that if these countries are to continue to demand power and fuel to run industries and vehicles, there will be a continuing demand for petroleum or some alternative fuel.

ESCAP suggested in 1978 that the increasing costs of petroleum products may increase the use of noncommercial energy forms in developing countries of the Far East.[15] The high usage of traditional fuels (fuelwood, bagasse, and the like) in the developing countries of the region (see Appendix Table B) also suggests, however, the magnitude of the potential for an increasing, or at best stable, dependence on commercial fuels in general in the next 20 to 30 years, as economic development programs are pursued. These programs generally imply increased mechanization and the consequent additional demand for commercial fuels. Given the current petroleum orientation of industrial, agricultural, and transport technology, a marked decrease in petroleum dependence in these countries would be an unrealistic expectation.

For practical purposes, therefore, such increase in the use of traditional fuels may not be expected to be significant, in view of their limited application in the major economic activities related to large-scale economic development programs.

TABLE 3.6
Asia-Pacific Consumption of Gasoline and Fuel Oils, 1970, 1978 (Selected Countries)

Country	Annual Growth Rate (%) GDP per Capita (real) 1970-78	Fuel Oils[a] Percent of Total Energy Used 1970	Fuel Oils[a] Percent of Total Energy Used 1978	Fuel Oils[a] Annual Growth Rate 1970-78 (%)	Gasoline[b] Percent of Total Energy Used 1970	Gasoline[b] Percent of Total Energy Used 1978	Gasoline[b] Annual Growth Rate 1970-78 (%)
South Asia and West Asia							
Afghanistan	2.7[c]	31	31	5	24	21	3
Bangladesh[d]	4.08	39[d]	31	10	3[d]	3	6
India	0.9	14	20	6	2	2	2
Nepal	0.1	38	36	-2	25	21	-4
Pakistan	1.5	31	24	-3	5	6	-2
Sri Lanka	3.46	46	49	-1	15	10	-7
Iran	5.58	29	31	5	6	8	6
Southeast Asia							
Burma	1.3	37	43	5	16	18	4
Indonesia	4.9	30	67	20	14	10	8
Malaysia	4.9	72	74	8	14	9	3
Philippines	3.18	59	68	6	27	17	-2
Singapore	7.08	84	61	4	12	8	4
Thailand	4.49	67	67	6	12	15	9

East Asia							
China	4.47	-	-	-	1	3[d]	6[d]
Japan	2.23	53	50	2	7	8	5
South Korea	8.15	42	52	11	4	3	2
North America							
Canada	3.89	28	23	1	17	17	3
United States	2.29	16	19	4	18	19	2
Oceania							
Australia	1.59	26	17	-1	18	17	3
New Zealand	1.66	25	25	4	30	22	1
Fiji	6.3	60	57	1	20	21	2
Papua New Guinea	1.78	44	64	16	32	20	6

Source: Based on data in United Nations, World Energy Supplies, Series J, nos. 19, 21 (New York: 1976, 1979, and International Monetary Fund, International Financial Statistics (Washington, D.C.: various issues).

[a] Includes distillate and residual fuel oils. Data start from 1972.

[b] Includes aviation gasoline but not jet fuel.

[c] GNP data from World Bank, 1980 World Bank Atlas (Washington, D.C.: 1980).

[d] Data to 1976 only.

Potential Availability and Alternative Commercial Sources

All but five of the countries shown in Table 3.7 are heavily dependent on imports for their oil supplies. This is reflected in the third column, showing the ratio of oil consumption to total production as well as in the fourth column.

In 1978 and 1979, the bulk of oil imports into the region came from the Middle East (see Table 3.8). For the countries outside the United States, from 75 to 90 percent of oil supplies were from that region (see Table 2.13, Chapter 2).

Dependence on any commodity becomes a problem only if there is some uncertainty with regard to its supply being able to meet demand. The scarcity of petroleum supply relative to demand at any given price bears on two issues: (1) physical access as a result of nonavailability of additional resources and reserves, and (2) monetary access. If price reflects the value that society has placed on petroleum as a commodity, petroleum will continue to be available to buyers who have the purchasing power to pay that price, assuming that price is the only determinant to physical access. There are other aspects related to petroleum supply that enter the scarcity issue when it is viewed from the structure of the market and control over supply, and these issues will be dealt with in other chapters. Supply interruptions not related to price, however, also bear on the issue of petroleum dependence and the accompanying economic implications. It will be useful at this point, therefore, to consider the region's prospects for developing domestic petroleum resources and for shifting to other indigenous energy resources.

Table 3.9 shows the energy resources of the countries of the region. The future outlook of import dependence on petroleum depends on the natural endowments of the countries of the region, their levels of consumption, and the ability of these countries to switch to fuels other than petroleum.

The bulk of oil reserves in the region are found in China, Indonesia, Iran, and the United States, although other countries like Canada, Australia, India, and Malaysia have significant reserves.[16] The rest of the region is, however, poorly explored, and to date most of the production comes from onshore areas (although most new finds in Asia are offshore). Grossling argues that over 50 percent of the world's undiscovered petroleum lies in the subsurface of countries outside the centrally planned economies.[17] Of this portion, about 18 to 20 percent is to be found in South and Southeast Asia Extended[18] and about 6 percent in China. Moreover, Grossling estimates that, using an index of U.S. = 1.00, the region comprising non-OPEC South and Southeast Asia Extended holds a rank of 1.01.[19]

Access to these resources will, however, depend on several factors. One important factor recognized by those in the business of exploring for petroleum is that the nature of the market determines the search for a material, and therefore the state of knowledge concerning that material. To carry this argument further, such market developments and state of knowledge also explain the

TABLE 3.7
Asia-Pacific Liquid Fuels Consumption, 1978

Country	Volume (mtoe)[a]	Ratio to Total Energy Consumption	Ratio to Total Oil Production	Oil Imports: As Percent of Total Oil Consumption	Oil Imports: As Percent of Total Energy Consumption
South Asia					
Afghanistan	0.39	0.68	39.00	97	66
Bangladesh	1.24	0.51	-	100	51
India	21.83	0.28	1.93	48	15
Nepal	0.08	0.80	-	100	80
Pakistan	3.59	0.40	7.48	87	35
Sri Lanka	1.00	0.91	-	100	91
Iran	23.17	0.54	0.09	-	-
Southeast Asia					
Brunei	0.11	0.08	1.01	-	-
Burma	1.03	0.74	0.71	-	-
Indonesia	23.43	0.84	0.26	-	-
Malaysia	5.67	0.88	0.59	-	-
Philippines	10.06	0.94	-	100	94
Singapore	3.90	1.00	-	100	100
Thailand	9.50	0.95	95.00	99	94
East Asia					
China	91.92	0.18	0.88	-	-
Hong Kong	5.18	1.00	-	100	100

TABLE 3.7 (cont'd)

Country	Volume (mtoe)[a]	Ratio to Total Energy Consumption	Ratio to Total Oil Production	Oil Imports: As Percent of Total Oil Consumption	Oil Imports: As Percent of Total Energy Consumption
East Asia (cont.)					
Japan	222.20	0.74	396.79	99.75	74
North Korea	1.26	0.04	-	100	4
South Korea	20.89	0.61	-	100	61
<u>Oceania</u>					
Australia	26.56	0.41	1.18	15	6
Fiji	0.19	1.00	-	100	100
New Zealand	4.09	0.51	6.20	84	43
Papua New Guinea	0.57	0.95	-	100	95
Tonga	0.01	1.00	-	100	100
<u>North America</u>					
Canada	77.91	0.49	1.06	6	3
United States	801.20	0.47	1.66	40	19

<u>Source</u>: Basic data from United Nations, <u>World Energy Supplies</u>, Statistical Series J, no. 22 (New York: 1979).

[a]mtoe = million metric tons oil equivalent.

TABLE 3.8
Asia-Pacific Imports of Crude Oil and Products by Geographical Region
(10^6 metric tons)

To / From	Year	Southeast Asia		Japan		Australasia		United States	
		Volume	Percent of Total Imports	Volume	Percent of Total Imports	Volume	Percent of Total Imports	Volume	Percent of Total Imports
Middle East	1979	86.7	88	205.2	74	14.0	75	104.6	25
	1978	79.6	87	198.1	75	12.1	71	113.6	28
Africa	1979	0.6	1	0.7	(a)	-	-	123.7	29
	1978	0.3	(a)	0.5	(a)	-	-	111.2	27
South Asia	1979	0.5	1	0.5	(a)	-	-	-	-
	1978	-	-	-	-	-	-	-	-
Southeast Asia	1979	-[b]	-[b]	58.2	21	4.4	24	25.5	6
	1978	-[b]	-[b]	52.1	20	4.7	27	27.4	7
Japan	1979	0.4	-	-	-	-	-	-	-
	1978	0.5	1	-	-	-	-	0.3	(a)
Australasia	1979	-	-	0.7	(a)	-	-	0.4	(a)
	1978	0.5	1	1.8	1	-	-	0.5	(a)
Centrally planned economies	1979	9.0	9	8.0	3	-	-	-	-
	1978	10.2	11	7.7	3	-	-	0.5	-
Other	1979	0.8	1	2.3	1	0.2	1	165.6	39
	1978	0.3	(a)	2.4	1	0.3	2	155.7	38
Total imports	1979	98.0	100	275.6	100	18.6	100	419.8	100
	1978	91.4	100	262.6	100	17.1	100	409.2	100

Table 3.8 (cont'd)

Source: British Petroleum Co., Ltd., B.P. Statistical Review of the World Oil Industry (London: 1978, 1979).

[a]Less than 1 percent.

[b]The B.P. series incorrectly shows no intra-regional trade. To leave the data in this table consistent, information from other sources is not incorporated.

---

reason for variations in estimates of petroleum resources over the past fifty or so years.[20] The estimates are really estimates of the "economic petroleum" base at any point in time.

Most of the search for petroleum has to date been focused on the economic resources (or the reservoirs that have been or will be produced in the foreseeable future under known technology and expected cost/price relationships); these are fields of "small" to "super giant" size, with the very small deposits untouched.

Obviously, improved technology or economic relationships (cost/price relationships) could encourage relatively more drilling in certain areas now considered uneconomic or beyond the reach of current technological developments, and would increase the range of economically recoverable reserves. A substantial portion composed of very small deposits will, however, not be economical to produce at any price.[21]

With or without less than adequate exploration, however, reserve additions at a level that will ensure adequate supplies over the next 50 years or so may not necessarily be forthcoming even if the resources are there.[22] Technical ability already exists for the extraction of the subset of economically recoverable resources. Other factors may, however, preclude exercising the option to extract them, given technical ability. These factors may be structural--social, political, economic, or institutional--in nature. (A discussion of these factors is beyond the scope of this chapter.)

Grossling's findings, among others, are considered significantly responsible for the World Bank's program to accelerate petroleum production in developing countries. In a January 1979 report, the World Bank noted that many countries were underexplored, and that India, the Philippines, Pakistan, and Vietnam warranted an increase in exploratory activity[23] (see Figure 3.1). Although the potential resources of these countries are by no means "high" when compared to the OPEC countries, these resources are significant in the context of their levels of consumption and future needs, especially during the transition period to alternative fuels.

The countries are also trying to tap their other indigenous resources (see Table 3.9 again). For example, some non-OPEC developing countries have significant gas resources (Bangladesh, Afghanistan, India, Malaysia, Brunei, China, Australia, New

TABLE 3.9
Energy Reserves and Resources in the Asia-Pacific Region

Country	Coal[a] Reserves[h] (10^6mt)	Coal[a] Geological Resources[i] (10^6mt)	Petroleum[b,c] Reserves[h] 1/1/81 (10^6b)	Petroleum[b,c] Potential Resources Recoverable[i] (10^6b)	Natural Gas[b,c] Reserves[h] 1/1/81 (10^9cu.ft.)	Natural Gas[b,c] Potential Resources Recoverable[i] (10^9cu.ft.)	Hydro Power[d] Assessed[j] (10^6mtce)	Uranium[e] Reasonably Assured Resources (10^3metric tons U)	Uranium[e] Estimated Additional Resources (10^3metric tons U)	Shale Oil[f] Resources (10^6b)	Geothermal Power[g] Installed Generating Capacity (MW)	Geothermal Power[g] Planned Capacity (MW)
	1	2	3	4	5	6	7	8	9	10	11	12
South and West Asia												
Afghanistan	n.a.	85	84[k]	D	2,600[k]	C	111	-	-	-	-	-
Bangladesh	519	1,649	-	D	8,000 (9,000)	C	40	-	-	-	-	-
India	33,700	56,799	2,580	C	12,000	C	1,722	(29.8) [0]	(23.7 [0]	-	-	1
Nepal	n.a.	-	-	E	-	E	2,680	-	-	-	-	-
Pakistan	n.a.	1,375	197	C	15,000 (21,000)	C	646	-	-	-	-	-
Sri Lanka	-	-	-	E	-	E	29	-	-	-	-	-
Iran	193	385	57,500	B	485,000	B	230	-	-	-	-	-
Southeast Asia												
Brunei	n.a.	1	1,710	D	7,300	D	-	-	-	-	-	-
Burma	n.a.	286	30	C	180	C	1,384	-	-	2,012.8	-	-
Cambodia	n.a.	n.a.	-	C	-	D	175	-	-	-	-	-
Indonesia	1,430	3,723	9,500	C	23,500 (56,000)	B	923	-	-	-	-	-
Laos	n.a.	n.a.	-	-	-	D	277	-	-	-	-	8,000[l]
Malaysia	n.a.	75	3,000	C	15,000 (27,300)	C	140	-	-	-	-	-
Philippines	150[l]	550[l]	20	C	10	C	121	(0.3) [0]	(0) [0]	-	3	892 (1988)[m]
Singapore	-	-	-	E	-	C	-	-	-	-	-	-
Thailand	n.a.	78	0.2[n]	D	8,000 (5,000)	D	412	-	-	817	-	-
Vietnam	n.a.	3,000	-	D	-	D	1,077	-	-	-	-	-
East Asia												
China	98,883	1,438,045	20,500	C	24,500	B	8,118	-	-	27,864.7	-	-
Japan	1,006	8,641	52	D	500	C	800	(7.7) [0]	(0) [0]	-	70	100
South Korea	386	921	-	C	-	C	61	(0) [3]	(0) [0]	-	-	-
Oceania												
Australia	27,353	262,134	2,360	C	30,000 (31,300)	B	151	(289) [7]	(44) [5]	251.6	-	-
New Zealand	144	790	173	E	6,100	C	339	-	-	251.6	202	165 (1981)
Fiji	-	-	-	E	-	E	n.a.	-	-	-	-	-
Papua New Guinea	-	-	-	n.a.	-	n.a.	748	-	-	-	-	-
North America												
Canada	9,381	115,352	6,400	C	87,300	B	n.a.	(167) [0]	(392) [264]	44,030	-	-
United States	177,588	2,570,398	26,400	B	191,000	A	4,185	(523) [120]	(838) [215]	2,000,220	522	1,621

[a]World Bank, Coal Development Potential and Prospects in the Developing Countries (Washington, D.C.: October 1979).

Table 3.9 (cont'd)

[b]Data on petroleum and natural gas reserves from Oil and Gas Journal, year-end issues. Figures in parentheses for natural gas reserves from government and other sources quoted in Ernest P. DuBois, "Major Gas Reserves of Southeast Asia and Australasia: An Overview," paper presented at the Fertilizer Raw Materials Resources Workshop, East-West Center, Resource Systems Institute, Honolulu, August 20-24, 1979.

[c]Data on petroleum and natural gas resources from John Albers et al., Summary of Petroleum and Selected Mineral Statistics for 120 Countries, Including Offshore, U.S. Geological Survey Professional Paper 817 (Washington, D.C.: U.S. Government Printing Office, 1973).

[d]United Nations, Economic and Social Commission for Asia and the Pacific, "Energy Resources in the Region: Progress in Energy Development," Document NR/WGMEPP/1, Report by the Secretariat to the Working Group Meeting on Energy Planning and Programming, Bangkok, August 15-21, 1978; U.S. data from World Energy Conference.

[e]International Atomic Energy Agency and The Organization for Economic Cooperation and Development Nuclear Energy Agency, Uranium: Resources, Production and Demand (Paris: Organization for Economic Cooperation and Development, 1977). ( ) indicates less than 80 US$/kgU. [ ] indicates 80-130 US$/kgU. See also Table 4.9 in this book.

[f]John R. Donnell, "Global Oil Shale Resources and Costs" in R. F. Meyer (ed.), The Future Supply of Nature-Made Petroleum and Gas (New York: Pergamon Press, 1977).

[g]H. Christopher H. Armstead, Geothermal Energy (New York: John Wiley & Sons, 1978).

[h]The term "reserves" refers to that subset of "resources" (see note i) that not only have been identified by geological or engineering methods but can also be exploited at a profit and can be legally extracted at the time of reporting.

[i]The term "resources" refers to concentrations of a mineral discovered, undiscovered, and surmised to exist in such form that extraction is currently or potentially feasible.

[j]The basis for the hydropower figures is the annual energy available at average river flow (load factor ranging from 30 to 80 percent) taken from the WEC data, converted by the United Nations by a factor of 0.123 tons of equivalent coal per 1,000 kilowatt-hours, and multiplied by a commonly used figure of 50 years' availability. The figures are also supplemented by UN-collected data.

[k]1978 data.

Table 3.9 (cont'd)

[l]Indonesian Government, National Committee, World Energy Conference, 1978 Energy Workshop.

[m]Philippine Government, Ministry of Energy, Ten-Year Development Program 1979-1988 (Manila: 1979).

[n]January 1, 1979 data from International Petroleum Encyclopedia, 1979 (Tulsa, Oklahoma: The Petroleum Publishing Co., 1979).

n.a. = data not available.

Legend: Petroleum and natural gas resources (10^9 b of oil or 10^{12} cubic feet of gas)

A = 1,000 - 10,000
B = 100 - 1,000
C = 10 - 100
D = 1 - 10
E = 0.1 - 1
- = zero or negligible

Zealand) and hydropower potential (Nepal and India). Others are well endowed with coal resources (e.g., India, China, Australia), while still others are looking more seriously at their oil shale resources (e.g., Thailand) and geothermal resources (e.g., the Philippines, Indonesia, and New Zealand).

At the moment, natural gas use is still low, even where reserves have been identified and proved. Mainly, this is because of the huge financial costs of installing the required infrastructure to transport and distribute the gas even in the domestic markets, and the inability of these countries to raise such financing for a combination of reasons. Examples are the natural gas in Bangladesh, Pakistan, and Thailand. Thailand has completed negotiations with one private company to develop its resources, but at this writing had not made progress with the second company.

The Philippines has been accelerating development of its program to develop alternative sources of energy. In a 1980 report, the government expected to reduce oil consumption to 83 percent by 1981, and to 74 percent by 1985. It would do this by increasing the use of coal, hydroelectric power, geothermal power, nuclear power, and nonconventional sources.[24]

Although many countries have alternative energy resources potentially available for development, many problems preclude their early availability. The most serious are finance related. Estimates of the required investment for coal, natural gas, oil and power are given in Appendix Table C. These estimates include the cost of exploration, production, and related infrastructure. Other problems are also associated with early access to coal and uranium; these are discussed more fully in Chapters 5 and 6. Even if alternatives became available quickly, other

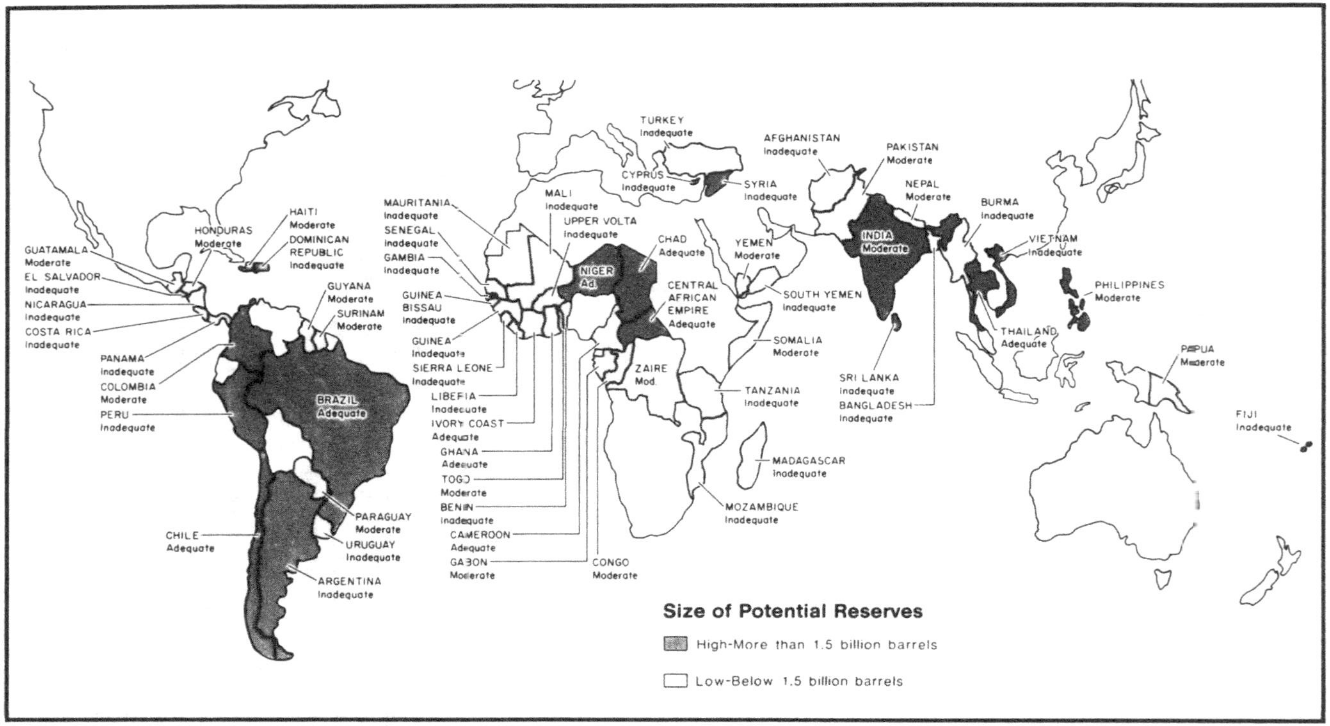

Figure 3.1. Petroleum reserves and exploration levels in less developed countries. Countries are listed with an estimate of the level of exploration activity. (Adapted from the New York Times, August 22, 1979, based on data from the World Bank.)

demand-related problems--principally the pervasiveness of oil-related technology--limit the ability to shift to nonliquid fuel alternatives.

Furthermore, notwithstanding international aid programs and promotional activities by host countries to increase energy exploration and development investments by foreign firms, mostly private, such development programs may be expected to suffer from the usual lags between investment and production. Also, until discoveries have been made and reserves proven, it is impractical to attempt to forecast how much of a country's energy consumption will be satisfied by domestic production. If countries cannot extract their own energy resources for whatever reason, the only viable--and more realistic--alternative is to count on some level of oil imports for the foreseeable future.

ENERGY DEMAND CONSTRAINTS FOR NOILDCS

Even during the relatively short time frame apparently left for the petroleum era, the main problem is not so much one of the existence of resources but <u>access</u> to those resources. The problem of access, to repeat, involves two types: (1) physical access, and (2) monetary access. The first refers to the supply side, and the second to the demand side. Any constraints affecting access from either aspect would determine the future role of oil and gas in achieving socioeconomic goals. The constraints on the supply side would affect both industrialized and developing countries insofar as availability and prices are thus influenced; constraints on the demand side would arise in both developing and industrialized countries, but probably less severely in the industrialized countries. The first constraint has been touched on in the previous section. This section will deal with the second.

The economic value of petroleum and the problems of obtaining sufficient supplies to meet demand, because of monetary or other constraints, justify the serious concern over a country's access to petroleum supplies. This concern increases with the level of a country's dependence on international petroleum, and as its balance of payments position deteriorates.

Although the net oil-importing industrialized countries must address the issue of their international payments positions in looking over their future prospects, these countries are economically stronger than any of the net oil-importing developing countries. They are, therefore, in a less precarious position over the long term insofar as maintaining the economic welfare of their citizens is concerned. The focus of the discussion that follows will, therefore, be on the adjustment options open to the Asia-Pacific NOILDCs, even while remaining aware of the external positions of the industrialized nations of the region.

The "Haves" and the "Have-Nots"

High oil prices have both desirable and undesirable implications, depending on which side of the resource picture a country is on. Table 3.10 shows the earnings of five countries in the region from their net exports of either oil or gas, and Table 3.11 shows the values of oil imports and the proportions of foreign exchange earnings from goods and services that net importers in the region must pay.

As Table 3.11 shows, petroleum exports account for a significant source of export earnings for two of three LDC exporters--Indonesia and Iran--with petroleum accounting for 97 percent of merchandise exports for Iran in 1978. These foreign exchange earnings have been the backbone of the economic development programs in both countries. In both cases, the contribution of petroleum exports to foreign exchange earnings have increased significantly between 1972 and 1978.

As for net importers, Table 3.12 shows the cost of oil imports in 1972 and 1978 from net importers in the Asia-Pacific region. The values are expressed in U.S. dollars, as well as in terms of proportions of merchandise exports and imports and proportions of total goods and services exports and imports. In certain NOILDCs of the region, oil import costs rose dramatically in the six-year period. Recognizing that these cost increases include volume increases, it is useful to compare the growth rate column in this table with that for oil consumption in Table 3.5. It is clear that import costs increased by from four to six times the rate of increase in physical consumption, some of which include domestic production (as in the United States).

Concern over the size of oil import costs appears to be justified both for the industrialized countries as well as for both the rapidly industrializing and the lower-income developing countries. In absolute amounts, five of the ten selected NOILDCs had oil import bills in the range of US$0.8 billion (Thailand) to US$2.2 billion (South Korea) in 1978. Even with Japan's GDP of US$1,000 billion, Japan's bill of US$25.7 billion is huge, since this represents over 20 percent of its export earnings. The US$42 billion import bill of the United States also explains some of the serious concerns, both within the U.S. and among other industrialized countries, because of the wide-ranging, international implications of U.S. energy policy.

The oil price increases in 1979 and 1980 doubled most of these amounts. In 1979, Bangladesh paid US$238 million, and the Philippines paid US$1.1 billion; the estimated 1980 bill of the Philippines was US$1.7 billion. In 1980, Thailand paid US$1.9 billion, and Korea paid US$5.7 billion. Japan's 1980 bill rose to US$57 billion.[25]

The rise in oil import bills is also reflected in the higher proportions they take of the foreign exchange earnings of nations in the region. This may be viewed in terms of the proportion of export merchandise earnings and in terms of earnings from the total export of goods and services and from other foreign-source incomes. The latter may be a more useful indicator of international purchasing capability, as some countries, such as the Philippines, India, Bangladesh, and Pakistan, have high "services and other incomes" exports sectors. (These countries have either a significant tourism industry or a large number of migrant labor

TABLE 3.10
Asia-Pacific Net Oil Exporters: Trade Accounts
(US$ millions, except where otherwise indicated)

Country	Total Exports		Oil or Gas Exports[a]		Exports as Percent of Total	
	1972	1978	1972	1978	1972	1978
Afghanistan	122	321	18	46	15	14
Indonesia	1,777	11,643	913	7,439	51	64
Malaysia	1,722	7,381	79	977	5	13
Canada	21,151	48,178	1,324	3,304	6	7
Iran	4,040	22,431	3,637	21,769	90	97

Source: Basic data from International Monetary Fund, International Financial Statistics Yearbook (Washington, D.C.: 1979).

[a]Oil for Indonesia and Malaysia; natural gas for Afghanistan; natural gas and crude oil for Canada; oil for Iran.

TABLE 3.11
Asia-Pacific Oil Imports and the External Accounts

Country	Oil Imports c.i.f. (US$10^6)			Oil Imports as Percent of				Oil Imports as Percent of			
				Merchandise Exports		Merchandise Imports		Goods and Services Exports		Goods and Services Imports	
	1972	1978	Growth Rates (%)	1972	1978	1972	1978	1972	1978	1972	1978
South Asia											
Bangladesh	n.a.	167	14.9[a]	n.a.	30.4	n.a.	12.1	n.a.	24.4	n.a.	9.8
India	189	1,060[b]	28.7	7.7	19.1[b]	8.5	18.7[b]	6.9	15.9	6.0	17.3[b]
Pakistan	33	408	41.9	5.1	27.7	5.0	12.5	4.4	22.0	2.8	9.9
Sri Lanka	6	153	54.0	1.8	18.1	1.6	15.8	1.6	15.3	1.4	13.7
Southeast Asia											
Philippines	140	907	31.1	12.8	27.2	10.2	17.7	9.5	18.7	8.4	14.4
Singapore	-64[c]	283[d]	31.0	2.9[c]	2.8	-	2.2	2.0[c]	2.2	-[c]	2.0[d]
Thailand	117	813	32.3	10.8	19.9	7.9	15.2	7.4	16.4	6.9	13.0
East Asia and Pacific LDCs											
South Korea	206	2,210	39.6	12.7	17.4	8.2	14.8	9.2	12.9	7.5	11.8
Taiwan	172	1,586	37.0	5.9	12.5	6.8	14.4	5.1	11.0	6.9	12.5
Fiji	16	57	21.2	20.0	28.4	1.0	16.1	10.6	25.8	5.0	14.0
Industrialized											
Australia	247	1,274	27.3	3.8	8.8	4.8	8.2	3.2	7.6	3.5	6.3
Japan	4,477	25,676	29.1	15.4	26.3	18.8	32.3	13.1	22.3	16.5	26.5
New Zealand	113	299	16.2	6.3	8.0	7.4	8.6	5.7	7.0	6.3	6.4
United States	4,300	41,598	37.8	8.7	22.9	7.3	19.0	5.6	18.9	5.4	18.2

Source: Basic data from International Monetáry Fund, *International Financial Statistics* (Washington, D.C.: monthly and annual) and *Balance of Payments Yearbook* (Washington, D.C.: annual) except for oil imports of Sri Lanka and New Zealand, which are from UN data; imports at c.i.f. values, exports at f.o.b. values.

[a] Growth rate from 1973.

[b] 19[illegible] data.

[c] Net exports.

[d] Net imports.

NOTE: Oil import cost data not available for Nepal and Papua New Guinea

n.a. = data not available.

TABLE 3.12
External Debt Position of Asia-Pacific NOILDCs[a]

Country	Current Account[b]		Total External Debt (US$10^6)		External Debt Service (US$10^6)		Debt Service as Percent of Goods Exports[c]		Debt Service as Percent of Goods and Services Exports		Percent of Debt from Private Creditors	
	1972	1978	1972	1978	1972	1978	1972	1978	1972	1978	1972	1978
South and West Asia												
Afghanistan	n.a.	n.a.	961	2,004	25.4	51.9	20.7	16.5	n.a.	n.a.	1.7	2.3
Bangladesh	-487.2[d]	-843.3	n.a.	4,274	n.a.	94.0	n.a.	17.1	n.a.	13.7	n.a.	4.9
India	-315.9	-1,739.0[e]	11,999	20,568	615.9	941.5	25.2	14.1	23.6	n.a.	4.2	1.6
Nepal	n.a.	-61.0	60	382	0.3	2.7	0.5	3.0	n.a.	n.a.	1.3	0.2
Pakistan	-285.5	-840.1	4,781	9,859	176.2	383.5	27.1	26.1	23.7	20.8	8.2	4.5
Sri Lanka	-49.3	-122.7	615	1,527	49.6	89.1	14.8	10.5	13.5	8.9	12.4	4.9
Southeast Asia												
Philippines	-101.0	-1,278.3	1,331	7,377	148.7	650.7	13.6	19.5	10.0	13.4	26.5	43.8
Singapore	-514.6	-743.7	504	1,315	33.0	302.5	1.5	3.0	1.0	2.3	24.4	61.6
Thailand	-80.3	-1,227.0	653	3,680	43.6	186.3	4.0	4.6	2.7	3.8	7.2	23.3
East Asia LDCs												
South Korea	-418.0	-1,145.6	3,925	18,146	406.2	1,795.2	25.0	14.1	18.2	10.5	49.0	54.8
Taiwan	-514.6	-591.3	1,589	4,348	134.9	632.7	4.6	5.0	4.0	6.8	23.5	53.5
Pacific Islands												
Fiji	-39.6	-41.1	46	132	1.2	9.4	1.5	4.6	0.8	4.2	12.2	15.3
Papua New Guinea	n.a.	-253.5	190	460	5.5	32.5	2.5	4.1	n.a.	n.a.	54.0	52.7

Source: Basic data from International Monetary Fund, *International Financial Statistics* (Washington, D.C.: monthly and annual); *Balance of Payments Yearbook* (Washington, D.C.: annual); and World Bank, *World Debt Tables* (Washington, D.C.: 1979).

[a] NOILDC = net oil-importing less developed country.

[b] Excludes unrequited official transfers, but includes private transfers.

[c] Exports from International Monetary Fund--goods only.

[d] 1973 data.

[e] 1977 data.

n.a. = data not available.

in overseas employment who remit earnings to their home countries, or both.) Depending on which indicator is used, oil imports in 1978 cost anywhere from an eighth to a third of the export earnings of most net oil-importing countries in the region. These proportions increased in 1979 and 1980 to at least a third of merchandise export earnings for all NOILDCs; that for Japan rose to about 40 percent.[26] Clearly, the burden becomes relatively more onerous the greater the need of a country to use its foreign exchange earnings to cover the costs of its socioeconomic development programs.

Foreign Exchange Constraints and Economic Development

The foregoing suggests the desirability of a closer analysis of the allocative impacts of rising oil prices for a net oil-importing country, especially the NOILDCs.

In the initial stages, a rise in price of oil has two main, short-run effects on an oil importer, assuming no excess demand exists: (1) substitution effects against oil, and (2) an income effect. The substitution effects may take the form of (1) shifts to other energy sources, domestically produced or imported at lower prices; (2) shifts to other factors of production, such as capital; and (3) reduction in the consumption of energy-intensive products as prices rise relative to those of less energy-intensive products. In the short run, however, the substitution effects are likely to be less than unity.

The income effect is felt in the drop of real income as a result of an increase in the price of oil. Assuming no substitution is possible, the same amount of funds will buy less oil. The same total allocation of funds will buy less of all goods. The income effect is felt both in the foreign sector and in the domestic sector: (1) in the balance of payments, because the effective purchasing power of foreign exchange earnings changes; (2) in the domestic sector, because of impacts on the cost of inputs, directly through oil import costs and indirectly through the repercussions through the system of higher oil input costs. The latter will result from higher-priced nonenergy imports reflecting passthroughs of energy-input costs. Another round of effects reflecting the higher domestic prices affects the price of export goods. If real incomes drop and real expenditures drop as a consequence, such high oil prices will have deflationary effects on both the oil-importing industrialized and the developing countries. For the NOILDC, however, such deflationary effects could result in a decline in demand for NOILDC exports to the industrialized country. The round of effects is partially summarized in Figure 3.2.

These considerations suggest a dilemma for the NOILDC. The NOILDC needs an increasing amount of commercial imported energy to pursue its economic development strategies. In the short term, as pointed out earlier, significant domestic substitutes are not expected to be available and technology limits its choice of other fuels. Furthermore, it is dependent on foreign sources for

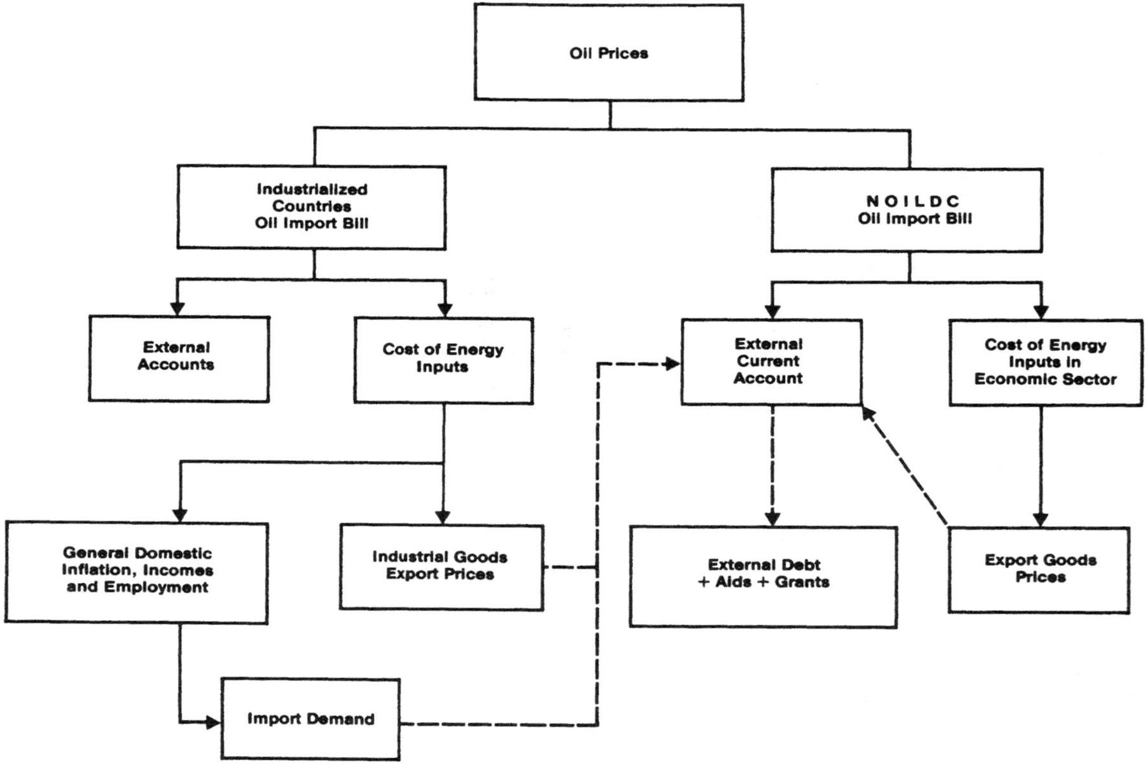

Figure 3.2. Paradigm of the impact of high oil prices on oil-poor developing countries.

development-related capital goods. A NOILDC with a given amount of foreign exchange must optimize its imports of oil and capital goods. Should it adopt a policy of maintaining its original level of oil imports and capital goods, it must either increase its earnings or resort to borrowing.

In response to the shock of drastic world oil price increases, an oil-importing country could adopt any of several policies. These could be in the form of: (1) deflationary measures, (2) import protection, (3) export subsidies, (4) devaluation, (5) development of nontraditional exports, (6) credit and tax measures, (7) foreign borrowing, and (8) revised investment plans and projects.[27] The first five would be aimed at reducing imports and increasing exports, as well as at reducing aggregate domestic demand.

The growth rates of the real value of export earnings of the NOILDCs have declined over the period 1973-1980 compared with the period 1967-1972.[28] Whereas the low-income countries' export earnings grew at an average rate of 3.4 percent during the 1967-1972 period, real growth was negative at -0.23 percent in the 1972-1979 period. In 1978, earnings dropped in real terms by -5.5 percent, and continued to drop at -1.2 percent in 1979. For the other net oil importers, real export earnings in 1973-1980 grew at 1.5 percent compared to 5.1 percent in 1967-1972. In 1980, the earnings of this group of countries dropped by -6.6 percent in real terms.

World Bank projections of export earnings by developing countries are not very encouraging. Exports of primary products (other than fuel and energy) are projected to grow at 3.3 percent annually during the period 1976-1990; for manufactured products, the rate projected is 9.0 percent annually, or an average for both of 6.1 percent. Both rates are lower than those that prevailed during the period 1960-1976.[29]

Of serious concern, then, is the continuing and significant negative current account balances of most of the NOILDCs (see Table 3.12). The sharp price increases in 1979 and 1980 would normally have to be offset over the long term by increases in the foreign exchange earnings of these countries. If this cannot be done, and should a policy be required to maintain expenditure levels unchanged, foreign borrowing would be necessary to permit the financing of a deteriorating trade balance. This would postpone the real adjustment by the country but would allow a country to avoid interruption in the growth process by using such debt to finance investment. Many Asian countries have used borrowing, mostly public, to increase national output, real growth, and exports.[30]

In response to what is sometimes referred to as "Oil Shock I" (a reference to the price increases of 1973-1974), certain NOILDCs encouraged the "export" of manpower to the capital surplus countries of the Middle East. Table 3.13 shows a comparison of overseas workers' remittances to their Asian home countries for the years 1972 and 1977. For several countries, such earnings have represented significant sources of nontraditional export earnings.

TABLE 3.13
Asia-Pacific NOILDC[a] Overseas Workers' Remittances, 1973-1977

Country	Workers' Remittances (10^6 US$)		As Percentage of Total (Goods and NFS[b])	
	1972	1977	1972	1977
Bangladesh	-0.2 (1973)	78.3	-0.1 (1973)	14.5
Fiji	1.0	n.a.	0.5	n.a.
India	118.3	616.5 (1976)	4.4	9.5 (1976)
Pakistan	130.3	866.3	17.8	60.6
Papua New Guinea	n.a.	-0.1	n.a.	...
Philippines	n.a.	207.8	n.a.	5.5
South Korea	19.5	100.4	0.9	0.8

Source: World Bank, World Tables, 2nd ed. (Washington, D.C.: 1980).

[a]NOILDC = net oil-importing less developed country.

[b]NFS = nonfactor services.

... = negligible.

n.a. = data not available.

NOTE: No data are available on the following countries: Afghanistan, Burma, Indonesia, Nepal, Singapore, Sri Lanka, and Thailand.

Table 3.12 shows, however, that most of these countries nevertheless ran negative current account balances by 1978.

Table 3.12 also shows the external public debt outstanding for 1972 and 1978 for the developing countries of South, Southeast, and East Asia.[31] This table also shows the payments on interest and principal on such debt (the debt service) for the same years, and the proportions of earnings from export goods and services that these payments take up. The country ratios of interest to us are those of export earnings, since the debt service must be paid in foreign currency. In Table 3.12 these are shown for both merchandise trade only and for trade of total goods and services; the levels are varied--with the lower income LDCs spending a relatively larger share of their merchandise export earnings making payments on their external debt.

A World Bank study showed that net oil-importing countries in the aggregate had a debt service pattern rising over the period 1973 to 1977.[32] In South and Southeast Asia, the percentages have declined from 1970 to 1977 as well as over the period 1975 to 1977.[33] These proportions may be somewhat misleading, however, with respect to the real burden of the debt over the long run. In 1978, for example, the Philippines had an outstanding public debt of over US$7 billion, and India had one of close to US$21 billion. It is true that some part of the external debt goes into increasing a nation's international reserves or to servicing some maturing debt. It is also true that countries can roll over some of their debt, and that international liquidity conditions can at certain times result in extremely low commercial rates (such as low Eurodebt rates in 1979 or so). Nevertheless, serious concerns have been voiced in the financial sector about the absolute debt levels and the debt service burden of certain "commercial" countries such as the Philippines and about the maturity structure of their debts. These concerns have been expressed in the light of prevailing and projected economic conditions.

Table 3.12 also shows how much of the debt of Asia-Pacific NOILDCs has been supplied by private creditors. Except for Afghanistan, there appears to be a considerable drop in the proportion of such loans during the period 1972-1978 in South Asia, with the reverse taking place in the middle-income countries of Southeast and East Asia. Private sector loans concluded in 1980 with the Philippines and Thailand,[34] despite the rather poor economic conditions prevailing in these countries, were seen as a reflection of the liquidity in the international market rather than of reduced concern over the future of debt payments. The trend towards concentration of commercial loans in the rapidly industrializing middle-income countries of the region, rather than being one of dispersion among both the lower and middle-income countries, confirms the notion that commercial lenders must follow certain guidelines in assuring that commercial lending risks are minimized.

Some analysts in the international organizations have tended to undervalue the severity of the external debt problems of the developing countries. It has often been pointed out that, despite the dire predictions following the 1973-1974 price increases, neither the economies or the international money markets have collapsed. Although on the aggregate, the debt service ratios of the developing countries do not appear to have deteriorated (mainly because such ratios are closely monitored by the International Monetary Fund), there is no doubt that individual countries do face problems of coping with their credit needs in the near term as well as in the long term. Such debts are a cause for concern not only among government officials but among private citizens worried about the economic--and political--futures of their respective countries. The implication of debt levels, insofar as economic conditions would affect the NOILDCs' creditworthiness and insofar as overall demand for credit may restrict liquidity in the 1980s, have been recognized in the 1980 _World Development Report_ by the World Bank. For although NOILDCs were able to cope with the 1973-1974 price increases for various reasons (including the decline in real terms of oil prices between 1974 and 1979 and the liquidity in international money markets), similar conditions mitigating the adverse effects of the recent price hikes and expected real increases in the coming years are not expected.

If these predictions turn out to be true, then the NOILDCs will find their efforts to achieve certain growth rates thwarted by two major constraints: (1) their limited indigenous petroleum resources, and (2) their limited foreign exchange sources to import such supplies.

Table 3.14 shows a partial picture of the import structure of certain Asian NOILDCs, specifically the shares of fuels imports and of machinery imports for the years 1973 and 1977. (At this writing more recent information is not available.) Although dramatic upward shifts in the share of fuel imports invariably occurred, the share spent on imports of machinery and transport remained fairly stable. This suggests that shifts may have occurred in the foods and raw materials sectors (including the imports of fertilizers and chemicals). To maintain, at the minimum, the same absolute levels of food, raw materials, and capital inputs, however, purchasing capabilities would have had to increase. Obviously, countries would not wish to reduce the imports of capital equipment because of their far-ranging implications for the development process. Table 3.12 indicates that, in the absence of such alternatives the logical approach to the dilemma of limited foreign exchange was then to increase borrowing.

One study shows elasticities of substitution of capital/energy and labor/energy to be 1.0 and 0.8, respectively, for manufacturing in certain European countries and in the United States.[35] We cannot, however, assume that the same capital/energy substitution elasticity will hold for manufacturing industries in NOILDCs, because such capital substitution would normally tend to be nonprocess related (that is, related to conservation measures). But if we can assume the same for NOILDCs, they have no choice that would reduce import costs. If they try to substitute more capital, which is imported rather than locally produced for most purposes, either way they will be affected by (1) the shortage of foreign exchange, or (2) the inflationary impacts of oil prices on imported manufactured goods. Again, a way out in the medium term is through foreign borrowing.

Also, substitution is a long-term phenomenon in the energy/capital relationship, for technical and economic reasons. For newly industrializing countries, capital equipment will be relatively new; it would be more costly to adopt more energy-efficient technology if this means discarding equipment that has not been fully depreciated or has not become nonfunctional. In the same way, depending on local conditions, it could be economically irrational to convert oil-using power generating plants to coal in the short to medium term. In the period under review, then, it must be assumed that, for the NOILDCs, capital and energy are in general complementary inputs, suggesting that financial constraints on one may reduce demand for the other.[36] The implication of such a reduction would be a magnified impact of lower energy use.[37]

The impacts of world oil price increases on the NOILDCs will vary according to the adjustment policies adopted by the affected country. The nature of such adjustment policies, however, and the

TABLE 3.14
Structure of Merchandise Imports of Asian NOILDCs[a]
(percentage shares of merchandise imports)

Country	Fuels		Machinery and Transport	
	1973	1977	1973	1977
Afghanistan	7[b]	8	14[b]	7
Bangladesh	n.a.	24	n.a.	13
Burma	6	n.a.	34	n.a.
India	10[b]	26	25[b]	19
Pakistan	8	16	24	28
Philippines	12	24	29	26
Singapore	13	26	28	26
South Korea	7	20	27	27
Sri Lanka	8	24	19	12
Taiwan	6	19	35	27
Thailand	10	22	32	30

Source: World Bank, World Tables, 1976 (Washington, D.C.: 1976); World Bank, World Development Indicators (Washington, D.C.: 1979); and World Development Report, 1980 (Washington, D.C.: 1980).

[a]NOILDC = net oil-importing less developed country.

[b]1972 data.

n.a. = data not available.

NOTE: Data on Nepal not available.

ability of a country to adopt them are highly dependent on the economic structure of such countries.[38] In this connection, the characteristics of the country's production function, but especially the relationship among the inputs (for example, of oil) and the flexibility of shifting inputs, determine the extent of the impact.[39]

Constraints on Growth Resulting from Oil-Related Inflation

Earlier, the far-ranging effects of reduced real incomes as a result of increased oil prices were pointed out.

There may be a tendency to underestimate the impacts of energy price increases on overall costs. Energy may represent a small component of the GNP in direct dollar units. The mutual dependence of the basic inputs of production through their indirect energy costs indicates, however, that the repercussions through the system may be more extensive. One conduit is the inflation multiplier.[40] Although the direct impact (or the "mechanical impact") on the cost of living index of inflation arising from oil price increases may be small, the wage price escalation triggered by this impact may result in a multiplication of the initial impact. A study of high energy prices on growth and inflation in Korea showed that a 10 percent increase in energy and other prices could slow down the manufacturing industry by -2.215 percent.[41]

Caution must, of course, be exercised in attributing to high oil prices the general inflationary spiral in a country. In analyzing the impacts of rising energy prices on the general inflation index, two critical factors must be considered: (1) the proportion of the increase in the overall index contributed by oil, and (2) the proportion of the individual consumption expenditure that is accounted for by energy price increases. A recent study points out, first, that sharply rising prices of all other goods and services in many oil-importing countries diminished the impact of the relative rise in oil prices; and, second, that some of the price increases were in fact a result of changes in the foreign exchange rates of the oil importing countries.[42]

Table 3.15a shows the real price index for imported oil computed by Dunkerley and Jankowski for four Asian countries. There are some methodological problems in their index, because the deflationary tool used (the domestic consumer price index, or CPI) already reflects oil price increases and therefore exaggerates the fall in the real price of oil. Notwithstanding such estimation problems, except for South Korea, the table shows that countries suffered serious deteriorations in their oil purchasing power. Table 3.15b shows that South Korea had a severe exchange rate depreciation that further raised the magnitude of the real increase in oil import costs. On the other hand, South Korea had the highest rate of inflation between the period 1974-1978 among the countries shown, so that, of the four developing countries shown in Table 3.15a, it appears to have suffered the least.

The role of exchange rate changes in determining the real price of imported oil also is circular in nature. Pressures on the foreign exchange are partially brought on by domestic inflationary conditions, which are partially caused by higher import costs. Changes in the exchange rate have, in certain Asian countries, been partially the result of inflationary pressures arising from a rise in the money supply, which in turn was a short-run policy response to the oil price shocks.[43] A further circular problem arises where the import of an input such as oil, for which the elasticity of substitution between imported and domestic factors of production is less than unity, results in a balance of payments deterioration.[44] The trade balance, as shown in a recent study, plays a crucial role in exchange rate determination.[45] These links between imported oil prices, domestic inflation, the balance of trade, and exchange rate determination indicate a problem with identifying the exact value of the real price of oil. To ignore this circularity is to understate the impacts on inflation of increased world oil prices.

TABLE 3.15a
Index of Real Price of Imported Oil[a]
(1974 = 100)

Country	1973	1974	1975	1976	1977	1978	1979	1980[b]
India	34	100	108	134	130	122	152	197
Japan	32	100	100	98	89	69	92	137
Philippines	37	100	109	114	116	113	119	155
South Korea	27	100	82	76	75	67	76	116
Thailand	35	100	104	108	108	102	124	164
United States	31	100	101	102	103	98	118	156

Source: Joy Dunkerley and John E. Jankowski, Jr., "The Real Price of Imported Oil," The Energy Journal 1, no. 3 (June 1980): Table 1. Data are from the International Monetary Fund, International Financial Statistics; and Organization for Economic Cooperation and Development, Energy Balances of OECD Countries and Workshop on Energy Data of Developing Countries, 1978.

[a]The indices were derived by converting the dollar price of Saudi Arabian (Ras Tanura) oil to national currencies at current exchange rates. The resulting price of oil expressed in national currencies was then adjusted for inflation in each country by deflating by the consumer price indices. The real price index was calculated from the latter set of numbers.

[b]Partially estimated.

TABLE 3.15b
Percentage of Real Price Change Due to Nominal Price Rise, Exchange Rate, and Inflation Effects[a]

Country	1973-1974				1974-1978				1978-January 1980			
	Total Percent Change in Real Price of Imported Oil	Percent Point Changes Represented by			Total Percent Change in Real Price of Imported Oil	Percent Point Changes Represented by			Total Percent Change in Real Price of Imported Oil	Percent Point Changes Represented by		
		Rise in $ Price of Oil	Rate of Exchange with US$	Inflation Rate		Rise in $ Price of Oil	Rate of Exchange with US$	Inflation Rate		Rise in $ Price of Oil	Rate of Exchange with US$	Inflation Rate
India	196	267.5	10.8	-82.3	22	30.8	1.3	-9.9	62	87.8	-4.4	-21.4
Japan	212	270.5	17.2	-75.7	-31	25.5	-31.5	-25.0	100	95.1	18.9	-14.0
Philippines	172	262.1	1.0	-91.2	13	4.0	12.5	-3.5	37	89.6	1.0	-53.5
South Korea	276	414.0	6.9	-144.9	-33	70.0	46.9	-149.9	73	97.3	28.5	-52.8
Thailand	187	259.4	-2.6	-69.8	2	31.1	-0.2	-28.9	61	89.5	0.6	-29.1
United States	226	261.5	*	-35.5	-2	36.8	*	-38.8	59	89.3	*	-30.3

Source: Dunkerley and Jankowski, "The Real Price of Imported Oil." Based on data from International Monetary Fund, International Financial Statistics (various issues).

[a]Partially estimated for some countries.

*Not applicable.

Percent change in the real price of oil is due to the inflation effect + price effect + exchange rate effect defined as:

Inflation effect = $(P_{NF} \div I_N) - P_{NF}$

Price effect = $(P_{ND} - P_{OD})\ [(E_{OF} + E_{NF}) \div 2]$

Exchange rate effect = $[(P_{OD} + P_{ND}) \div 2]\ (E_{NF} - E_{OF})$

Where: P_{OFR} = Real price of "old" oil in the foreign currency: $P_{OFR} = P_{OD} \cdot E_{OF} \div I_O$

P_{NFR} = Real price of "new" oil in the foreign currency: $P_{NFR} = P_{ND} \cdot E_{NF} \div I_N$

P_{OD} = Old $ price of oil.

P_{ND} = New $ price of oil.

E_{OF} = Old exchange rate.

E_{NF} = New exchange rate

P_{NF} = New foreign currency price of oil.

I_o = CPI of base year.

I_N = CPI of the new period related to the base year.

The issue of inflation transfer from industrialized to developing countries also gets obscured at times in the discussions, where people argue that inflation only gets transferred internationally through monetary policies. Such arguments ignore the effect of an open economy and the weight of the import price index on the total price index of a country. To a debtor country, world inflation has meaning insofar as its capacity to import capital goods and other important production inputs like energy is affected. The capacity to import is a function of the behavior of export volumes and prices, of capital flows (loans, aids, grants, and investments), and of import volumes and prices. An UNCTAD group argues that the burden of debt can be judged only relative to the current account as a whole, and that the terms of trade ratio--or the movements of export and import prices--are of critical importance in determining whether or not a debtor country benefits from the impacts of inflation on the debt burden.[46]

Table 3.16, taken from the UNCTAD study, shows the debt burden for several Asian NOILDCs for the years 1973 and 1975. The benefit index column shows that by 1975, despite inflation but because of the nature of the current account balances of the countries concerned, the debt burden of four countries--Afghanistan, the Philippines, Sri Lanka, and Thailand--worsened. Because inflation, oil prices, the current account balance, and the debt burden are all so closely intertwined, the main cause of the problem is difficult, if not impossible, to identify. It is clear, however, that the only winners are those who are able to find ways to reduce the impacts of the increased oil prices. Uncertainty in future terms of trade because of the world market performance of LDC exports implies a further deterioration in the economic welfare of NOILDCs seeking to mitigate such impacts through increased export earnings.[47]

CONCLUSION: AGENDA FOR THE FUTURE

The foregoing has shown that the implications for both industrialized and developing countries of rising oil prices are severe. Because of their weaker economic structure, however, the allocative implications for the NOILDCs are more serious. The problems the Asian NOILDCs could face in the next two decades may be gleaned from a projection of their projected foreign exchange requirements to meet oil import bills in the years 1990 and 2000, as shown in Table 3.17, should certain conditions prevail.

Projected Oil Import Costs

These projections are based on relatively conservative GDP per capita and population growth rates,[48] and incorporate--with a few exceptions--the income elasticity estimates shown in Table 3.3. They are also based on rather pessimistic changes in oil import dependency levels, as well as very pessimistic growth rates in world oil prices.[49] Furthermore, for simplicity and because development targets are assumed to hold very high priorities, so

TABLE 3.16
Inflation and the Debt Burden in the Asia-Pacific NOILDCs[a]

Country	1973		1975	
	Benefit Index[b]	Terms of Trade Ratio[c]	Benefit Index[b]	Terms of Trade Ratio[c]
Afghanistan	0.5	1.3	-1.5	-0.7
India	-0.2	0.6	1.3	2.1
Pakistan	-0.5	0.3	0.0[d]	0.8
Philippines	1.4	2.3	-4.0	-3.0
South Korea	-0.2	0.7	2.6	3.5
Sri Lanka	-0.2	0.7	-7.9	-7.1
Thailand	1.9	2.9	-3.3	2.4

Source: UNCTAD, "Some Aspects of the Impact of Inflation on the Debt Burden of Developing Countries," World Development 7, no. 2 (February 1979): 135-143.

[a]NOILDC = net oil-importing less developed country.

[b]Benefit index = ηP_x, P_m - $[1-(r + \alpha)D/P_x.X]$.

[c]Terms of trade ratio = $\frac{P_x X}{P_m M}$.

[d]Less than 0.05.

Where: D = total outstanding debt
X = volume of exports
M = volume of imports
η = elasticity
P_x = price of exports
P_m = price of imports
r = average rate of interest on D
α = average rate of amortization on D

TABLE 3.17
Projected Asia-Pacific NOILDC Oil Import Costs, 1990 and 2000[a]

NOILDC	Imports (10^6 barrels)		Import Costs (US$$10^6$) Case I[b]		Case II[c]	
	1990	2000	1990	2000	1990	2000
South Asia						
Bangladesh	41.29	78.30	2,643	9,553	3,386	17,383
India	232.10	345.17	14,854	42,111	19,032	76,628
Nepal	1.39	1.55	89	189	114	344
Pakistan	55.58	68.27	3,557	8,329	4,558	15,156
Sri Lanka	22.60	29.85	1,446	3,642	1,853	6,627
Southeast and East Asia						
Philippines	198.23	255.50	12,687	31,171	16,255	56,721
Singapore	80.02	183.56	5,121	22,394	6,562	40,750
Thailand	349.42	717.74	22,363	87,564	28,652	159,338
South Korea	872.38	1,980.79	55,832	241,656	71,535	439,735
Pacific						
Fiji	5.21	8.28	333	1,010	427	1,838
Papua New Guinea	11.14	12.62	713	1,540	914	2,802

Source: See appendix tables of this Chapter for basic data and assumptions.

[a] NOILDC = net oil-importing less developed country.

[b] Case I: Base of US$30 per barrel in 1980 projected at 7.5% annual growth to 1990 and 7.0% growth to 2000 = US$64/b in 1990 and US$122/b in year 2000.

[c] Case II: Base of US$30 per barrel in 1980 projected at 10% growth = US$82/b in 1990 and US$222/b in year 2000.

that price is ignored by policy planners, the price elasticities have been ignored in the projections.

Though attempting to project future events is always a hazardous task, this exercise was done to show what problems the Asian NOILDCs could face if very few or no changes occur to reduce existing dependence on oil imports. For India and the Philippines, given their prospects for increasing domestic oil production and their efforts to shift to coal, dependence levels below those of 1978 were used. For all other countries, 1978 import dependence levels were used.

Under the Case II price assumptions, India's import bill could rise to US$20 billion by 1990 and to about US$77 billion by the year 2000. These projections are not out of line when comparing the Case I projections of about $12.7 billion for the Philippines in 1990 and of about US$1.5 billion for Sri Lanka with actual or projected 1980 import bills of these two countries. In 1980, the estimated oil import bill of the Philippines was US$1.7 billion, amounting to 31 percent of export goods earnings.[50] Sri Lanka's trade deficit in 1980 was expected to rise to around $800 million, largely resulting from its oil import requirements.[51]

Prospects for Adjustment

Efforts to find solutions to the problems arising from increased oil prices have not always been easy, whether in the industrialized countries or in the NOILDCs. Such efforts by governments have, in some instances, been stalemated by consumer actions that caused governments to adopt pricing policies that misrepresent the true costs of energy products. In general, in the name of equity or redistribution of income, policies have been adopted to shield certain segments of society from the burden of increasing energy prices. As a result, both demand and supply responses within each country concerned have been distorted.[52]

Nevertheless, even after the 1973-1974 price increases, most NOILDCs were able to cope. The middle-income NOILDCs of the Asia-Pacific region were able to sustain growth rates close to those attained in the 1960s. Growth in the low-income NOILDCs in South Asia were, however, lower (dropping from 4.1 percent on the 1960s to 3.4 percent in the 1970s). NOILDCs were able to adjust to the huge imbalances in their balance of payments in the 1970s, through various means. These were: (1) the recycling of OPEC surpluses by the private banking system, (2) official development assistance, (3) increases in NOILDC exports of goods and services to the industrialized countries, (4) through workers' remittances from overseas employment, mainly in the OPEC countries with large balance of payments surpluses, and (5) through fiscal, monetary, and exchange rate policy changes.

Unfortunately, it is not clear that the first three adjustment measures will be as easily available in the 1980s. The borrowing record in the previous decade has shown that both debt and debt service are concentrated in a few developing countries, primarily the larger and more dynamic ones. Most middle-income countries were able to borrow from private sources. The low-income countries, with their limited financial resources and limited skilled human resources, have been more restricted in their borrowing capacity. Because of such poor creditworthiness, most of them had to rely on ODA, that is, concessional loans and grants. Since availability of the latter is limited, borrowing in the low-income countries has grown more slowly than in the middle income group.

In the 1970s, the middle-income NOILDCs have been able to borrow from the private banking sector because of the liquidity in the Eurocurrency market, in effect borrowing a share of the OPEC surpluses to finance their oil deficits. Even for these larger borrowers, however, external borrowings had grown much more slowly by 1980. The management of a country's external debt involves matching its borrowing capacity to its economic performance This is not an easy task because, as the increases in oil prices show, some of the major determinants of economic performance--and therefore of borrowing capacity--are beyond the control of the economic managers of a borrowing country. For example, exports to the industrialized countries currently constitute two thirds of their market.

The most important aspect of a country's economy that determines its ability to borrow is its future ability to pay. Its future ability to pay is largely determined by its export performance. This is not just an academic statement, nor are fears of the implications of poor export performance in the borrowing market unfounded. The deterioration of the Turkish economy and lending cutoffs by foreign creditors in 1979 are sufficient reminders of this basic economic fact. Thus, if the growth in the value of export earnings from NOILDCs is not expected to match the rate at which oil prices are expected to grow, the external debt as an adjustment option becomes less clear in the forthcoming years.

The increasing importance of private sources of financing has been accompanied by much harder overall terms. These included shorter periods of maturity, the use of variable rates of interest (so that changes in interest rates affected debt service on past loans as well), and the like.[53] The tendency for the latter and for selective lending may become stronger in the 1980s, despite expectations that a slowdown in economic activity in the industrialized countries will encourage banks to seek lending opportunities among the LDCs.[54] While the economic slowdown may favor lending to LDCs in general, the weaker of the set will be faced by prudential constraints that are expected to be more serious in the 1980s than in the years following "Oil Shock I."[55]

Furthermore, the fact remains that borrowing involves risk. Even if, as some studies suggest,[56] default by one NOILDC may not have any significant impact on the international market, for the affected country or bank such a default could be disastrous. Thus the risk is not borne by each country in isolation: Any spectacular default would affect the creditworthiness and growth prospects of other middle income countries.[57] With the huge debts and debt service obligations of many NOILDCs, it seems rational for lenders to begin to question the capacity of some countries to withstand the costs of still larger debts.

Agenda for the Coming Decades

Over the long term, then, the NOILDC will have to review its supply and demand conditions. On the supply side, the NOILDC can look to domestic energy alternatives, not only to relieve the pressure on its balance of payments but to reduce the inflationary impacts of high oil prices. It will also have to seek further increases in export earnings from nontraditional sources.

On the demand side, a reassessment of its sectoral requirements for energy (including the type of industries that have been planned for economic development purposes) and a review of its energy pricing and energy related taxation policies that influence sectoral use of energy will be required.[58]

An earlier section described the indigenous commercial resources of the countries in the region, as well as efforts by NOILDCs to develop such resources. The rise in oil prices from US$12.70 in 1978 to US$32.00 in 1980 has improved the economic tradeoffs with reference to both oil and nonoil resources.

There are, furthermore, enormous potentials for the utilization of renewable resources: (1) the use of biomass, both in traditional forms (wood and agricultural waste as solid fuel) as well as in the synthesis of biomass into liquid and gaseous fuels; and (2) the use of small-scale hydro (mini-hydro), solar, and wind power units. Moreover, in addition to the use of natural gas and coal in the conventional way, these fossil fuels could be converted to liquid forms to reduce a resource-rich country's dependence on imported liquid fuels if such an approach will save foreign exchange. This policy is being considered for Bangladesh and Pakistan. The main issues in these approaches would, of course, be their economic and social costs, and the related trade-offs.

The outward orientation of Asian developing countries has been viewed as a partial explanation of their reduced vulnerability to the impacts of "Oil Shock I," and the world recession of 1974-1975.[59] Outward orientation is associated with high export and import shares, allowing a country to reduce imports without seriously and adversely affecting the functioning of the economy. Such reductions would, however, be associated only with nonessential imports. Oil imports are essential imports, and reductions can only be acceptable if usage in the productive sectors is reduced either through a no-growth policy in energy-using activities (while achieving conservation or efficiency in use), or if domestic resources can replace imports. The previous discussions show that borrowing allowed import levels to be sustained or raised, in the absence of these options.

For the net oil-importing country pursuing economic development programs that are based on modernization (that is, mechanization) objectives, the cost of security of supply for the relevant period may be more important than the actual cost of imports or of developing more expensive but indigenous resources.

That is, development of smaller, more costly indigenous resources may be economically justified in the short term if such costs are weighed against the long-term economic costs of interruptions in imported energy supply for financial or other reasons. As development of indigenous resources requires some lead time (if the resources are present) and if shifts to other fuels are not immediately feasible, the payment of high oil import bills in the short term may be the only rational response. For the long term, however, a country that has some indigenous energy resources but limited foreign exchange resources could consider development of indigenous resources if the constraints of competing development-oriented demands on its foreign exchange pose other long-term growth problems.

The interrelated issues of demand and security of supply may be expressed in terms of the following equation:

$$NB(D,S) = TB(D) - C(D,S) - G(D,S)$$

where NB = net benefits, D = demand, S = security of supply, C = direct supply costs, G = economic costs of demand-supply gap. A government may wish to maximize the security of supply (S) to maintain economic development programs, or minimize the shortfall (G) to reduce the costs of demand-supply gaps or interruptions in supply.

The World Bank has a program for financing energy projects in developing countries during the next five years.[60] The Asian Development Bank has also taken an interest in helping nations of the region address their energy issues; it engaged in a survey of the main problems faced by the developing member countries in 1980 and has moved energy loans to a top priority.[61] It has been suggested elsewhere that petroleum exploration in developing countries be considered and debated against a broader perspective of the future needs and potential of the world community as a whole.[62] This is not necessarily a practical approach to an urgent problem. Resolutions of international approaches take time, as evidenced by the Law of the Sea negotiations.

Development of indigenous energy resources is, however, only a partial answer. For the long term, plans will have to be made to take into consideration the overall impacts of the altered structure of energy prices. Besides addressing some of the immediate needs related to oil imports, it will mean designing international commercial policies that will enable the NOILDCs to make the best possible use in the longer term of comparative advantages. Although income transfer through concessional loans is commendable, these are short-term solutions just as are all doles and loans. The best guarantee that NOILDCs will be able to face up to the problem of growth in the face of increasing energy prices is a guarantee that their economy and their exports will grow. Meeting the needs calls for innovation on the part of the NOILDCs themselves but requires at the same time more marketing opportunities. Loans must be used efficiently to increase output that will lead to exports, as well as serious and sincere efforts on the part of the industrialized countries to reduce obstacles to the flow of such exports. Furthermore, the role of foreign investment in funding indigenous resources, development programs, and increasing export-oriented output cannot be overemphasized. But an increase in that role from current levels requires an appropriate investment climate both in the receiving country and in the investor's home country.

In summary, the major issues are these: Is an accelerated indigenous energy production program the answer to coping with rising oil prices? To what extent can overall energy and trading policies in the industrialized countries help mitigate the plight of the NOILDCs? Will the needy developing countries remain sufficiently creditworthy to be attractive lending prospects? What approaches are available to the NOILDCs to reduce their vulnerabilities? What happens to these countries if their export/import conditions do not keep pace with the rise in oil prices? Fortunately, as one economist put it, "It is just as well that human ingenuity keeps a step ahead of economists' predictions."[63] That is our main hope!

NOTES

1. The term NOILDC is used in this paper strictly to refer to net oil-importing developing countries. It has a more restricted coverage than the term nonoil developing countries used in publications of the International Monetary Fund or the term non-OPEC in those of the World Bank. The latter terms include non-OPEC but net oil exporters like Mexico and Malaysia. See for example, R. C. Williams et al., International Capital Markets: Recent Developments and Short-Term Prospects, Occasional Paper 1 (Washington, D.C.: International Monetary Fund, September 1980).

2. The term value is used here to refer to the "total utility" yielded by energy. This is often referred to in economics as its value in use. Its scope is broader than what the term importance would encompass.

3. The term commercial is used in the energy literature generally to refer to forms traded in large markets, and the term noncommercial to those not falling under the former category. Strictly speaking, of course, the term is incorrectly used. The Oxford English Dictionary defines the word commercial as "trading." Those of us who come from developing countries know that firewood is traded in every sense of the word. For the purposes of this chapter, however, the terminology commonly used in the literature will be applied.

4. These sector categories are defined as follows: primary--agriculture, forestry, and fishing; secondary--manufacturing, mining, and construction; tertiary--commerce/trade, transport, and services.

5. See Corazon M. Siddayao, "The Demand for Energy and Economic Growth: Some Measurement Issues," working paper in progress.

6. See for example, U.S. Central Intelligence Agency, Office of Economic Research, The World Oil Market in the Years Ahead

(ER79-10327U), August 1979, and studies surveyed in Corazon M. Siddayao (with the assistance of Thammanun Pongsrikul), Survey of Energy Demand Elasticity Estimates in Developing Countries, Technical Memorandum TM-80-7 (Honolulu, Hawaii: East-West Resource Systems Institute, 1980).

7. See note 5.

8. For the issue on conversion factors and purchasing power parity see Irving B. Kravis, Alan Heston, and Robert Summers, International Comparisons of Real Product and Purchasing Power (Baltimore, Md.: Johns Hopkins University Press, 1978).

9. See Ali Ezzati and Frank J. P. Pinto, "An Analytical Framework to Assess the Energy Outlook in Developing Countries," in Workshop on Energy Data of Developing Countries, December 1978, vol. 1, Summary of Proceedings and Technical Papers (Paris: Organization for Economic Cooperation and Development, 1979).

10. In a few cases, data were available for shorter periods.

11. The elasticity concept is a device to indicate the degree of responsivenes of changes in one variable to changes in another. Arithmetically, it may be expressed as the ratio of percentage changes in two variables. The values of the coefficients are designated as elastic, unit elastic, or inelastic if they are greater than 1, equal to 1, or less than 1, respectively.

12. More detailed information on specific product usage for the years 1973 and 1978 is given in Table 2.14.

13. This category includes both distillate fuel oil and residual fuel oil. Distillate fuel oil includes light fuel oils such as No.1 and No. 4 diesel oil and fuel oil, and kerosene, which are used primarily for home heating, diesel engine fuel (including railroad engine fuel and fuel for agricultural machinery), and electric power generation. Residual fuel oil is the heavier oil that remains after the distillate fuel oil and lighter hydrocarbons are boiled off in refinery operations. It includes No. 5 and No. 6 oils, heavy diesel, and Bunker C oil which are used for electric power generation, industrial and commercial space heating, ship fuel, and various industrial uses.

14. This category includes aviation gasoline but not jet fuel.

15. See United Nations, Economic and Social Commission for Asia and the Pacific, "Energy Resources in the Region: Progress in Energy Development" (NR/WGMEEP/1), Report by the Secretariat to the Working Group Meeting on Energy Planning and Programming, Bangkok, August 15-21, 1978.

16. The term resources refers to concentrations of a mineral discovered, undiscovered, and surmised to exist in such form that extraction is currently or potentially feasible. The term reserves refers to that subset of resources that not only have been identified by geological or engineering methods but can also be exploited at a profit at existing or expected price levels and can be legally extracted at the time of reporting.

17. See Bernardo F. Grossling, "A Long-Range Outlook of World Petroleum Prospects" (Paper prepared for the Subcommittee on Energy of the Joint Economic Committee, Congress of the United States, March 2, 1978).

18. "Southeast Asia Extended" includes the countries in the Pacific.

19. See Bernardo F. Grossling, "The Petroleum Exploration Challenge with Respect to the Developing Nations," in *The Future Supply of Nature-Made Petroleum and Gas, Proceedings of UNITAR-IIASA Conference*, ed. R. F. Meyer (New York: Pergamon Press, 1977), pp. 57-96. In this scale, OPEC holds a rank of 0.93.

20. See J. Moody, "Where Oil and Gas Stand in the Energy and Ecology Dilemmas," *Oil and Gas Journal* 76, no. 35 (August 28, 1978), pp. 185-190; Corazon M. Siddayao, *The Supply of Petroleum Reserves in Southeast Asia: Economic Implications of Evolving Property Rights Arrangements* (Kuala Lumpur: Oxford University Press, 1980); and revisions in the estimates of U.S. petroleum reserve and resource estimate in G. L. Dolton et al., *Estimates of Undiscovered Recoverable Resources of Conventionally Producible Oil and gas in the United States--A Summary*, U.S. Geological Survey (Open File 81-192), 1981.

21. See figures 4, 5, and 7 in P. W. J. Wood, "New Slant on Potential World Petroleum Resources," *Ocean Industry* 14, no. 8 (August 1979): 55-70.

22. In a separate study, the author made a detailed study of the contractual framework and the relationship of negative influences on drilling decisions to the supply of petroleum reserves in South Asia. See Siddayao, *The Supply of Petroleum Reserves*.

23. World Bank, *A Program to Accelerate Petroleum Production in Developing Countries* (Washington, D.C.: World Bank, January 1979), p. 18.

24. Philippines Ministry of Energy, *Five-Year Energy Program, 1981-1985* (Manila, 1980), p. 56.

25. All data are from the International Monetary Fund, *International Financial Statistics* 34, no. 5 (May 1981): 1-441.

26. International Monetary Fund, *International Financial Statistics* (monthly).

27. Some of these are outlined in Bela Balassa, "Policy Responses to External Shocks in Developing Countries: A Research Strategy," in *Seminar on The Implications for Third World Countries of the Economic Slowdown in Industrialized Countries* (Paris: OECD Development Centre, December 11, 1978).

28. See International Monetary Fund, *World Economic Outlook* (Washington, D.C.: International Monetary Fund, May 1980), Appendix C.

29. World Bank, *World Development Indicators* (Washington, D.C.: World Bank, 1979), Table 3.

30. See for example, the analysis in Burnham O. Campbell, "Recycling," unpublished manuscript, June 1981.

31. See Appendix Table D for more details.

32. World Bank, *World Development Report* (Washington, D.C.: World Bank, 1979).

33. See Corazon M. Siddayao, "Regional Oil Dependency: Rising Oil Import Bills, the External Accounts, and NOILDC Development Policies," *Energy* 6 (1981).

34. "Loans for Thailand and Philippines Prosper, Despite Countries' Economic, Political Woes," Asian Wall Street Journal, 29 July 1980, p. 5.

35. James M. Griffin and Paul R. Gregory, "An Intercountry Translog Model of Energy Substitution Responses," American Economic Review 66, no. 5 (December 1976).

36. See Edward A. Hudson and Dale W. Jorgenson, "U.S. Energy Policy and Economic Growth, 1975-2000," Bell Journal of Economics and Management Science 5, no. 2 (Autumn 1974): 461-514; Ernst R. Berndt and David O. Wood, "Technology, Prices, and the Derived Demand for Energy," Review of Economics and Statistics 7, no. 1 (Spring 1975): 259-268; and J. R. Magnus, "Substitution between Energy and Non-Energy Inputs in the Netherlands 1950-1976," International Economic Review 20, no. 2 (June 1979): 463-484.

37. See William W. Hogan, "Capital Energy Complementarity in Aggregate Energy-Economic Analysis," Energy Modelling Forum Paper EMF 1.10 (Stanford, Ca.: Stanford University, 1978).

38. Some early research on this issue has been done, and the initial results are reported in Louka T. Katseli-Papaefstratiou and N. P. Marion, "Adjustment to Variations in Imported Input Prices: The Role of Economic Structure," Economic Growth Discussion Paper No. 360 (New Haven, Conn.: Yale University, August 1980).

39. See G. Fishelson, "The Effects of Restricted Energy Imports," Energy Economics 2, no. 3 (July 1980): 166-171.

40. This is a term used by the French that appears to be quite descriptive. See Jean Waelbroeck, "Energy and International Trade Issues " (Paper presented at the Workshop on Energy and the Developing Nations, sponsored by the Electric Power Research Institute and Stanford University, Palo Alto, Ca., March 18-20, 1980).

41. Chong Kiew Liew, "The Impact of Higher Energy Prices on Growth and Inflation in an Industrializing Economy: The Korean Experience," Journal of Policy Modelling 2, no. 3 (September 1980): 389-407.

42. See Joy Dunkerly and John E. Jankowski, "The Real Price of Imported Oil," Energy Journal 1, no. 3 (June 1980): 113-118.

43. This was reported at a seminar by Professor Maxwell Fry, Department of Economics, University of Hawaii, March 27, 1981.

44. See the study of Louka T. Katseli-Papaefstratiou, "Transmission of External Price Disturbances and the Composition of Trade," Journal of International Economics 10, no. 3 (August 1980): 357-375.

45. See Carlos Alfredo Rodriguez, "The Role of Trade Flows in Exchange Rate Determination: A Rational Expectations Approach," Journal of Political Economy 88, no. 6 (December 1980): 1148-1158.

46. See UNCTAD, "Some Aspects of the Impact of Inflation on the Debt Burden of Developing Countries," World Development 7, no. 2 (February 1979): 135-143; and Patrick Guillaumont, "The Impact of Declining Terms of Trade and Inflation on the Export Proceeds and Debt Burden of Developing Countries," World Development 8, no. 3 (October 1980): 763-768.

47. See J. Eaton, "The Allocation of Resources in an Open Economy under Uncertain Terms of Trade," *International Economic Review* 20, no. 2 (June 1979): 391-403.

48. The interested reader may refer to Table 2.4 for a comparison of projections.

49. Details are presented in Siddayao, "Measurement Issues."

50. International Monetary Fund, *International Financial Statistics* (monthly).

51. "Sri Lanka Meets Mixed Success in Move toward Free Economy," *Asian Wall Street Journal*, 30 October 1980, p. 3.

52. See Corazon. M. Siddayao, *Petroleum and Coal Pricing Policies*, report prepared for the Regional Energy Survey, Asian Development Bank, Manila. In press.

53. For a discussion of the problems related to borrowing, see Bahram Nowzad, "Managing External Debt in Developing Countries," *Finance and Development* 17, no. 3 (September 1980): 24-27.

54. Williams et al., *International Capital Markets*, p. 5.

55. Ibid., p. 9.

56. See note 30.

57. J. A. Holsen and Jean L. Waelbroeck, "The Less Developed Countries and the International Monetary Mechanism," *American Economic Review* 66, no. 2 (May 1976): 171-176.

58. See note 52.

59. See Bela Balassa, *The Process of Industrial Development and Alternative Development Strategies*, Essays in International Finance No. 141 (New Jersey: Princeton University, December 1980).

60. World Bank, *Energy in the Developing Countries* (Washington, D.C.: World Bank, 1980).

61. See Asian Development Bank, *Regional Energy Survey* (Manila: Asian Development Bank, 1981).

62. See Hasan S. Zakariya, "Petroleum Exploration in Developing Countries: The Need for a Global Strategy Based on Public Policy," *OPEC Review* 5, no. 1 (Spring 1981): 8-32.

63. Ashok V. Desai, "The Effects of the Rise in Oil Prices on South Asian Countries, 1972-1978," unpublished paper prepared for the National Council of Applied Economic Research, New Delhi, p. 19.

APPENDIX TABLE A.1
Per Capita Consumption of Commercial Energy: 1965-1978 (kce)

Country	1965	1966	1967	1968	1969	1970	1971	1972	1973	1974	1975	1976	1977	1978
South and West Asia														
Afghanistan	30	32	41	42	28	34	50	n.a.	n.a.	n.a.	n.a.	n.a.	n.a.	n.a.
Bangladesh	n.a.	n.a.	n.a.	n.a.	n.a.	n.a.	n.a.	n.a.	32	27	29	33	39	43
India	174	174	170	181	180	181	186	193	156	158	168	172	178	n.a.
Nepal	9	9	10	12	18	15	9	15	12	11	11	11	11	n.a.
Pakistan	n.a.	n.a.	n.a.	n.a.	n.a.	n.a.	n.a.	177	169	183	184	182	170	172
Sri Lanka	118	119	123	135	166	153	129	138	131	117	112	107	114	109
Iran	395	412	386	486	666	1,020	980	993	1,098	1,245	1,317	1,453	1,542	n.a.
Southeast Asia														
Burma	49	53	53	49	55	58	62	61	53	59	53	56	60	n.a.
Indonesia	112	103	106	118	110	120	138	128	136	151	184	223	290	278
Malaysia[a]	357	416	439	446	466	486	454	686	663	697	667	742	733	738
Philippines	218	235	246	280	290	301	307	287	314	300	316	327	342	339
Singapore	749	922	1,001	1,135	1,252	1,402	1,670	1,466	2,006	1,842	2,025	2,358	2,433	2,461
Thailand	130	126	173	200	167	247	309	264	290	275	284	306	327	327
Cambodia	46	54	53	52	78	54	24	24	n.a.	n.a.	n.a.	n.a.	n.a.	n.a.
Laos	35	45	59	31	57	98	71	82	83	n.a.	n.a.	n.a.	n.a.	n.a.
South Vietnam	149	318	452	443	526	530	433	425	407	210	n.a.	n.a.	n.a.	n.a.
East Asia (LDCs)														
China	456	493	341	443	469	548	562	572	598	631	653	675	761	837
South Korea	435	513	582	593	655	815	858	796	956	995	1,090	1,115	1,240	1,359
Pacific Islands (LDCs														
Fiji	n.a.	n.a.	n.a.	365	398	473	478	491	432	556	505	404	507	466
Papua New Guinea	76	85	103	119	124	138	167	253	267	278	278	291	293	n.a.
Industrialized Countries														
Australia	4,659	4,696	4,938	5,116	5,293	5,440	5,562	5,640	5,826	6,110	6,127	6,510	6,648	6,622
Canada	7,074	7,301	7,562	7,959	8,389	8,708	8,997	9,612	9,660	9,802	9,802	10,011	9,994	9,930
Japan	1,817	1,962	2,304	2,574	2,946	3,342	3,478	3,557	3,922	3,830	3,621	3,775	3,806	3,825
New Zealand	2,437	2,647	2,594	2,596	2,679	2,895	2,911	3,178	3,389	3,470	3,299	3,555	n.a.	n.a.
United States	9,176	9,684	9,770	10,272	10,745	11,020	11,146	11,617	11,799	11,357	10,784	11,497	11,574	11,374

Source: United Nations, World Energy Supplies, Series J, nos. 19, 21, 22 (New York: 1976, 1978, 1979).

[a]Data for Malaysia from aggregate energy consumption (million tons coal equivalent) divided by population.

n.a = data not available.

APPENDIX TABLE A.2
Ratio of Commercial Energy Consumption per Capita
to GDP per Capita: 1965-1978[a]

Country	1965	1966	1967	1968	1969	1970	1971	1972	1973	1974	1975	1976	1977	1978
South and West Asia														
Afghanistan	0.264	0.283	0.360	0.352	0.236	0.286	0.419	n.a.	n.a.	n.a.	n.a.	n.a.	n.a.	n.a.
Bangladesh	n.a.	n.a.	n.a.	n.a.	n.a.	n.a.	n.a.	n.a.	0.260	0.200	0.219	0.227	0.271	0.284
India	1.415	1.430	1.321	1.398	1.336	1.280	1.314	1.403	1.115	1.154	1.154	1.189	1.180	n.a.
Nepal	0.075	0.072	0.083	0.101	0.148	0.123	0.077	0.127	0.104	0.092	0.092	0.090	0.089	n.a.
Pakistan	n.a.	n.a.	n.a.	n.a.	n.a.	n.a.	n.a.	1.153	1.062	1.131	1.155	1.120	1.059	1.024
Sri Lanka	0.624	0.613	0.611	0.659	0.765	0.675	0.578	0.614	0.514	0.428	0.402	0.375	0.399	0.364
Iran	0.528	0.516	0.447	0.515	0.649	0.906	0.802	0.714	0.724	0.782	0.826	0.838	0.926	n.a.
Southeast Asia														
Burma	0.412	0.475	0.502	0.428	0.476	0.489	0.517	0.504	0.449	0.501	0.439	0.447	0.461	n.a.
Indonesia	0.758	0.696	0.718	0.738	0.659	0.686	0.764	0.665	0.652	0.690	0.822	0.956	0.190	1.076
Malaysia	0.676	0.764	0.793	0.793	0.753	0.776	0.697	0.990	0.881	0.880	0.853	0.885	0.829	0.792
Philippines	0.763	0.812	0.834	0.927	0.940	0.959	0.954	0.874	0.901	0.833	0.844	0.843	0.853	0.830
Singapore	0.724	0.710	0.815	0.824	0.811	0.811	0.875	0.690	0.862	0.755	0.808	0.887	0.859	0.807
Thailand	0.588	0.523	0.688	0.756	0.603	0.868	1.035	0.873	0.826	0.826	0.815	0.831	0.852	0.802
Cambodia	0.608	0.698	0.666	0.640	0.966	0.761	0.367	0.400	n.a.	n.a.	n.a.	n.a.	n.a.	n.a.
Laos	0.522	0.648	0.782	0.398	0.720	1.212	0.862	0.990	1.041	n.a.	n.a.	n.a.	n.a.	n.a.
South Vietnam	0.993	2.213	3.243	3.486	3.941	3.762	3.044	3.054	2.909	n.a.	n.a.	n.a.	n.a.	n.a.
East Asia (LDCs)														
China	1.856	1.808	1.330	1.755	1.718	1.769	1.740	1.734	1.643	1.711	1.689	1.781	1.887	1.889
South Korea	1.605	1.727	1.900	1.757	1.724	2.019	1.980	1.764	1.855	1.815	1.870	1.708	1.754	1.754
Pacific Islands (LDCs)														
Fiji	n.a.	n.a.	n.a.	0.567	0.600	0.627	0.624	0.578	0.445	0.504	0.420	0.344	0.422	0.373
Papua New Guinea	n.a.	0.231	0.275	0.316	0.319	0.333	0.373	0.565	0.545	0.589	0.583	0.636	0.624	n.a.
Industrialized Countries														
Australia	0.971	0.973	0.973	0.967	0.965	0.939	0.941	0.946	0.939	0.974	0.967	1.018	1.021	1.007
Canada	1.459	1.426	1.450	1.477	1.503	1.523	1.486	1.513	1.408	1.346	1.349	1.296	1.284	1.272
Japan	0.738	0.724	0.758	0.756	0.787	0.815	0.830	0.784	0.805	0.827	0.805	0.825	0.815	0.780
New Zealand	0.631	0.673	0.675	0.668	0.662	0.699	0.696	0.740	0.751	0.752	0.717	0.777	n.a.	n.a.
United States	1.529	1.535	1.520	1.542	1.593	1.665	1.642	1.619	1.579	1.575	1.530	1.551	1.503	1.430

Source: United Nations, World Energy Supplies, Series J, nos. 19, 21, 22 (New York: 1976, 1978, 1979).

[a] Ratio based on GDP data and per capita energy consumption from previous table. Commercial energy consumption per capita in kilograms of coal equivalent. GDP per capita in US$ at 1975 prices.

[b] Ratio for China uses GNP per capita. GNP data from U.S. Central Intelligence Agency, National Foreign Assessment Center, China: The Continuing Search for a Modernization Strategy (Washington, D.C.: April 1980).

n.a. = data not available.

APPENDIX TABLE B
Asia-Pacific Traditional Fuels as a Proportion of Total Fuel Consumption[a] (percent of total)

	Percentage of Fuelwood[b]
More than 90 percent (traditional fuels)	
Burma	85
Laos	87
Nepal	96
75-90 percent	
Bangladesh	63
Papua New Guinea	66
50-75 percent	
India	28
Indonesia	62
Malaysia	25
Sri Lanka	55
Thailand	34
Less than 50 percent	
China	9
Fiji	2
Pakistan	22
Philippines	1
South Korea	8

Source: W. Knowland and C. Ulinski, "Traditional Fuels: Present Data, Past Experience and Possible Strategies" (Washington, D.C.: Agency for International Development, 1979) as presented in Harrison Brown and Kirk R. Smith, "Energy for the People of Asia and the Pacific," *Annual Review of Energy* 5 (1980): 173-240.

[a]Total energy equals commercial energy plus the primary energy of traditional fuels.

[b]Official estimate of fuelwood as a percentage of commercial energy plus fuelwood.

APPENDIX TABLE C
Asia-Pacific Developing Countries: Estimated Investment Needs in the Energy Sector for Commercial Energy Development, 1978-1985 and 1985-1990 (US$ million, at current prices)

	1978-1985					1985-1990				
	Coal	Natural Gas	Oil	Electricity	Total	Coal	Natural Gas	Oil	Electricity	Total
I. Net Energy Exporting Countries										
Indonesia	800	1,604	14,280	1,368	18,052	958	3,280	17,974	5,205	27,417
Malaysia	-	438	1,854	3,303	5,595	-	930	2,228	1,493	4,651
Subtotal	800	2,042	16,134	4,671	23,647	958	4,210	20,202	6,698	32,068
II. Net Energy Importing Countries										
A. Low-Income Group										
Afghanistan	10	15	-	41	66	9	20	-	116	145
Bangladesh	10	1,090	36	1,320	2,456	-	1,256	50	2,268	3,574
Burma	5	10	525	275	815	-	10	619	573	1,202
Laos	-	-	-	-	-	-	-	-	627	627
Nepal	-	-	10	437	447	110	2,211	13,050	6,168	21,539
Pakistan	469	2,560	5,260	4,293	12,582	-	-	-	830	830
Sri Lanka	1	-	36	1,286	1,323	-	-	-	-	-
Subtotal	495	3,675	5,867	7,652	17,689	119	3,497	13,719	10,582	27,917
B. Lower Middle-Income Group										
Papua New Guinea	-	-	-	431	431	-	-	-	1,022	1,022
Philippines	392	-	5,704	9,395	15,491	665	-	4,702	9,698	15,065
Thailand	153	3,092	50	6,449	9,744	405	1,170	100	8,907	10,582
Subtotal	545	3,092	5,754	16,275	25,666	1,070	1,170	4,802	19,627	26,669
C. High & Upper Middle-Income Group										
Fiji	-	-	-	335	335	-	-	-	387	387
Singapore	-	-	-	2,408	2,408	-	-	-	4,317	4,317
South Korea	-	-	-	18,809	18,809	-	-	-	27,201	27,201
Taiwan	-	599	-	11,144	11,743	-	906	-	19,568	20,474
Subtotal	-	599	-	32,696	33,295	-	906	-	51,473	52,379
Total	1,840	9,408	27,755	61,294	100,297	2,147	9,783	38,723	88,380	139,033

Source: Asian Development Bank, Regional Energy Survey (Manila: March 1981), Table 9.12.

APPENDIX TABLE D
Asia-Pacific Developing Countries: External Debt Position, 1972 and 1978

Country	Total Outstanding (US$ millions)		Total Private Creditors (US$ millions)		Debt Service (US$ millions)		Debt Service/Export Goods[a] (%)		Debt Service/Export Goods and Services[b] (%)	
	1972	1978	1972	1978	1972	1978	1972	1978	1972	1978
South and West Asia										
Afghanistan[c]	961.20	2,003.70	16.60	45.80	25.40	51.90	20.74	16.51	n.a.	n.a.
Bangladesh	n.a.	4,274.40	n.a.	210.70	n.a.	94.00	n.a.	17.11	n.a.	13.73
India[d]	11,999.00	16,425.00	508.50	421.00	615.90	751.70	26.90	13.89	22.48	11.28
Nepal[c]	60.10	381.60	0.80	0.70	0.30	2.70	0.52	3.31	n.a.	n.a.
Pakistan	4,780.70	9,858.60	390.20	447.80	176.20	383.50	28.13	27.32	23.73	20.78
Sri Lanka	614.80	1,527.40	76.20	74.40	49.60	89.10	15.67	10.54	13.49	9.20
Iran	5,981.20	11,571.00	2,235.10	8,048.60	860.90	889.50	21.71	3.65	20.10	3.18
Southeast Asia										
Burma	299.10	1,323.00	69.40	186.60	23.70	55.10	19.35	20.05	17.75	18.04
Indonesia	5,128.80	18,554.00	735.10	5,844.00	126.00	1,381.00	7.03	12.53	6.86	12.21
Malaysia	1,020.00	4,359.70	385.10	2,224.90	51.20	703.70	3.05	9.84	2.73	8.76
Philippines	1,330.90	7,376.60	352.90	3,233.30	148.70	650.70	13.09	19.03	10.05	13.43
Singapore	504.40	1,314.60	123.10	809.60	33.00	302.50	1.62	3.17	1.05	2.30
Thailand	653.20	3,680.30	46.80	858.40	43.60	186.30	4.17	4.61	2.74	3.75
East Asia (LDCs)										
Taiwan	1,588.60	4,348.10	372.80	2,327.40	134.90	632.70	4.53	5.03	3.98	4.37
South Korea	3,925.20	18,146.00	1,924.60	9,935.40	406.20	1,795.20	24.23	14.16	18.24	10.51
Pacific Islands (LDCs)										
Fiji	45.90	132.00	5.60	20.20	1.20	9.40	1.81	5.26	0.79	2.55
Papua New Guinea	189.50	459.90	102.40	242.50	5.50	32.50	n.a.	4.52	n.a.	4.11

Source: International Monetary Fund, Balance of Payments Yearbook (Washington, D.C.: December 1979); and International Financial Statistics (Washington, D.C.: 1979, 1980).

[a]Debt service as percentage of export of goods.

[b]Debt service as percentage of total credits (good and services).

[c]Figures for exports of goods taken from International Monetary Fund, International Financial Statistics (Washington, D.C.: various issues).

[d]Data for India 1972 and 1976, and for Iran 1972 and 1977.

n.a. = data not available

4
Nuclear Power in the Asia-Pacific Region

Kirk R. Smith

The global nuclear age was born in Asia in August 1945, and--though the center of research, development, and implementation of military and civilian nuclear energy has remained in the North Atlantic--the role of Asia and the Pacific in global nuclear affairs has continued to be significant. The Pacific Islands have experienced more atmospheric nuclear detonations (over 200) than any other area in the world during the American, British, and French testing programs, and they have also been prime candidates for international nuclear spent-fuel storage and waste disposal sites. Japan, with one of the largest civilian nuclear power capacities in the world, is undertaking a rapid expansion program. South Korea and Taiwan have given high priority to nuclear power as a principal component of their ambitious economic development plans. China and India are the only two developing countries who have openly joined the nuclear weapons club. Australian uranium and Indian thorium comprise a considerable fraction of world nuclear fuel resources. To understand the role of nuclear energy in Asia today, therefore, it is important to recognize that nuclear technology has had a long period of development in the region and has helped shape present Asian political and economic patterns.

ASIA'S NUCLEAR HISTORY

Nuclear technical capability was first developed for war and from the beginning was perceived to bring power both for energy production and for weapons to those who obtained it. This perception led the United States to attempt to maintain its monopoly on nuclear technology during the years immediately after World War II. In Asia, this period of denial hampered the development of nuclear technical capability. India, however, began nuclear research as early as 1944, set up its Atomic Energy Commission in 1948, and, in 1955, Indian scientists designed and built India's first research reactor with fuel elements from the United Kingdom.[1]

The entry of the Soviet Union and the United Kingdom into the nuclear weapons club, to be closely followed by France, demonstrated the hopelessness of the U.S. denial policy. These events led to the second phase of international nuclear relations, initiated by the United States in 1953 with its proposal to the United Nations for the "Atoms for Peace Plan." This was an ambitious program to harness U.S. resources for the transfer of peaceful nuclear technology worldwide. In return, however, prohibitions were provided in the plan to discourage the development of military applications of nuclear energy. In addition, by the late 1950s, American industry was showing considerable interest in an international nuclear market.[2]

Helped along by the declassification of information that occurred as part of the Peaceful Uses of Nuclear Energy Conferences in Geneva (1955, 1958), the result of the Atoms for Peace Plan was the widespread transfer of nuclear technology for use in agriculture, medicine, and, to a lesser extent, power production. A number of Asian developing countries were able to take advantage of the plan with India, Taiwan, South Korea, and Pakistan benefiting most.[3] Thousands of students from these countries received training in nuclear sciences and engineering in U.S. laboratories and universities. Of the more than 40 research reactors operating in Asia today, most were built before 1965 during the initial burst of enthusiasm about the Atoms for Peace Plan (see Table 4.1).[4]

TABLE 4.1
Asian Research Reactors

Country	Number	Year of First Operation
Australia	3	1958
China	several	1958
India	5	1956
Indonesia	2	1964
Japan	22	1961
Malaysia	(1)	(1981)
Pakistan	1	1965
Philippines	1	1963
South Korea	2	1962
Taiwan	5	1961
Thailand	1	1962
Vietnam	1	?

Source: Modified from Albert Wohlstetter et al., Swords from Plowshares (Chicago: University of Chicago Press, 1979).

As part of the Atoms for Peace Program, in 1955 the United States proposed and offered to sponsor an Asian Regional Nuclear Center in Manila. A study team from the Brookhaven National Laboratory of the U.S. Atomic Energy Commission (USAEC) made recommendations for the design of the center based on its two-month tour in 1956. This constituted the first study of the potential of nuclear energy in LDCs. A major finding of the survey team is one that has been repeated often by studies in later years:

> . . . many countries are deluding themselves as to the immediate application of atomic energy to their problems, especially the linking of atomic energy development to their long-term power needs . . .[5]

The study emphasized the importance of developing Asian medical and agricultural applications of nuclear technology, with power production as a longer-term objective possible only after the development of basic scientific and engineering infrastructure in the Asian countries. In spite of the formal offer by the United States in late 1956 to fund the Center, however, negotiations were never completed.

The largest of the U.S. bilateral programs in this period was with Japan. The initial Agreement for Cooperation was concluded in 1958, not long after implementation of the 1956 Japanese Atomic Energy Act. The sale of what was to become the first of more than 20 Japanese research reactors was a part of the agreement. During the 1960s, the relationship intensified to the point that, in addition to extensive transfers of knowledge resulting from Japanese participation in U.S. training programs, Japan received the majority of its nuclear technology from U.S. manufacturers through license arrangements and loans from the U.S. Export-Import (EXIM) Bank. Japan was also the first country to allow its nuclear facilities to be visited by international inspectors, an action in support of the often-beleaguered U.S. nonproliferation policy. In the 1970s, however, this close Japanese-U.S. tie became strained by changes in the economic relationship between the two countries, shifts in U.S. nuclear export and nonproliferation policies, and the impact on Japan of the evolving global energy situation.[6]

One outgrowth of the Atoms for Peace Plan was the creation in 1957 of the International Atomic Energy Agency (IAEA). In addition to promoting peaceful uses of nuclear energy, the IAEA was to be charged with safeguarding nuclear technology from being used militarily as soon as it was ready to take on the responsibility. This safeguards role was not implemented quickly, however, largely because of opposition by the Soviet Union until the early 1960s.

In this period, the Soviet Union had initiated its own nuclear technology transfer program with the People's Republic of China and by 1959 had helped China build a research center, develop its uranium resources, and train thousands of scientists and engineers. The cooperation ended in that year, however, when relations deteriorated with China for a number of reasons, including China's evident intention to develop nuclear weapons. The first Chinese test occurred in 1964, by which time the Soviet Union had agreed to the need for developing an international safeguards scheme for nuclear technologies. Safeguards negotiations culminated in the 1968 Nonproliferation Treaty (NPT). The IAEA safeguards program was significantly strengthened by the addition of NPT functions and the transfer of a number of bilateral safeguards programs to the agency by member countries.

In 1980, however, most of the population of Asia still lived in countries that had not ratified the NPT--most notably, North Korea, China, India, and Pakistan--although those countries with the largest nuclear power[7] programs were all signatories--Japan, South Korea, and Taiwan.

In the late 1960s and early 1970s, there was considerable optimism about the future of nuclear power both in developed and developing countries and relatively little expressed concern over proliferation issues. Japan and India, for example, both mapped out ambitious nuclear development plans during this period. In 1974, however, two events occurred that have greatly affected the future of nuclear power: the oil embargo and the Indian nuclear test. Energy security has taken greater precedence in national planning and concern over the proliferation of nuclear weapons has once again become dominant in international and bilateral nuclear technology transfer.

In addition to these events, several longer-term trends affecting nuclear power development became significant by the mid-1970s. By that time, it was becoming apparent that the relative cost advantage of nuclear power was worsening rapidly in spite of the price rises of other fuels. In combination with the downward shifts in energy/GNP elasticities and economic growth rates experienced globally in the 1970s, the worsening of nuclear economics led to a decreased number of annual reactor orders after a peak in the mid-1970s and many cancellations of previous orders (Figure 4.1). Also in the mid-1970s, French and German exporters of nuclear technologies began to play a significant role in a world market traditionally dominated by American and Canadian industries. These political events and industry trends, together with rising public concerns about safety and waste disposal, led to a new phase of global nuclear development beginning in the mid-1970s.

Asian nations--notably India, China, and Japan--have played important roles in the postwar history of nuclear energy. In the newest phase, these countries retain their significance, and Taiwan and Korea will come to play increasingly important parts.

NUCLEAR POWER PLANTS IN ASIA: PRESENT AND PROJECTED

In Asia, nuclear power plants are at present a phenomenon principally of the North Pacific. As Table 4.2 indicates, Japan dominated the picture in 1981 with nearly 85 percent of the total civilian nuclear electric capacity in Asia which, at 15 gigawatts (GW) was about equal to the Soviet program although far behind the

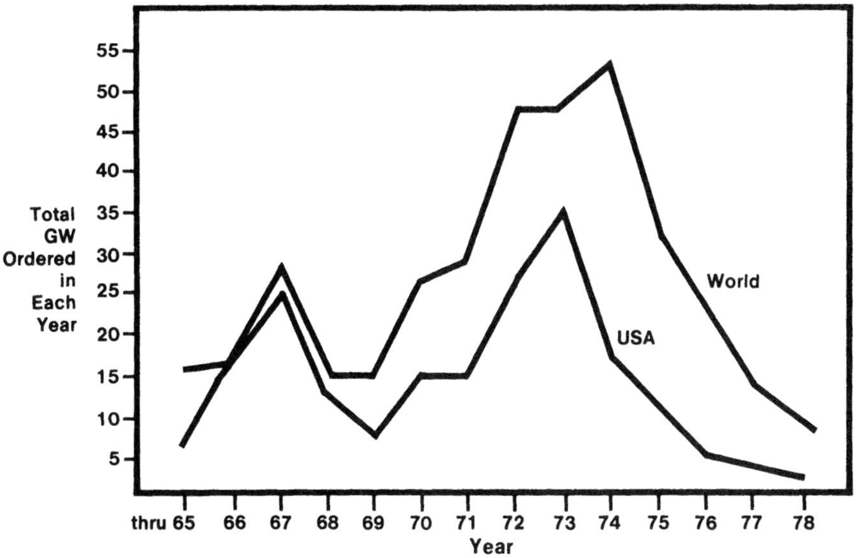

Figure 4.1. History of nuclear power plant orders. (Source: R. Krym and J. P. Charpentier, "Nuclear Power Development: Present Role and Medium-Term Prospects," IAEA Bulletin 22, 2 [April 1980]: 11-12.)

---

world leader, the United States at 52 GW installed capacity. Taiwan and South Korea, though late starters, have ambitious expansion plans well along in implementation. By the late 1980s, therefore, Japan's share will probably drop and the share of Taiwan and South Korea will rise to more than one quarter. India is planning a number of fairly small units that will not soon add to a large total capacity. Although Pakistan has announced plans for a large nuclear program, there is little evidence that its goals will be achieved during the next decade. The Philippines has one reactor under construction, but much delayed because of Filipino concerns about cost overruns, safety, and waste disposal.

Three other Asian countries seem to be moving toward the introduction of nuclear power in the next decade. China has a significant technical capacity as a result of its nuclear weapons program and has been negotiating with the French for two pressurized water reactors (PWRs) and planning to build a small heavy water moderated reactor (HWR) of domestic design (see Chapter 5 for details of the Chinese nuclear power program). Bangladesh and Indonesia are also planning to introduce nuclear power plants during this period although no contracts had been signed by 1980.

No other Asian or Pacific country has committed itself to a nuclear power project in any but general terms, although a number have small research reactors (see Table 4.1). New Zealand and

TABLE 4.2
Nuclear Power Plants in Asia (January 1981)

Country	In Operation (Under Construction)					Advanced Planning Stages
	Capacity Net MW	PWR	BWR	HWR	GCR	
		Numbers				
East Asia						
Japan (1966)	14,560 (9,480)	9 (5)	12 (6)	1	1	3,300 MW (1 PWR, 2 BWRs, 1 LMFBR)
South Korea (1978)	565 (6,835)	1 (7)		1		3,600 MW (4 PWRs)
Taiwan (1978)	1,210 (3,715)	(2)	2 (2)			1,700 MW (2 BWRs)
(China)	(a)					(2,225 MW)[b] (3 PWRs, 1 HWR)
Southeast Asia						
Philippines (1984)	(620)	(1)				--
(Indonesia)						(600 MW)[b] (1 HWR)
South Asia						
India (1969)	800 (880)		2	2 (4)		420 MW (2 HWRs)
Pakistan (1972)	125					600-900 MW (1 PWR)[c]
(Bangladesh)						(300 MW)[b] (1 PWR)
Total	17,260 (21,530) MW	10 (15)	16 (8)	3 (5)	1	12,800 MW

Source: "World List of Nuclear Power Plants," *Nuclear News*, February 1981, pp. 75-94; "Power Reactors," *Nuclear Engineering International*, July-August 1980, pp. 64-72.

PWR = pressurized water reactor; BWR = boiling water reactor; HWR = heavy water reactor; GCR = gas-cooled reactor.

Table 4.2 (cont'd)

[a]China has a small nuclear electric generating capacity as part of one of its research HWRs and reportedly has two plutonium production reactors.

[b]Feasibility studies have been done, but no bids have been let (Nuclear Engineering International, May 1980, p. 5; November 1980, p. 7; December 1980, pp. 4-5; January 1981, p. 5). Since these facilities would be the first nuclear power plants in these countries, their timing and prospects are more uncertain than the expansion program in other countries.

[c]Government approval has been given for a PWR and a design contract let to a Spanish firm to compare 600 and 900 MW units (Nuclear Engineering International, May 1981, p. 4).

Singapore have explicitly stated that they will not proceed with nuclear power at all for the next few years. Australia, Sri Lanka, Malaysia, and Thailand have tentative schedules for nuclear power programs commencing with one or two plants in the 1990s. Hong Kong is negotiating with China to share the financing and electricity for PWRs potentially to be built nearby in Guandung Province (see Chapter 5).

As elsewhere in the world, the development of nuclear power in Asia has been slower than was anticipated a few years ago. In 1974, for example, IAEA calculated that the potential nuclear power capacity for Asia in 1990 would be more than 110 GW; in 1977, this estimate had been lowered to 75-90 GW. The combination of higher nuclear plant costs, lower growth in electricity demand, longer construction times, and international and domestic political concerns about nuclear fuel-cycle hazards have tended to outweigh the advantage to nuclear power of the rising costs of energy alternatives during the 1970s to the extent that the official projections for 1990 shown in Table 4.3 seem unlikely to be realized. Indeed, it would seem that at most, only about one half of the projected capacity (45 gross GW) could be in place in 1990, given the amount in operation or under construction today. Nor does it seem likely that the official projection for 2000 will be accurate. Table 4.3 also lists the results of three studies of international nuclear prospects for the year 2000 published in late 1979 or 1980. All indicate substantially lower projections, and the latest one (by Interdevelopment, Inc.), which included considerations of the particular conditions in each country such as construction times and electricity demand, indicates that less than half of the official projections will be realized. In reflecting on the accuracy of the projections in Table 4.3, however, it should be remembered that most of the analyses were done before the precipitous decline in uranium prices that occurred in 1980-1981.

Other than China, it seems unlikely that any other Asian country beyond the four listed in Table 4.3 will operate more than

TABLE 4.3
Asian Nuclear Power Projections (GW)

Country	"Official"[a] 1990	"Official"[a] 2000	INFCE[b] 2000	ICGNE[c] 2000	Interdevelopment[d] 2000	Built or Under Construction 1981
Japan	53	110	100	76	79	24
South Korea	13	48	43	8	12	7.4
Taiwan	8	25	-	10	11	4.9
India	6	10	4.7	11	2.3	1.7
Total Asia	90	225	160	130	100	38

[a]Japan: Ministry of International Trade and Industry (MITI), Interim Report on Energy, Advisory Committee for Energy, Japanese Government (Tokyo: August 1979); Nucleonics Week, December 4, 1980, p. 2.
Korea: J. H. Cha, "Nuclear Power Program in Korea," Workshop on the Future of Large-Scale Energy Systems, East-West Center, Honolulu, 1979.
India: Nuclear Engineering International, September 1980, p. 3.
Taiwan: Taipower and Its Development (Taipei: Taiwan Power Company, May 1980).

[b]International Nuclear Fuel Cycle Evaluation, "Working Group 1 Report: Supply" (Vienna: International Atomic Energy Agency, 1980).

[c]Thomas Connolly et al., World Nuclear Paths, International Consultative Group on Nuclear Energy (New York: Rockefeller Foundation, 1979).

[d]Study by Frank Faris, Interdevelopment Inc., Arlington, Va., reported in Nucleonics Week, May 22, 1980, p. 1.

two or three nuclear power plants by 2000.[8] The total for the People's Republic is difficult to forecast (as discussed in Chapter 5) but could possibly reach 15 GW, although 10 GW is more likely and less than 5 GW is possible.[9]

Grid Size

It has often been pointed out that a barrier to the introduction of nuclear power in many developing countries is the size of electricity grids. This is sometimes called the "capacity paradox" because nuclear power plants are usually considered to experience considerable economies of scale, at least up to 1.0-1.5 GW. Although a number of factors affect systems operation, it is fair to say that units large enough to be a significant fraction of total grid capacity tend to reduce system reliability substantially and increase system inefficiency. In a study done in 1974, the IAEA calculated the maximum plant size that could be accommodated according to total size of the electric grid. For a 600-megawatt(MW) plant, the smallest size readily available on the international market, this study indicated that the electric grid should be at least 7000 MW, whereas for a 900-MW plant, the grid should be at least double this amount (see Figure 4.2). By this measure, South Korea and Taiwan both reached the threshold in about 1978, when they each introduced their first 600-MW power plants (see Table 4.4). The 125-MW Kanupp plant near Karachi came on line in 1972 at a time when the grid surrounding Karachi met the 700-MW criterion of the IAEA (Figure 4.2). The 7000-MW level could be reached by 1990 at a 9 percent growth rate but would require strong interconnection among the five major regional grids in Pakistan. The western region of India, which includes Bombay and received the first Indian power reactor (220 MW) in 1969, contained sufficient capacity at the time. Rajasthan in the north, however, which received a power plant in 1973, did not, and significant problems have resulted. The strengths of the interregional grids in India have not been great and systems stability remains a problem, especially in Rajasthan.

If this criterion is applied in the rest of Asia, it seems there are few places where 600-MW or larger plants will fit in the 1980s (see Table 4.5). The grid on Luzon in the Philippines, although connecting more than 70 percent of the nation's capacity would not reach the required 7,000 MW until 1990 at an 8 percent growth rate even though the 620-MW PWR near Manila was originally planned for activation in 1982 (now 1986). The Javanese grid would have to grow at more than 15 percent annually to reach 7,000 MW by 1990. The Thai electric grid does not yet connect all major regions of the country and would have to grow at 7 percent annually to reach 7,000 MW by 1990.[10] The region being considered for a 300-MW PWR in Bangladesh will not reach the required 2,500 MW this century. An interconnected Malaysian system would have to grow at 12 percent annually to reach 7,000 MW unless strong interties were made with neighboring countries. Hong Kong and Singapore, which both have serious siting problems, could reach 7,000 MW by 1990 at 7 and 13 percent growth rates,

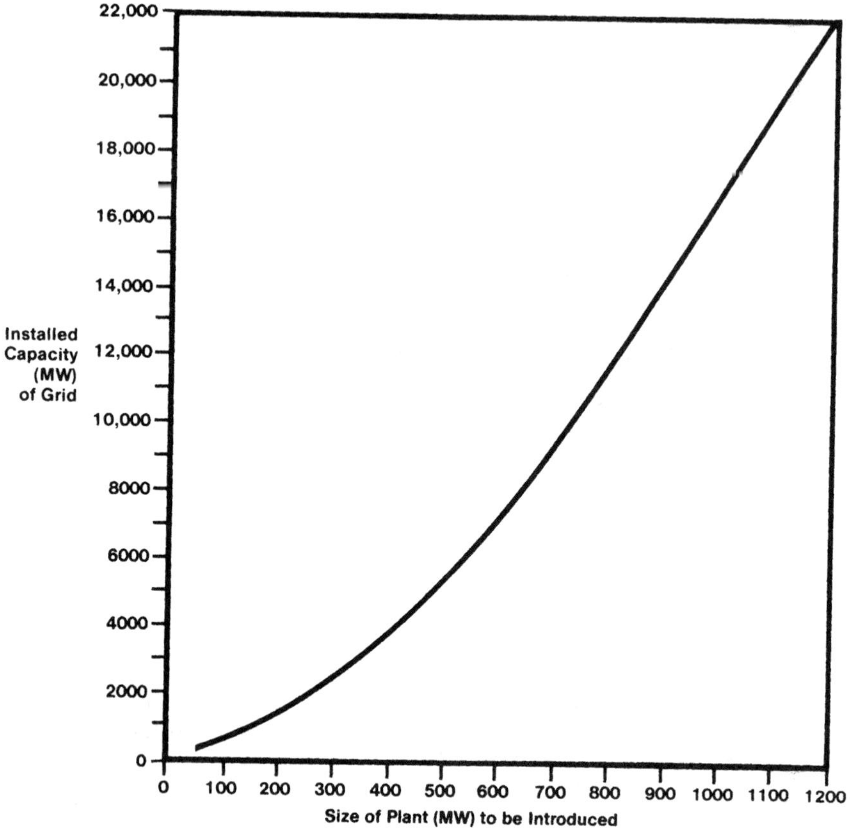

Figure 4.2. Electric grid size required for efficient absorption of new plants. (Source: International Atomic Energy Agency, Market Survey for Nuclear Power in Developing Countries [Vienna: 1973].)

---

respectively, although both have potentials for interties with neighbors. As shown in Table 4.5, no other Asian country could possibly reach the required 7,000-MW grid size until well beyond the year 2000 unless growth rates exceed 13 percent. Table 4.4 shows that such growth rates have rarely been sustained in the recent past. New Zealand has no plans for nuclear power but could reach the required capacity on the North Island by 1990 at an annual growth rate in electric generating capacity of a few percent.

Because several factors affect reliability and because transmission systems technology is also changing, the capacity paradox curve shown in Figure 4.2 should be viewed as only a rough guide to nuclear projections. Nevertheless, the future of nuclear power in Asia and Oceania depends on the growth rate of electricity demand, the extent to which national and international grid systems develop, and the potential development of economic small

TABLE 4.4
Installed Electrical Capacity in Asian Countries with Present and Potential Nuclear Power Programs (MW)

Country	Year	Fossil Steam	Diesel	Gas Turbines	Nuclear	Hydro	Total	Percent Nuclear	1972-1978 Average Annual Growth Rate (%)
Japan	1965	24,495	225	80	15	16,280	41,095	0	6.9
	1978	91,840[a]	n.a.	n.a.	8,870	26,690	127,400	7	
South Korea	1965	515	40	0	0	215	700	0	10.2
	1978[b,c]	5,330	80	210	585	710	6,915	8.2	
Taiwan	1965[d]	570	20	55	0	630	1,275	0	13.9
	1978[d]	5,020[a]	n.a.	n.a.	1,270	1,390	7,680	16.5	
China	1973	21,600[a]	n.a.	n.a.	0	5,400	27,000	0	13.2
	1978	40,080[a]	n.a.	n.a.	0	10,020	50,100	0	(1973-1978)
Thailand	1965	260	155	0	0	145	560	0	9.5
	1978[e]	1,700	35	165	0	910	2,810	0	
Philippines	1965	455	340	0	0	290	1,085	0	7.4
	1978[f]	1,905[a]	585	0	0	755[g]	3,245	0	
Indonesia	1965	125	175	0	0	260	560	0	9.9
	1978	1,110[a]	n.a.	n.a.	0	550	1,660	0	
India	1965	5,350	405	135	0	3,855	9,745	0	6.9
	1978	16,710[a]	n.a.	n.a.	640	9,450	26,800	2.4	
Pakistan	1965	565	115	15	0	380	1,075	0	3.1
	1978	1,230[a]	n.a.	n.a.	140	865	2,235	3.6	
Bangladesh	1965	-	-	-	-	-	-	-	6.8
	1978	860[a]	n.a.	n.a.	0	110	970	0	
Australia	1965	6,155	170	0	0	2,085	8,410	0	4.3
	1978	17,225[a]	-	-	0	5,600	22,825	0	

Source: 1965 figures are from United Nations, Electric Power in Asia and the Pacific (Bangkok: Economic and Social Commission for Asia and the Pacific, 1959); 1978 figures are from United Nations, World Energy Supplies 1973-1978, Series J, no. 22 (New York: 1979) except as otherwise stated.

Table 4.4 (cont'd)

[a]Total capacity of all thermal units. Breakdown by type of thermal unit is not available.

[b]Republic of Korea, Statistical Handbook (Seoul: annual).

[c]Arthur D. Little, Inc., U.S./Republic of Korea Energy Assessment: Electric Sector Evaluation, Cycle I (draft).

[d]Republic of China, Taiwan Statistical Data Book (Taipei: annual).

[e]Electricity Generating Authority of Thailand, Annual Report (Bangkok: annual).

[f]Philippine Government, Ministry of Energy, Ten-Year Energy Program 1980-89 (Manila: 1980).

[g]Includes geothermal capacity.

n.a. = data not available.

---

nuclear units. Such units are under development in several European countries and Japan, for example.[11] A number of other considerations also affect the future, including commercial arrangements and technical and regulatory infrastructures. These are perhaps easiest to discuss in the context of each major country.

South Asia

The three poorest countries in the world with populations of more than 80 million are in South Asia. Two have embarked on nuclear programs and the other is considering one.

India. In ordering its first two power reactors on a turnkey basis from the American General Electric Company in 1963, India obtained highly concessionary loans from the U.S. EXIM Bank and the U.S. Agency for International Development (USAID). Bechtel Corporation was the architect/engineer and contractor, and General Electric provided the generators for the boiling water reactors (BWRs). Partly because of the uncertainty in obtaining long-term access to enrichment services necessary for light water reactor (LWR) fuel, India turned to heavy water reactors for all additional plants. The next plant was a CANDU (Canadian Deuterium [HWR] reactor) in which the reactor was provided by Canadian General Electric and the architect/engineer services by Canadian companies. Construction, however, was done by Hindustan Construction Company. The reactor for the fourth Indian nuclear power plant, which began operation in 1980, was supplied domestically. All additional power plants are being constructed with more than 90 percent Indian participation.[12] The Indian

TABLE 4.5
Approximate Growth Rates Required to Reach 7,000 MW Total Capacity

Country	National Capacity in 1978 (MW)	Growth Required to Reach 7,000 MW	
		by 1990	by 2000
Afghanistan	335	26	14
Bangladesh	970	16	9
Brunei	147	35	18
Burma	450	24	13
Cambodia	52	46	24
Fiji	120	37	19
Guam	302	27	15
Hong Kong	2,971	7	4
Indonesia	1,660	13	7
Laos	55	45	23
Macau	102	38	20
Malaysia	1,575	12	7
Mongolia	300	27	15
Nepal	64	44	23
New Zealand	5,400	2	1
North Korea	5,000	3	1
Pakistan	2,235	9	5
Papua New Guinea	285	28	15
Philippines	3,245	7	4
Samoa	8	68	34
Singapore	1,470	13	7
South Korea	6,915	-	-
Sri Lanka	421	24	13
Thailand	2,810	8	4
Tonga	4	78	35
Trust Territory	70	43	22

Source: United Nations, *World Energy Supplies 1973-1978*, Series J, no. 22 (New York: 1979).

government's policy has been to reduce import dependence by developing indigenous capacity in all major industries.[13] Thus, although 10 years behind the original schedule set up in the early 1960s and at some loss of economic efficiency, India has been able to obtain an indigenous power plant industry that includes several major Indian corporations. Although the next power plants being designed are relying on the 220-MW reactor used in the two plants now under construction, it is planned to move to a 500-MW HWR design after about 10 reactors.

Indian reactors have not performed as well as expected. There have been problems with equipment, including leaks in the cooling system, as well as problems in the late 1970s in obtaining enriched uranium from the United States for the boiling water reactors. The result is that the two BWRs have a cumulative capacity factor of about 50 percent. This is not far from the global average for all BWRs (see Table 4.6). These low capacity factors, whether close to global norms or not, mean that nuclear electricity has been significantly more expensive than the original projections which were usually based on 70 to 80 percent capacity factors. RAPP-1, the first Indian heavy water reactor, has a cumulative capacity factor of 40 percent compared to the global average of over 70 percent. Because RAPP-1 has operated since 1973, this reflects problems that have extended well beyond the expected low performance during reactor start-up. One problem is that the grid in Rajasthan was too small and too poorly integrated with the rest of the northern region to accommodate a 220-MW power plant adequately and another is excessive leakage of heavy water.

In 1980, for example, Rajasthan industry was subject to 100 percent power cuts sometimes lasting weeks, partly because the RAPP-I power plant was down so often and RAPP-II had been delayed. India as a whole suffered a 19 percent power deficit in 1980 that was exacerbated by the poor grid interconnections and low capacity factors of all thermal plants (42 percent).[14] The planned strengthening of regional power grids under the coordination of the five Regional Electricity Boards (REBs) is an important priority for India's energy program.[15] At present, however, the REBs operate only to share infrequent surpluses in the separate state electric grids and are not effective in sharing deficits. Worry about system stability is apparently one of the principal reasons that the Indians turned down an offer of a 1000-MW pressurized water reactor from the Soviet Union in 1980.[16]

Safety has also become a concern in the siting and operation of Indian reactors. In 1980, for example, a controversy developed about the siting of two reactors under construction in Narora, Utter Pradesh because of an active earthquake fault 50 km away.[17] Accusations were made that political considerations had prevailed over the recommendations of the Bhabha Atomic Research Center Environmental Group, which had the responsibility to clear sites for safety.[18] In 1980, an Indian Government Committee on power recommended that India create an agency modeled after the U.S. Nuclear Regulatory Commission (NRC) to establish and enforce safety standards independently from the Department of Energy.[19]

TABLE 4.6
Capacity Factors of Asian Nuclear Power Plants (percentages)[a]

Country	PWR	BWR	HWR	GCR
Japan	49[b]	59	50	62
South Korea	47	-	-	-
Taiwan	-	59	-	-
India	-	50[c]	40	-
Pakistan	-	-	35	-
Global average	54	52	73	59

PWR = pressurized water reactor; BWR = boiling water reactor; HWR = heavy water reactor; GCR = gas-cooled reactor.

[a]Cumulative weighted averages as of September 1980 (*Nucleonics Week*, October 23, 1980, pp. 11-17; *Nuclear Engineering International*, December 1980, pp. 40-41).

[b]During the year immediately following the accident at the Three-Mile Island PWR (March 1979), the capacity factor at Japanese PWRs dropped to 39 percent (see Figure 4.3).

[c]In the late 1970s and 1980, the Indian BWRs were kept at lower power levels partly because of U.S. delays in shipments of enriched fuel.

India's nuclear power program has been plagued by delays, some of which are common worldwide. Others, such as systems integration difficulties, particularly poor reactor performance, and a strong government push to develop indigenous manufacturing capacity, have characterized the Indian program. In addition, the unevenness of the Indian economy and small failures in government planning have led to frustrations such as the delay in the Narora plant construction in 1980 because of the lack of locally available cement.[20] Further, there is growing doubt about the suitability of a large-scale electric power program for India's largely rural population and increased concern about safety issues. All of these factors have led not only to delays but to what is called "fatigue" that plagues the professional staff of the Indian nuclear establishment.[21] The hopeful days in the early 1960s, when Dr. Homi Bhabha, the founder and most articulate proponent of the Indian nuclear power program, set out India's nuclear power goals as an answer to its development needs, seem far away as India faces the 1980s.

<u>Pakistan</u>. In 1954, the American Atoms for Peace exhibit toured Pakistan as part of the U.S. global effort to expand the peaceful applications of nuclear energy. The Pakistan Atomic Energy Commission (PAEC) was established shortly thereafter and many Pakistanis were sent abroad for training. Although a number of these scientists and engineers seemed to have gone down the brain drain in the early years, by the mid-1970s many had returned to Pakistan thanks to changing technical job markets and an aggressive recruiting program by the PAEC.[22]

Pakistan's only nuclear plant, ordered in 1965, was built on a turnkey basis by Canadian General Electric with concessionary loans through the Canadian government. As described later, Canada's assistance and participation ended in 1976 when it imposed an embargo on spare parts, heavy water, and fuel because of proliferation concerns. This, in combination with heavy water leaks, has led to a very low lifetime capacity factor for the 125-MW HWR. For example, it was out of service for more than half of 1980 and had only come back up to half power by early 1981.

Since the mid-1970s, official announcements have made it seem that Pakistan has been on the verge of starting construction on its second reactor, but as of early 1981, no bids had yet been solicited. Site work was started in 1978 at Chashma, Punjab, but financing problems have apparently remained the stumbling block. The expected cost is $1,700/kW.[23] Most announcements have indicated that a light water reactor will be built, although some Pakistani observers believe that the country would be better served by moving directly to an indigenous heavy water reactor design which could eliminate dependence on uranium enrichment.

Pakistan has had problems in recent years with a new sort of brain drain, the movement of skilled and semiskilled workers to the oil-exporting countries of the Middle East. Substantially higher salaries have drawn reactor operators from Pakistan to work in chemical plants in countries to the west. This has left Pakistan short of trained and experienced reactor operators, a situation worrisome to PAEC officials in light of the evident importance of reactor operator competence revealed by the Three-Mile Island accident. This is one example of a wide variety of infrastructure problems that are potentially troublesome in Pakistan's nuclear power expansion program. The simple transport of large reactor components from the port of Karachi to Chashma would require substantial modification of roads or trainbeds, for example.[24]

Pakistan has traditionally projected nuclear power as the residual in its power program. The gap between projected electricity demand and supply of planned conventional power sources has been projected to be filled with nuclear power. In its first projection in 1955, the gap for West Pakistan was expected to be 600 MW by 1980 and 5300 MW by 2000. Present projections illustrate what might be called the "Pakistan paradox," often observed in other countries as well. Even though actual capacity has fallen considerably <u>below</u> past projections, present projections have become even higher.[25] Thus in 1980 the official projection for 2000 was some 16 GW, nearly three times higher than before, in spite of the fact that the 1980 capacity was some four times lower than previously projected.[26] Present Pakistani projections are clearly not realistic and must be viewed as idealized development goals and, possibly, means to help justify the construction of Pakistani fuel cycle facilities for other purposes.

Bangladesh. Cut off from most of Pakistan's technical capability by gaining its independence in 1971, Bangladesh was one of 14 developing countries to be surveyed for nuclear power in 1972 in an IAEA program funded by the international banks, USAID, USAEC, and West Germany.[27] Based on this study and others done in the mid-1970s by Belgian, Canadian, and Japanese consultants, Bangladesh has proposed to introduce a reactor in Rappour (Western Bangladesh) because it is often isolated from the main fossil and hydro plants in the eastern areas by flooding of the Brahmaputra River. The small size of the western grid (less than 100 MW) precludes large power plants, however. The French government has apparently offered concessionary loans totaling about a third of the cost of a 300-MW LWR based on French submarine reactor technology.[28] Reportedly, the cost would exceed $2,000 per kW, or nearly 10 percent of the annual GNP. In spite of possibilities of Saudi Arabian loans in the near future, it seems unlikely that so much capital would be devoted to a project requiring such a large foreign exchange component and placing so much grid capacity in a design that had never before been built. Even taking into account the long-planned East-West grid interconnector, the necessary national capacity will probably not be reached until the end of the century (see Figure 4.2).

East Asia

In sharp contrast to South Asia, East Asia has three of the most rapidly growing economies in the world. Each has made a significant commitment to nuclear power.

Japan. The first power reactor purchased by Japan was a British natural-uranium gas-cooled reactor, the 160-MW Magnox, ordered in 1960 and put into operation in the Japanese nuclear research and development center at Tokai Mura in 1966. This sale startled U.S. nuclear manufacturers, who had expected to be dominant in the export market to Japan. Consequently, U.S. government and industry both worked hard to establish a strong position in the Japanese market through long-term enriched fuel contracts and offers of technology transfer. By the mid-1960s, the Japanese had decided to rely solely on U.S. light water reactors for the immediate future, partly because of these changes in the U.S. position and partly because of the uncertain future of the British reactor export industry. Ironically, however, the Tokai Mura Magnox has consistently outperformed Japan's U.S.-style LWRs, although the difference in ages and sizes makes comparison questionable (see Figure 4.3).

The rush to capture the Japanese market led to some practices that, in hindsight, might seem overenthusiastic and unnecessary.

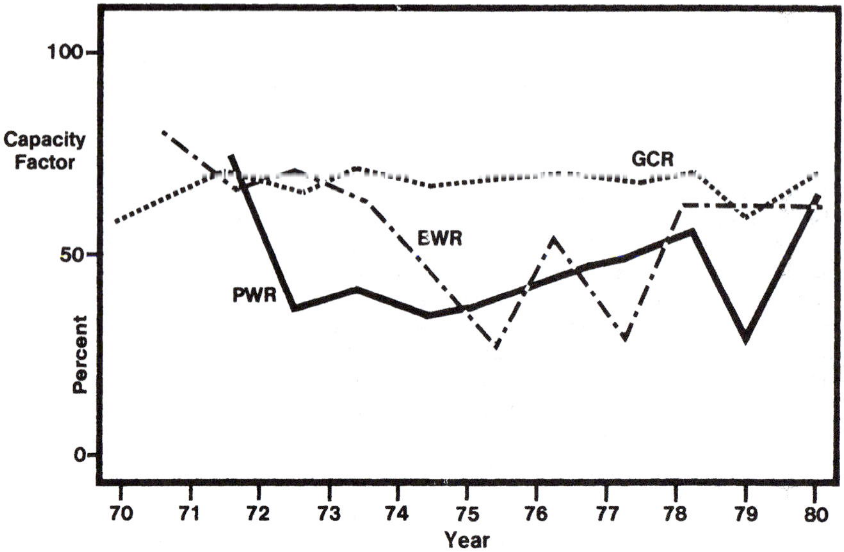

Figure 4.3. Trends in capacity factors for Japanese nuclear power plants. PWR = pressurized water reactor; BWR = boiling water reactor; GCR = gas-cooled reactor. (Source: "Power Plant Performance," Nuclear Engineering International, various issues.)

The U.S. EXIM Bank funded many Japanese nuclear power projects at terms more liberal than the international standard, well past the time at which Japan could be reasonably considered a developing country, and even beyond the time when the United States began to accrue large trade deficits with Japan--1967. This might be justified as a government subsidy to the sometimes financially troubled U.S. nuclear industry as well as an investment in the energy security of an ally. An obvious question, however, is whether it was necessary, given the probability that Japan would have chosen and been able to pay full prices for U.S. LWRs anyway.[29]

Another important characteristic of U.S. reactor sales to Japan during this period was the high component of technology transfer involved. Under licensing arrangements with the two largest U.S. reactor manufacturers, Westinghouse and General Electric, several Japanese companies rapidly developed sufficient capability to design, fabricate, and construct complete nuclear power stations. Table 4.7 illustrates this transition. Note that Japanese practice was to secure direct U.S. industrial participation, particularly when changing to larger models of each type of LWR. The growth in the size of U.S. LWRs, however, seems to have leveled off at about 1300 MW. The Japanese have indigenous industrial capacity at near this size since construction began on the 1067 MW Fukushima Daina-1 in 1976. Since the three large plants went into operation in 1979, as shown in Table 4.7, U.S. direct participation in the Japanese market has been much less prominent though still important in commercial arrangements to help Japanese companies compete within Japan. In 1980, Bechtel, for example, signed an agreement oriented toward increasing Hitachi's chances of competing successfully with Toshiba, the present leader in installing BWRs in Japan. Thus, for the foreseeable future, U.S. participation in the Japanese market will consist of selling relatively minor technical consulting services and collecting license fees. In spite of the large license fees collected, the U.S. nuclear industry now apparently questions the overall economic wisdom of having helped the Japanese develop industrial capacity so rapidly.[30] Total Japanese reactor construction capacity is about 8 GW per year, although, as in most of the world, this capacity seems likely to be underutilized in the near future and the three vendors--Hitachi, Toshiba and Mitsubishi--were only just beginning to achieve marginal profits in 1980. Losses in previous years totaled more than one-half billion dollars.[31]

In 1975, the government embarked on a program to standardize the design, licensing, and construction of light water reactors to increase reliability, which has been notoriously low (see Table 4.6 and Figure 4.3), and to decrease the cost and time of construction. The installation time has been increasing rapidly. A Japanese study in 1980 indicated that Japanese utilities now require an average of 15 years to proceed from the first siting proposal to local governments through to operation. This figure compares with an average of 10 years only a few years earlier.[32] Construction delays, however, were not the major reason, increasing only from 4.5 to 6 years.[33] Japanese light water reactors have also been plagued with a series of minor but debilitating operational problems that have led to quite low capacity factors in many cases.[34] Clogging of coolant channels, stress corrosion of pipes, cracked pump impellers, fuel pin leaks, and abnormal turbine wear are examples of these problems.[35] The first of these caused the virtual complete shutdown of the Mihama-1 340-MW PWR from 1974 to 1980, leading to the tenth worst cumulative capacity factor in the world.[36]

The Ministry of International Trade and Industry (MITI) has established a schedule of standardization for LWRs written not only to stabilize size and design characteristics but to incorporate Japanese modifications to those U.S. BWR and PWR designs and operational procedures that the Japanese feel have led to their past troubles.[37] The U.S. view is that many of these small, but disrupting, technical problems have been caused by the Japanese delay in adapting U.S. reactor system designs to local engineering regulations and practices. In Europe, for example, such adaptation has allowed the French and Germans to avoid these sorts of problems.[38] Part of the standardization effort has been funded by a national tax on electric power which in 1979 amounted to 0.4 mill per kilowatt-hour (kWh).[39] Additional portions of these tax

TABLE 4.7
Evolution of Japanese Nuclear Industrial Capacity[a]

Years of		Size in		National Base of Major Plant Contractors and Suppliers[b]					
Order	Operation	MW/Type	Utility	Main Contractor	Architect-Engineer	Reactor Manufacturer	Fuel Contractor	Turbines	Comments
1960	1966	160 Magnox	Japco	UK	UK	UK	UK	UK	Turnkey; carbon-dioxide cooled, graphite moderated reactor
1965	1970	342/BWR	Japco	US	US	US	US	US/J	Turnkey
1966	1971	440/BWR	Tokyo	US	US	US	US/J	US/J	Turnkey
1966	1970	320/PWR	Kansai	US/J	J/US	US	US/J	J	
1966	1972	470/PWR	Kansai	J	J	J	US	J	
1967	1973	760/BWR	Tokyo	J/US	US	US/J	J	US/J	Step increase in BWR size
1967	1974	440/BWR	Chusoka	J	J	J	J	J	Shimare, first all-Japanese nuclear power plant
1967	1976	515/BWR	Chubu	J	J	J	J	J	Hamaoka-1, first all-Japanese turnkey nuclear power plant
1968	1979	165/Fugen	PNC	E/J	-	E	J	J	Prototype light-water cooled, heavy-water moderated reactor
1969	1974	780/PWR	Kansai	US	J/US	US/J	US	J	Step increase in PWR size
1969	1975	780/PWR	Kansai	J	J	J	US	J	
1970	1983	500/BWR	Tohoku	J	J	J	J	J	
1970	1975	529/PWR	Kyushu	J	J	J	J	J	Turnkey
1970	1976	760/BWR	Tokyo	J	J	J	US	J	

1970	1978	814/BWR	Chubu	-	J	J/US	J	J	
1971	1979	1120/PWR	Kansai	US/J	US/J	US	US	J	Step increase in PWR size
1971	1979	1120/PWR	Kansai	J	US/J	US	US	J	
1971	1978	760/BWR	Tokyo	J	J	J	J	J	
1971	1978	760/BWR	Tokyo	J	J	J	J	J	
1971	1977	538/PWR	Shikoku	J	J	J	J	J	Turnkey
1971	1978	1056/BWR	Japco	US	US	US/J	J	US	Tokai-2, step increase in BWR size and last U.S. turnkey project in Japan
1972	1976	781/PWR	Kansai	J	J	J	US	J	
1972	1979	1056/BWR	Tokyo	US/J	US	US/J	US/J	US	Fukushima Daiichi-6, last LWR with significant U.S. participation
1975	1982	538/PWR	Shikoku	J	J	J	J	J	
1976	1982	1067/BWR	Tokyo	J	J	J	J	J	Fukushima Daina-1, first all-Japanese project over 1 GW

Source: "World List of Nuclear Power Plants," Nuclear News, February 1981, pp. 75-94; "Power Reactors," Nuclear Engineering International, July-August 1980, pp. 64-72; Petroleum News, November 1980, pp. 14-15.

PWR = pressurized water reactor; BWR = boiling water reactor; HWR = heavy water reactor; GCR = gas-cooled reactor.

[b]For all power plants ordered after 1972, there has been no significant foreign participation except for license fees.

[c]J = Japan; US = United States; UK = United Kingdom; E = Europe.

revenues go toward low-interest loans to utilities that purchase nuclear equipment made in Japan and to facilitate siting.

Partly as a result of the problems with light water reactors, the Japanese have been considering other reactor types that might act as insurance should the Japanese breeder program become delayed.[40] The 165-MW Fugen represents the most developed of these efforts. It is an advanced thermal reactor (ATR) prototype cooled by light water but moderated by heavy water and is capable of using a variety of fuel mixtures at relatively high efficiency. In competition with Fugen is adoption of the CANDU heavy water reactor by purchase from the Canadians. The CANDU would offer only a few of the fuel cycle advantages of an ATR although it would add flexibility to Japan's present all-LWR system. In 1979 and 1980, the Japanese seemed to be on the verge of purchasing a CANDU, but by early 1981 the relatively good performance of Fugen had apparently persuaded the Japanese to rely on their own design. Plans are being made to complete the design of a 600-MW demonstration ATR to be built in the late 1980s.[41] There remains substantial controversy within the Japanese nuclear establishment about the need for ATRs in the overall nuclear strategy, which is to move from LWRs to breeder reactors.

The Japanese have also been pursuing a number of other avenues in nuclear energy development that may eventually have significant commercial importance. In 1980, the U.S. General Atomic Company and the Japan Atomic Energy Research Institute signed an agreement to cooperate on Japan's very high temperature reactor (VHTR) program. A 50-MWth (thermal megawatts) reactor designed to deliver helium gas at 1000°C is being studied.[42] Commercial versions of such reactors could provide direct heat for steel making and other industrial processes.

In 1980 MITI announced plans to develop a 200 to 300 MW light water reactor designed both for electricity generation and steam production at 300°C.[43] Such reactors not only might have many potential applications in Japan if present siting restrictions were made more "flexible," but might enable Japan to expand its nuclear industry by exporting to developing countries.[44] To compete effectively with the existing supplier nations, such an export business would have to be supported by a much-expanded Japanese fuel cycle industry.[45] Many observers think that Japanese industry is not ready for exports and should cooperate directly with its traditional American partners--Westinghouse and General Electric--in any initial export plans.[46]

In the early 1970s, Japan built the 36-MWth _Mutsu_, a nuclear-powered merchant ship. This ship, which is under the jurisdiction of the Japan Nuclear Ship Development Agency (JNSDA), has been very controversial in Japan and has aroused severe public opposition in fishing communities. She developed a radiation leak on her maiden voyage in 1974, and after protracted negotiations among JNSDA, the Science and Technology Agency, and the Nagasaki Fisheries Association, the ship put in for repairs at the shipyard in Sasebo. It appears, however, that the repairs will not be completed before the agreement runs out in late 1981. It may be necessary to build a special port to handle the ship because of public concerns.[47] Her future and the future of the nuclear ship program remains clouded, although the Japanese Diet has reaffirmed the nuclear ship program.

The regulation of Japanese nuclear plants has sometimes been called "terroristic."[48] Frequent regulatory shutdowns and lengthy annual inspections, which are common occurrences, are reflected in low capacity factors. This is so in spite of the relatively small regulatory staff at MITI--about 50 in 1980. Although there were no licensing requirements, most observers felt that Japanese reactor operators had better training than their American counterparts before the Three-Mile Island accident. The regulatory apparatus was called into question in 1981, when it was revealed that contaminated water had been released from the 340-MW BWR Tsuruga plant south of Fukui. Along with allegations of a coverup by the utility to avoid government and public retribution, the resultant boycott of seafood from the region confirmed some of the worst fears of fishing communities about nuclear power plants.[49] It seemed likely that this incident would exacerbate the already considerable local resistance to plant siting in many areas as it has made finding a shipyard for the <u>Mutsu</u> more difficult than before.[50]

In late 1978, the Nuclear Safety Commission (NSC) began operation. It was designed to start the process of separating regulation and promotion, previously roles both played by the Japanese Atomic Energy Commission. The NSC oversees the work of MITI in regulation but does not implement safety regulations. It plays a crucial role in licensing procedure and has initiated local public hearings on reactor safety, held after the MITI public hearings, which also cover a range of nonsafety topics.[51]

Responding to the Three-Mile Island accident was the Nuclear Safety Commission's first real test. The accident precipitated inquiries from many local authorities, including pressure for shutdowns, and set the stage for a renewed series of antinuclear demonstrations around the country.[52] NSC responded by requiring MITI to undertake strict inspections of Japanese PWRs. At the time, only one PWR happened to be in operation, the others being shut down as part of routine procedure. This reactor, Ohi-1, was shut down for safety checks, and minor changes were made in the emergency systems of all PWRs. There were delays in granting permission to restart some of the PWRs, which are reflected by the low capacity factors illustrated in Table 4.4 and Figure 4.3. Besides a number of studies that are producing research, regulatory, licensing, design and operational changes, concerns over Three-Mile Island led to a gap in the granting of new construction licenses after the incident (March 1979) until late 1980. Other outcomes include MITI's plans to install inspectors in every plant and to develop elaborate local emergency response plans.[53] Finally, NSC also issued reassurances to the Japanese public that a Three-Mile-Island-type accident could not occur in Japan because of design differences and more stringent regulation.

Japan's famous nuclear allergy may be older but does not seem to be as strong as the allergies in some other OECD countries. One result, however, is that locating suitable and acceptable sites is extremely difficult, a problem exacerbated by Japanese geography as well as the Three-Mile Island accident and Tsuruga leakage. As can be seen in Table 4.7, about half of Japan's power plants are owned by two utilities, Kansai and Tokyo. This circumstance has decided advantages for quality assurance over the U.S. situation, in which many small utilities have one reactor but do not have large technical backup staff or long experience in nuclear plant operations. Indeed, Tokyo Electric Power Company is the largest privately owned utility in the world and has considerable economic and political influence in Japan as well as technical expertise.[54] This concentration means that Japan has been moving de facto into what has been termed "nuclear parks." These are emerging at the research headquarters--Tokai, at Fukushima south of Sendai, and in the region around Wakasa-wan Bay on the Sea of Japan. At Shimokita at the north end of Honshu, two utilities are planning to develop jointly a 20,000-MW site with the first reactor in 1990. To help alleviate the siting problem and to respond to seismic safety concerns, the Japanese government has been conducting feasibility studies for underground siting.[55] The reliability of this option is uncertain, but the emergence of more autonomous local decision making has led to a substantial reduction in available coastal sites.[56] It has also led to what are essentially transfer payments from urban to rural dwellers in the form of subsidies to local areas that do accept power plants, subsidies that result in higher electricity prices in the cities. To counter complaints that in the past these payments reached relatively few people and thus were more like bribes, in 1981 the government initiated two new subsidies--a progressive subsidy on utility rates and a direct subsidy to those prefecture governments from which a significant amount of electricity is "exported."[57] Disputes over the level of compensation have been a principal reason why the Onagawa plant north of Sendai had its operation date moved from 1975 to 1984.[58]

Siting is one of the important categories of problems that lead most observers to agree that MITI's projections for nuclear power shown in Table 4.3 are unrealistic. Simply summing the plants in operation, under construction, and in advanced planning status as of 1981 gives less than 30 GW by 1990 instead of MITI's 53 GW (see Table 4.2). Should a number of orders be placed early in the 1980s, as much as 35 GW might be feasible,[59] but the average time from proposal to operation would have to shorten substantially. Should MITI's targets be reached, nuclear power would account for about 25 percent of electricity in 1990 and 30 percent in 2000.

<u>Taiwan</u>. In contrast to many other countries, Taiwanese electrical demand during the 1970s grew substantially faster than had been projected in 1970 (see Table 4.4). Taiwan Power Company plans to expand nuclear capacity aggressively during the rest of the century to provide most of the new electrical capacity projected as a requirement for a continuing large GNP growth rate. To reach the goals set for 2000 (see Table 4.3), it will be necessary to build about one reactor per year. The eventual goal is

that about half of the island's electricity will be generated by nuclear power.[60]

As part of this expansion, serious consideration is being given to establishing a complete nuclear industry, although this would require a large expansion of present capabilities.[61] For the 1980s, purchases will be made from U.S. and, potentially, European vendors. Long-term financing may be a problem, however, as the requests to EXIM Bank for loans to Taiwanese nuclear projects in 1981 were close to 20 percent of all EXIM funds available.[62] Indeed, new financing arrangements will be required if U.S. manufacturers are to bid successfully for a large part of the planned rapid expansion of Taiwanese and Korean nuclear capacities.[63]

The Taiwanese nuclear program has seemingly run into very few serious problems. The construction times of its two BWRs were shorter than similar reactors in the United States (less than seven years). The Taiwan Power Company hopes to lower construction times to four years during the 1980s. Siting remains a problem, to a large extent because of the high seismicity of most of the island. Four sites at the northern and southern tips have been chosen for the first eight reactors and design modifications have been incorporated into imported U.S. equipment as results of this problem. In general, however, Taiwan has relied on U.S. safety criteria and regulatory approaches.[64] Taiwan has also been successful in expediting its reactor purchases by requiring that U.S. vendors receive government export approval before submitting bids. In 1981, this policy was approved by the U.S. Nuclear Regulatory Commission (USNRC) along with an agreement not to allow intervenor hearings during the export license procedure for Taiwan.[65] Taiwan, on the other hand, is planning to begin holding public hearings on power reactor siting decisions in 1982. These will undoubtedly slow the siting process.

<u>South Korea</u>. Taiwan and South Korea have the most aggressive nuclear programs in the world. As shown in Table 4.8, in terms of GNP, the construction programs in these two countries are two times as ambitious than the French program, which is the next largest. Their commitment to nuclear expansion stands out by a factor of nearly 4 among countries below the median global GNP per capita (approximately $2,500 in 1980). As indicated earlier, these expansion plans may be difficult to finance. This is true not only because of the capital limitations of lending institutions but also because of the magnitude of the loans in relation to the national economies. The marginal cost of the 120 MW per billion dollars of GNP (Table 4.8) under construction in South Korea probably represents nearly 20 percent of one year's GNP. If South Korea's official expansion plans were to keep on schedule, the total cost of reactors under construction in the early 1990s may exceed the 1980 GNP. As described in Chapter 3, there may be too many other claims on funds to allow such a heavy loan commitment to nuclear power. For this reason, and because of the general slowdown in economic growth, South Korea has been

TABLE 4.8
Nuclear Capacity Per Billion Dollars of Annual Gross National Product[a]

Country	Operating	Under Construction
Argentina	6 MW	21 MW
Belgium	15	36
Bulgaria	27	27
Canada	24	28
Czechoslovakia	12	43
Finland	38	17
France	24	60
Germany, Federal Republic of	12	13
Hungary	0	43
India[b]	5	9
Italy	5	10
Japan	14	9
Pakistan	6	0
Philippines	0	22
South Africa	0	38
South Korea	10	122
Spain	7	45
Sweden	37	57
Switzerland	22	10
Taiwan	37	115
United Kingdom	23	18
United States	22	38
U.S.S.R.	14	8

Source: "World List of Nuclear Power Plants," Nuclear News, February 1981, pp. 75-94; World Atlas (Washington, D.C.: World Bank, 1980); Annual Report 1980 (Taipei: Taiwan Power Company, 1981).

[a]GNP in 1979. Nuclear power plants operating or under construction as of December 1980.

[b]Asian countries are underlined.

reassessing its nuclear expansion plans beyond the 13 reactors listed in Table 4.2.

Although South Korea's present official schedule to the year 2000 is widely believed to be too ambitious, it is one of the few countries that has exceeded the nuclear projections done in the early 1970s. In 1973, for example, the IAEA projected that South Korea would have 9800 MW in 1990, which was a factor of 2 above official South Korean plans at the time.[66] As shown in Table 4.2, however, it seems likely that South Korea will have at least 11,000 MW on line in 1990. The official projections shown in Table 4.3 assume a consistent rapid annual growth in electricity demand and an even greater increase in the nuclear capacity--11 and 25 percent, respectively. This would bring nuclear power to 78 percent of the total electricity production by the year 2000.[67] Such high growth rates in electrical capacity and the ability to pay for it depend on maintaining high economic growth throughout the rest of the century--a very unlikely prospect, based on recent economic performance. The Korea Electric Company (KECO), the sole utility in the country, relies on internal funds, common stock, and loans from the Korean government, international banks, and bilateral arrangements with export financing banks located in countries where KECO purchases equipment. Apparently, about 50 percent of the capital required for overall system expansion in the 1980s will come from foreign loans. The first eight nuclear reactors, however, were financed nearly entirely from foreign sources.[68]

Such a large commitment to power stations that normally operate to meet baseload demand requires careful attention to system reliability and efficiencies. Where sites are available, pumped-storage hydroelectric facilities are one way to use the energy from base load plants to follow daily demand variations. Korea installed its first pumped-storage plant in 1979 (200 MW) and plans to have more than 2000 MW by 1990.[69]

An additional critical element in such rapid growth is skilled and professional staff. The Korean Energy Research Institute (KERI, formerly the Korean Atomic Energy Research Institute) estimates that the total personnel required by the end of the fifth Five-Year Plan (1986) will be 2.5 times that available in 1981. Although this problem is well recognized in Korea, major efforts will have to be made to retrain and attract qualified personnel.[70]

The scheduled lead time for nuclear plants is 12 years, during which it is planned to award the contract after 2.5 years and begin excavation after 4.5 years.[71] These lead times will also be affected by the availability of suitable sites, and the 44 plants planned for the year 2000 would require nearly 20 sites. Several factors, however, limit the number of suitable sites--seawater is generally used for cooling, coastal fishing industries are important, many coastal areas belong to the military, and all sites must be located below 37 degrees latitude for security in a north-south conflict.[72]

During the first phase of nuclear capacity additions, South Korea relied heavily on the regulatory programs of supplier countries (the United States, France, and Canada) to assure safety. Studies in recent years have recommended, however, that South Korea should develop a "strong, formalized, and self-sustaining Korean nuclear regulating program."[73] Standards, review procedures, codes, and personnel would all require strengthening in quantity and quality. The major institutions involved are all intertwined in ownership and management, and the boundaries of responsibilities are confused. Perhaps more seriously, the Korean Atomic Energy Commission (KAEC) like the old USAEC, is responsible for both promotion and regulation; difficult objectives to pursue simultaneously. The KAEC was originally modelled after the organization in Japan which, however, has now split these roles into different agencies.

The present system is burdened by administrative regulations, partly as a result of the lack of technical personnel. The Nuclear Regulatory Bureau (NRB), which functions as secretariat to the KAEC, for example, had only 20 persons with bachelor's degrees in 1980; turnover is high because of low salaries.[74] Part of the problem in attracting competent personnel to the NRB and other parts of the government's nuclear apparatus is extensive competition from private industry. South Korea is planning to increase steadily the local fraction of nuclear projects from approximately 30 percent in 1980 to 90 percent in 1990 and nearly 100 percent in the mid-1990s. The combined labor requirements of the utility (KECO), component manufacturers, research and development (KERI), the architect and engineering company, Korea Nuclear Engineering Services, Inc., and government is very high. Recent analyses indicate that the long-term interests of the country would better be served by lowered goals in both capacity growth and growth in localization. Such rapid growth may jeopardize safety and reliability as well as distort financial and labor markets, and defining the appropriate degree of localization has become subject to considerable debate within the South Korean nuclear establishment.[75]

South Korea has chosen the pressurized water reactor as its principal reactor type, not wanting to spread its technical personnel any thinner than necessary by trying to develop expertise in all three major reactors being exported today (PWR, BWR, and HWR). Accordingly, the decision to build Wolsung 1, a 630-MW Candu, may now seem overambitious. Although scheduled for completion in 1982, uncharacteristically poor project organization may result in some delay in this heavy water reactor. The first six PWR contracts were awarded to Westinghouse, but apparently the South Koreans felt the need to diversify the source of supply and the four new orders announced in 1980 were split between Westinghouse and the French company, Framatome. Although the two French reactors will be Westinghouse-style PWRs built under license, by choosing Framatome, the South Koreans gain access to different financing channels and different fuel suppliers, and hope to gain access to the considerable French nuclear experience in breeder reactors.

Southeast Asia

Nuclear power has had little impact on Southeast Asian countries, one of which--Indonesia--is the largest nonnuclear country in the world. Only one power reactor order has been placed by a country in this region.

Philippines. In 1973, the IAEA forecasted that the Philippines would have 9600 MW of installed capacity in 1990 and that nuclear power plants would account for 50 percent. A few years later, plans for nuclear plants had been scaled down to 1240 MW for 1990.[76] By 1980, official plans called for this total electrical capacity by 1985[77]--a 14 percent annual growth rate--although a study done by the World Bank in 1980 projected that this capacity would not be reached until nearly 1990.[78] In both cases, however, nuclear power will account for only 7 percent. Instead of six to eight reactors as originally projected by the IAEA, only one will be built--the 620-MW Westinghouse PWR under construction on the Bataan Peninsula across the bay from Manila.

The building of this power plant has generated one of the most intense controversies in the history of nuclear power. It brought to focus a number of issues surrounding the suitability of nuclear power to conditions in developing countries and the relations between supplier and consumer countries.

Philippine Nuclear Power Plant I (PNPP-I) is extremely expensive. Although originally budgeted at less than $1,000 per kW, the approved bid was $1,700 in 1976. Although this seemed significantly higher than other reactor contracts at the time (a 50 percent larger Westinghouse PWR was sold to Spain at two thirds the total cost), Westinghouse has steadfastly maintained that the extra cost was necessary because of the lack of local infrastructure. Nevertheless, the $1.1 billion represented about 30 percent of the external debt of the country at the time.[79] In 1980, the Philippines negotiated additional loans to pay cost overruns, which bring the total cost to $1.9 billion or $3,065 per kW. As shown in Chapter 3, Table 3.12, the Philippines has one of the region's largest foreign debt service ratios, of which this one nuclear project now accounts for about 20 percent.

The Luzon electrical grid is expected to grow at about 7 percent per year. From a systems stability viewpoint, therefore, the 620-MW reactor will be a decade too early if it comes on line in 1986 as now planned. The scheduled addition of 300 MW of pumped hydro in 1983 may help alleviate potential problems, however.[80]

In 1979, shortly after the Three-Mile Island accident, President Marcos suspended construction on PNPP-I in part to reexamine the safety of the plant, which was then about 20 percent complete.[81] By late 1980, when the construction ban was lifted, the total cost of safety changes had reached approximately $100 million. Of this, less than 10 percent were changes responding directly to Three-Mile Island, the rest being changes in response to new studies of the earthquake and volcanic hazards at the site.[82] These hazards were, and continue to be, the central focus of opposition by intervenor groups. Indeed, the controversy over the Bataan reactor led to public protests and legal battles on both sides of the Pacific and has become one of the principal rallying points for a number of anti-Marcos groups. Although intervenors, to be successful, must be based locally, this controversy has shown that international trade in nuclear technology is accompanied by an international trade among national public interest movements.

The U.S. export license for the reactor was issued in mid-1980 after four years of deliberations by the U.S. government. The U.S. State Department originally approved the application in 1977, but then asked the USNRC to delay final action because of domestic and Philippine concerns about safety. The construction ban in 1979 was partly directed toward pressuring the USNRC to issue the license. In late 1979, the USNRC decided to recommend approval. These deliberations were complicated by President Carter's executive order in early 1979 extending U.S. health, safety, and environmental criteria, although not the National Environmental Policy Act, to the final site of nuclear reactors exported by U.S. firms. In early 1980, the USNRC interpreted the executive order to mean that only impacts directly on the United States and on the global commons such as the ocean would be considered in reactor export proceedings. This eliminated the need for detailed safety reviews requiring site visits, which could have been considered interference with national sovereignty. In 1981, intervenor suits were still pending over the NRC's refusal to consider environmental impacts on the Philippines and on the 32,000 Americans living at military bases near the reactor site.

In his report to the World Bank on the South Korean nuclear program, S. Levy strongly recommended that

> a domestic reference plant in the foreign vendor's country should be specified at the purchase stage to be able to keep track of all the changes and new safety requirements being imposed in that country.[83]

Unfortunately, the Philippine reactor has no such reference facility because such plants were never built in the United States (the only one planned, in Puerto Rico, was canceled). This is a particularly troublesome omission because it places a greater burden on indigenous technical expertise to assure safety. Much more so than in South Korea, the Philippines is not yet able to field the necessary quantity and quality of trained personnel.

What to many Filipinos may have seemed a modern addition to national prestige in 1975 was generally considered a politically and financially expensive white elephant by 1980.[84] Although plans for further reactors have been shelved, some observers believe that the relative cost of the major alternatives--hydroelectricity, geothermal, and coal in the Philippines--may rise to the extent that nuclear power will be financially attractive by the end of the 1980s. In that case, the Philippine experience with PNPP-I may be judged as a necessary, if expensive, entry fee.

Other Southeast Asian Countries. Although figuring prominently in early projections of Asian nuclear expansion, it seems unlikely that any other plants will be built in Southeast Asia this century. Thailand, for example, although professing an interest in nuclear power in the early 1970s, briefly tested bitter public opposition and now finds its natural gas and hydropower reserves to be much less expensive to develop. Malaysia, with extensive gas and oil deposits, has not experienced high growth rates in electrical demand. It has been moving slowly into nuclear expertise by ordering a research reactor for completion in the early 1980s. Singapore presents nearly impossible siting problems, and a joint project between Singapore and Malaysia seems remote under present strained relations.

Indonesia is likely to be the next Southeast Asian country to build a full-scale nuclear power plant. It has expressed particular interest in heavy water reactors because of the freedom from enrichment facilities they enjoy. A joint research effort with Italy on HWRs was announced in 1980, along with a German contract for construction of a 25-MWth research and test reactor, and a pilot HWR fuel fabrication plant.[85] There are reports that the country hopes to build a 600-MW HWR in central Java;[86] by the late 1980s, the Java grid would be large enough to accommodate such a plant.[87] Most observers believe, however, that the financial and technical infrastructural barriers to nuclear power will persuade Indonesia to seek coal and gas alternatives until the next century.[88]

FUEL CYCLE

The nuclear fuel cycle is more international than perhaps any other major fuel cycle. Relatively low transport costs and high political visibility often lead to tortuous and extended paths through half a dozen or more countries from mine to reactor to waste disposal. Thus, it is valuable to examine the stages of the fuel cycle on a regional rather than national basis.

Uranium Supplies

For the rest of the century, the most important countries in the Asian region from the standpoint of uranium supply and demand will be Australia and Japan. They are likely to represent roughly 10 percent of global uranium production and consumption, respectively, during this period, although there are great uncertainties in the forecasts of global nuclear power growth. At present, only small amounts of uranium are mined in Japan; India produces a few hundred tons a year; and China seems to have an annual production of one or two thousand tons. Small deposits have been reported in Indonesia, South Korea, Pakistan, the Philippines, Thailand, and recently, in Taiwan,[89] but the bulk of the resources, as well as the production in the region, are found in Australia, which contains about 8 percent of the uranium resources and 18 percent of reserves outside the USSR, China, and

Eastern Europe (Table 4.9). If, as shown in Table 4.3, total Asian nuclear capacity were to reach 130 GW by year 2000, the uranium requirement would be about 19,000 tons per year at steady state. Actual requirements that particular year would be greater since many reactors would just be starting up and LWRs usually begin operations with a three-year load. Utilities also prefer to order somewhat in advance to build up a buffer stockpile. Nevertheless, this amount is apparently well within the potential production capacity of Australia, and the region could be self-sufficient in uranium for many years.

Australia partook in the global uranium boom in the 1950s when uranium was being purchased by the United States and the United Kingdom for military purposes. These contracts lapsed entirely by 1964, however, and all production stopped in 1971, at which time a small stockpile had been created. Several large new discoveries were made in the 1970s, but production did not resume until 1976 because of conditions in the international uranium market and domestic political opposition. In the 1970s, the government took on greater regulatory and ownership roles in the industry, but by 1980 the Liberal Party government of Malcolm Fraser had reduced these government roles substantially.[90]

Environmental concerns both about damage from mining and about the operation of the global nuclear fuel cycle combined with a growing Aboriginal rights movement in the 1970s to move Australia toward reevaluation of its uranium mining and export policies. A major inquiry was undertaken and the results finally accepted by the two major political parties.[91] Basically, the report recommended a phased and moderate rate of development with considerable attention to Aboriginal rights and environmental concerns. Not all of these recommendations have been implemented to date. In concert with other major OECD uranium producers (the United States and Canada), however, Australia has imposed safeguards restrictions on uranium sales, requiring, for example, prior approval in each case for reprocessing.[92] This policy had led to difficulties in obtaining contracts with Japan and European nations. In 1980, however, the government proposed to give generic approval to certain customers for reprocessing, a step welcomed by potential customers.

Production in 1980 was approximately 2,000 tons, with capacity totalling 3,500 tons under construction. Present deposits could supply a much larger capacity by 1990 if markets could be found, although in 1981 firm contracts amounted to only a commitment for 2,500 tons per year in 1990. The 1980 changes in safeguards requirements, however, were apparently leading to additional contracts that might as much as double the commitments for 1990.[93]

At the beginning of the 1980s, the conditions of the international uranium market did not bode well for development of new production capacity. The price of uranium was decreasing and, at \$25 per pound of U_3O_8, was lower than it had been since World War II in real terms except for a brief period in the early 1970s.[94] At the same time, there was an excess of production capacity in the world and substantial stockpiles of uranium in consumer countries (outside the United States and Canada). Stockpiles

TABLE 4.9
Nuclear Fuel Cycle in Asia

Country	Uranium (tons metal)		Enrichment (1,000 separative work units)		Reprocessing (tons metal)	
	RAR[d]	1980 Production	Present Capacity	Planned Additions	Present Capacity	Planned Additions
Australia	292,000[e]	1,950[i]	-	(1,500)[n]	-	-
China[a]	100,000?	1,000-2,000	~80	?	?	?
India	29,800	~250[j]	-	-	30	200
Japan	7,700	2	70-80[m]	1,000-2,000[o]	210	1,100[s]
Pakistan	small	~10[k]	-	p	-	-
Philippines	300	-	-	-	-	-
South Korea	(f)	-	-	-	-	-
Taiwan	(g)	-	-	-	(0.01)[r]	-
World[b]	1,850,000[h]	50,000[l]	30,000	20,000[q]	3,000	9,000[t]
1 GW plant-year[c]	145		115		30	

[a] Kim Woodard, <u>International Energy Relations of China</u> (Stanford, Ca.: Stanford University Press, 1980).

[b] Not including the USSR, Eastern Europe, or China.

[c] A LWR with a 30-year life and 0.2% tails assay. A HWR would use no separative work and about 120 tons of natural uranium per year (Thomas Neff and Henry Jacoby, "The International Uranium Market," MIT-EL-80-014, Cambridge, Mass.: Massachusetts Institute of Technology, 1980). The energy equivalent at 70% capacity factor is 2.4 million tons coal equivalent, or 11.5 million barrels oil equivalent. Gloval average capacity factors have been less than 60%, however see Table 4.6.

Table 4.9 (cont'd)

[d]"Reasonably Assured Resources" at a production cost of less than $80/kg of uranium in 1978 dollars according to NEA/IAEA (Uranium: Resources, Production and Demand, Paris: OECD, 1979). The NEA/IAEA lists Indonesia, Pakistan, and Thailand as having "moderate potential" for discovery of uranium resources (NEA/IAEA, World Uranium Potential, Paris: OECD, 1978).

[e]Reserves reported by companies total more than 350,000 tons (R. T. Madigan, "Minerals and Energy--Key to Development," paper presented at the Australian National Committee 12th General Meeting, Los Angeles, Ca.: Pacific Basin Economic Council, 1979; NUEXCO, "Monthly Report on the Uranium Market," no. 153, Menlo Park, Ca.: Nuclear Exchange Corporation, May 1981).

[f]Three thousand tons uranium at $80-130/kg and 41,000 tons thorium at presently subeconomic grades (Robert Hall, "Assessment of Thorium Resources in the Republic of Korea," World Bank, 1980).

[g]There are recent signs of small uranium deposits in Taiwan (Nuclear Engineering International, November 1980, p. 7).

[h]World total of RAR up to $130/k, U = 2,590,000 tons total of RAR, and estimated additional resources = 5,040,000 tons.

[i]The Nabarlek mine at 1,270 tons and Mary Kathleen at 680 tons (Nuclear Engineering International, January 1981, p. 9). Estimates of production in 1990 range from 5,000-23,000 tons (Neff and Jacoby, "International Uranium Market").

[j]India is planning to expand its present uranium production capacity from 1,000 tons/day of ore. India has about 320,000 tons of thorium reserves.

[k]Apparently Pakistan is producing enough uranium to partially fuel its 125-MW HWR (Nuclear Engineering International, October 1980, p. 4).

[l]Neff and Jacoby, "International Uranium Market."

[m]Nucleonics Week, October 9, 1981, p. 10.

[n]Considerable discussion but no definite plans for commercial-scale capacity by the early 1990s.

[o]A commercial-scale centrifuge facility is planned for 1990 (Nuclear Engineering International, January 1981, p. 6). In addition to the demonstration-scale centrifuge plant under construction, approximately 2 kSWU in chemical enrichment will be operating by 1983 (Nuclear Fuel, April 27, 1981, p. 4).

[p]There have been frequent rumors that Pakistan has been constructing a clandestine centrifuge plant for producing weapons-grade uranium

Table 4.9 (cont'd)

(*Nuclear Engineering International*, June 1979, p. 27; Ashok Kapur, "Nuclearizing Pakistan: Some Hypotheses," *Asian Survey* 20, 5 [May 1980]: 495-516). See, for example, the BBC television production *The Islamic Bomb*, 1980.

[q]P. Bauder, A. Hornscastle, G. Lurf, and M. Stephany, "Competing for the Non-USA Enrichment Markets," *Nuclear Engineering International*, October 1980, pp. 37-41; Thomas Neff and Henry Jacoby, "Supply of Assurance in the Nuclear Fuel Cycle," *Annual Review of Energy* 4 (1979): 259-311.

[r]Apparently dismantled in 1976 under pressure from the United States (Gene I. Rochlin, *Plutonium, Power and Politics*, Berkeley: University of California Press, 1979).

[s]*Nuclear Engineering International*, June 1980, p. 9.

[t]Rochlin, *Plutonium*, chapter 3.

---

totaled about three years' supply in 1980 and will total as much as seven years' supply by 1990 even if no more uranium contracts are made. Presently committed expansion plans in uranium production capacity will lead to about a 50 percent excess of export supply over demand in 1990. If all countries were to stretch to reach their maximum attainable capacities (23,000 tons per year, in the case of Australia), the excess would be a factor of 3 or 4.[95]

The largest customer in the region has contracted for only 12 percent of its supply from Australia through 1990. Japan has a policy to diversify its sources, and the largest single supplier during this period will be Canada at 35 percent. Unless new contracts are signed, however, Japan will not be importing uranium directly from South Africa, France, or the United States after 1985, increasing its reliance on Niger, the independent company Rio Tinto Zinc of London, and possibly, Australia. There is little impetus for signing new contracts, since in 1980 Japan held stocks equivalent to more than a 10-year supply for its operating nuclear capacity, and under current contracts and the most likely capacity growth (30 GW in 1990), might hold as much as a 15-year supply in 1990. The economic cost of holding stocks is high. A 15-year supply might lead to additional electrical generation costs of 25 to 50 percent, depending on the carrying charges and the degree to which the uranium is held in enriched or fabricated forms.[96]

The cost of this much energy security is high, especially since most observers believe that uranium contracts will remain a buyer's market throughout the decade and beyond. There will be many opportunities for purchases and for joint ventures. Japan, for example, has equity interest in Niger production, and South Korea has been investing in the Gabon uranium industry.[97] Indeed, there is so much uranium held in stocks by consumer countries that significant amounts may come on the market for resale. Iran, for example, has apparently been reselling uranium originally purchased from Namibia and South Africa.[98]

Although South Korea has concluded small contracts with Canada and Australia, and Taiwan has a modest contract with South Africa, these two countries have not yet become major actors in the uranium market. This is largely because their nuclear programs are relatively recent and initial cores and reloads traditionally are supplied by reactor vendors for new reactors.

Enrichment

Uranium, like petroleum, must be processed in special facilities before it can be used as an energy source. Relative to petroleum, the political and geographic distribution of present and potential exporters of significant amounts of uranium is large. In contrast to petroleum, processing capacity is restricted to a relatively small number of facilities. In 1980, this processing, called enrichment, was available for foreign use in significant amounts from only three suppliers--the U.S. Department of Energy (nearly 70 percent of total commercial capacity); the Eurodif facility in France (20 percent) owned by France, Belgium, Spain, Italy, and Iran (which is trying to dump its 10 percent share); and the Soviet Union, which has been offering enrichment services to non-Soviet bloc countries equivalent to 10 percent of the total available in these countries.[99]

As shown in Table 4.9, significant amounts of new enrichment capacity are under construction. In 1990, it is estimated that the USDOE's share will have dropped to about 60 percent, Eurodif will retain its 20 percent share, and there will be two new European facilities--Coredif, involving the same countries as Eurodif, and Urenco, involving the United Kingdom, West Germany, and the Netherlands. Under present contracts, the Soviet share will drop to 3 percent. Brazil, South Africa, and Japan are planning to build commercial-scale plants for domestic use in this period as well.[100]

It seems that enrichment services, like uranium itself, will be available in abundance during the next decade. Present contracts will result in significant stockpiles of enriched uranium by 1990, at which time there will still be unused capacity. The cumulative global surplus of enriched uranium in 1980 was about six years' forward supply, a surplus likely to remain until the end of the decade. By 1990, global demand will probably reach only 80 to 85 percent of enrichment capacity. Japan has brought nearly all of its enrichment services from the USDOE, although a contract with Eurodif began in 1981. In Japan, demand for enrichment may be less than half the amount for which it has contracted by 1990. Indeed, Japan could well enter the 1990s with five to eight years of cumulative enrichment supplies. Taiwan and South Korea have received all their enrichment services from the USDOE through contracts with U.S. firms associated with reactor purchases, although it is likely that Korea will purchase Eurodif enrichment as part of its 1980 deal to buy two French PWRs.

In this market climate, Japanese plans to expand the country's pilot-scale enrichment facility to supply roughly half of its needs by 1990 must be attributed to a heavy emphasis on energy security rather than economic efficiency.[101] It would seem difficult to compete in a market characterized by excess capacity in the large U.S. and European facilities even with the greater energy efficiency and flexibility promised by the centrifuge plants being planned in Japan. On the other hand, development of indigenous fuel-cycle facilities is recommended by some observers so that Japan can compete effectively in the reactor export market.[102]

Australia, too, has announced tentative plans to build enrichment facilities, perhaps with European or Japanese participation.[103] Although there is some incentive to capture more of the value added in uranium sales, Australia has no concrete plans for a significant domestic nuclear power capacity. In present international market conditions, the financial risk of such an undertaking would be great.

Part of the interest of countries such as Japan, Brazil, and South Africa in domestic enrichment capacity and the Australian belief in the potential profitability of alternatives to U.S. and European suppliers lies in the worry that political considerations might interrupt services from present suppliers. India, for example, has experienced delays amounting to several years in deliveries of enriched uranium from the United States because of U.S. concerns over safeguards.[104] Late in the Carter administration, Congressional approval for these shipments was belatedly granted, but in mid-1981 the Indians were still unhappy with the actual schedule of shipments.[105] Another worry of countries depending on foreign enrichment is that they might be subjected to economic pressures as a result. In the mid-1970s, for example, the United States tried to impose enrichment contracts that were interpreted by many foreign observers to be designed to secure commercial dominance in reactors and fuel cycle services.[106] The message to others who depend on foreign enrichment contracts is clear, however. It is not enough only to find sufficient uranium supplies; reliable sources of enrichment must also be tapped.

Enrichment can be avoided entirely by employing reactors that use natural uranium directly. Heavy water reactors, the most popular of which is the CANDU, offer this advantage in addition to using about 15 percent less uranium and being somewhat less complicated to build. Their higher initial capital cost makes them less economic, however, unless uranium prices rise significantly. Their advantages have led India to choose HWRs as the basis of their nuclear program and may lead Pakistan, Indonesia, and China in a similar direction.

Reprocessing

Unlike enrichment, reprocessing is not necessary for the light water reactor fuel cycle, although it does offer the advantage of recycling plutonium and unused uranium from spent fuel back to reactors for increased total fuel efficiency. Reprocessing costs are such that the financial advantage lies with once-through systems (no reprocessing) when uranium and enrichment costs are low. Since uranium supplies and enrichment services are expected to be in oversupply until at least the mid-1990s, reprocessing does not promise to be economically attractive during this period. Although there is considerable debate over the costs of reprocessing, the principal arguments for and against reprocessing do not lie directly with the relative economies. The argument being favored by the United States during the Ford and Carter administrations and to some extent adhered to by a few U.S. allies (principally Australia and Canada) was that reprocessing should be discouraged because it increases the chances of proliferation by making plutonium available in international and domestic commerce. The principal argument in favor of reprocessing is that it provides the required experience as well as a source of the initial plutonium cores for breeder reactors. Consequently, several European countries and Japan argue that reprocessing is necessary no matter what its present economics. The International Nuclear Fuel Cycle Evaluation Program (INFCE) conducted from 1978-1980 addressed this issue directly without coming to definitive conclusions.[107] Since the U.S. negative position was not strongly supported, however, the international view has tended toward the Japanese and European positions. The Reagan administration also is more sympathetic to reprocessing and the breeder than its predecessors.

Before obtaining operating licenses, Japanese nuclear utilities must make arrangements for the spent fuel to be generated by each reactor. Since Japan has only a small demonstration-scale reprocessing plant (built with French technology at Tokai Mura), spent fuel is being sent to Windscale, U.K. and La Hague, France for reprocessing. Contracts for 4,600 tons of spent fuel have been signed for the 1980s--roughly enough in combination with Tokai Mura to supply Japanese needs. The plutonium and fission products are being held in Europe for eventual shipment back to Japan. Japan agreed to hold back on plans to build a larger reprocessing plant until after INFCE. Plans seem to be proceeding for an 1,100 ton/year plant by 1990, although Japan also seems to have agreed to explore nonreprocessing options.[108]

One of these options is an international spent-fuel storage facility in the Pacific. Under the Carter administration, a feasibility study was begun for storing spent fuel on a relatively remote and uninhabited Pacific Island under U.S. control. It seems clear that the Pacific Island political and intervenor group opposition would become intense should such a facility actually be proposed. Although events such as the Israeli raid on Iraq's reactor may alter its position, the Reagan administration is not likely to push Pacific spent-fuel storage for Japanese fuel as strongly, given its increased attention to domestic reprocessing and development of the breeder. By 1990, on the other hand, if no disposal options have yet become viable, a Pacific spent-fuel facility may become an important part of U.S. efforts to help assure that plutonium is not recycled in Taiwan and South Korea

India has a small reprocessing plant at Tarapur near Bombay. The smaller reprocessing plant at Trombay used to separate the plutonium for India's nuclear explosion in 1974 has been closed down since 1975. The United States has not given permission for U.S. enriched fuel to be reprocessed in India. In 1981, however, India announced that it might reprocess anyway unless promised shipments of enriched uranium under the 1963 U.S.-India cooperative agreement were forthcoming. India, which has had to restrict the output of its two BWRs because of delays in enriched fuel shipments, has sought a "polite end" to this agreement.[109]

Waste Disposal

Nuclear waste disposal has yet to become as much of a bottleneck in Asian nuclear development as it has in the United States. There are signs, however, that it may. As in the United States, spent-fuel storage space has become critical at Tarapur in India and limited at some sites in Japan.[110] Outside China, only India and Japan have any reprocessing waste as yet. India has apparently been keeping it in underground concrete vaults, while Japan plans to store it at Tokai Mura until a waste management plan is deployed in the 1990s.[111]

Countries in the region with the largest nuclear programs seem to have a fairly limited potential for permanent disposal of high-level wastes. Only India and China have large regions with the prerequisite conditions of low population density, low rainfall, and low seismicity. Australia has a number of regions with these general characteristics but does not have the political climate to offer waste disposal services in spite of some proposals to this effect.[112] Part of the Australian public opposition to uranium sales is apparently due to fears that Australia may one day be called upon as a disposal site for the resulting waste.[113] The first power reactors in Taiwan and South Korea only began operation in 1978, and thus high-level nuclear waste issues are not likely to become acute unless there are still no accepted international management strategies by the late 1980s. The lack of an apparent long-term program for waste management is one of the reasons stated by the Philippine government for suspending the construction of its pressurized water reactor.[114] Japan seems to have neither the physical nor political conditions necessary for land-based waste disposal. The Japanese utilities are particularly sensitive to changes in the nuclear waste situation because of the licensing requirement that a plan be in place to handle all spent fuel over the life of each reactor. Thus, they are vulnerable through regulatory as well as physical links to political or technical restrictions that may develop in the transport, reprocessing, storage, and final disposition of Japanese spent fuel, which in the 1980s will be reprocessed in Europe. Subseabed and other international disposal options for high-level waste will probably receive more attention and support from these countries during the 1980s as nuclear waste issues become increasingly public and as spent-fuel storage pools grow crowded.

Japan intends to store returning high-level waste from Europe along with that from domestic reprocessing at Tokai Mura until permanent disposal techniques are perfected and tested. In the interim, however, it is finding difficulty in disposing of the much less dangerous low-level wastes from the nuclear fuel cycle. In 1980, a Japanese team toured Pacific Island nations to try to gain acceptance of a plan to dump these wastes 500 miles north of the Marianas.[115] In spite of receiving universally negative responses and sending initial diplomatic messages that the project had been scrapped, in early 1981 there were indications that the plan seemed to be going ahead.[116] Available land sites, which in Japan are basically restricted to existing nuclear facilities, were apparently not adequate. Dumping and leaking of low-level wastes in coastal waters has been a major concern of Japanese fishing communities and organizations, which are perhaps the most vocal and effective of the antinuclear groups in Japan.[117]

Problems surrounding transport of spent fuel and waste may also prove to be obstacles to foreign reprocessing and waste disposal options in the future. In 1979, for example, a British ship carrying Japanese spent fuel to Europe was barred from refueling in Honolulu Harbor by the Governor of Hawaii and the spent-fuel ships have been rerouted around Africa as a result.[118] Because of the shallow approaches to the docks at the Japanese reactor sites, these ships cannot be larger than about 3,000 tons and therefore carry about 100 tons of spent fuel.[119] At least 50 separate shipments will be needed to meet present contracts through 1991. The Pacific Nuclear Transport, Ltd. fleet, which is owned by the British and French national nuclear fuel companies and six Japanese utilities, will consist of four to six specially constructed spent-fuel ships by the mid-1980s.[120] The likely Japanese nuclear capacity in 1990 would result in an annual steady-state production of approximately 1,000 tons of spent fuel. If the commercial-scale Japanese reprocessing plant is not operating by then, about ten one-way shipments a year will be required. Conceivably, these ships could also be used to bring the high-level waste back to Japan, although a considerable amount of processing of the waste would be necessary in order to ship it safely in spent-fuel casks. Local concern about these shipments, as evidenced in Hawaiian, Panamanian, and European ports, may be the forerunner of a time when local authorities attempt to ban nuclear materials on international routes in spite of international agreements about acceptable safety precautions. In the late 1970s, for example, many local jurisdictions in the United States attempted to ban domestically routed radioactive material until stopped by federal court action.[121]

BREEDER REACTORS

The long-term future of nuclear power in this region, as elsewhere, depends on the future of the breeder reactor. Although there are significant uncertainties about the availability of uranium resources at low and moderate costs, essentially the only way in which these resources can be stretched to allow nuclear energy to become a long-term and widespread energy source is to take advantage of the great fuel efficiency offered by breeder reactors. Thus, to evaluate the future of nuclear power in the region, the present status of breeder development should be considered.

In Asia, two breeder concepts are being pursued. The liquid-metal fast breeder reactor (LMFBR), which converts fertile uranium to plutonium, is being developed in both Japan and India. India hopes to modify the system so that it can convert fertile thorium into the nuclear fuel uranium-233 as well and thus tap India's abundant thorium deposits.

The Japanese experimental 75-MWth fast reactor, Joyo ("Eternal Light") has been operating at O-Arai near Mito since 1978.[122] It is to be followed by the 280 to 300 MW prototype, Monju at Tsuruga, planned originally for 1978.[123] As of early 1981, construction had not yet begun and the earliest start-up had slipped to 1987.[124] Much of the delay to date had been due to local environmental opposition by the Fukui prefectural government. There is speculation that the spill of radioactive water at the Tsuruga plant in 1981 would result in increased public opposition to nuclear projects in the area and further delay of Monju. The cost of Monju is estimated to be nearly $2 billion, of which a federation of utilities will pay 20 percent.[125] A study has tentatively identified a 1000-MW design for the first commercial-scale breeder, but no concrete schedule had been established by 1981.[126]

India is building a small experimental fast reactor at Kalpakkam near Madras. This 15-MW facility is based on French technology and assistance and, though originally planned for 1976, will begin operating in 1982 at the earliest. Official Indian plans calls for 250 to 500 MW fast breeder designs by the late 1980s[127] and as many as 1,000 MW of breeders in place by 2000.[128]

Although much has been written in official discussions about the need to utilize Indian thorium resources, technical progress to date has apparently been minimal.[129] There are a number of unresolved questions about the necessary reactor and fuel-cycle modifications of the French-style fast-reactor technology being pursued by India so that it can use thorium efficiently and reliably.

Although there are many arguments in its favor,[130] the decision whether and when to employ the liquid-metal fast breeder reactor brings to focus many of the major problems surrounding nuclear power:

1. Development is expensive and lengthy, and the technology is complex.
2. The capital cost of the facilities will be high even compared to conventional nuclear plants.
3. The fuel cycle requires the circulation of weapons-suitable material.

4. The reactors have safety characteristics that may not lead to overall greater hazard but are different enough from present reactors to require lengthy and expensive evaluation.

India's investment in the high-technology and long-term commitment required by a breeder program is viewed by many observers to be a considerable economic and political risk because uranium and enrichment costs seem likely to remain low for some years, proliferation fears are likely to revive quickly upon entry of the next developing country into the nuclear weapons club, and developed country expertise in breeders is likely to remain far ahead for many years.[131] Even Japan might well be better served by adapting European or U.S. breeder exports on a schedule tailored to Japanese energy demand and the international market conditions affecting LWRs rather than embarking on an expensive and lengthy separate development program. Indeed, in spite of the optimistic official timetable, the breeder programs in both countries have now slipped more than a decade. The economic, technical, and political uncertainties can be expected to cause a further slippage so that it may be beyond the end of the century before an indigenous commercial-scale breeder is employed in Asia. It has been argued, on the other hand, that inconsistent and frequently changing policies of supplier countries are pushing Japan and India toward indigenous nuclear development.[132] Most of the policies singled out for criticism are those relating to safeguards designed to stop the proliferation of nuclear weapons.

PROLIFERATION

Energy security, an objective served by many actions including developing supplies and increasing efficiency of use, is valuable for the increase in national economic security it brings. Nuclear power is unique among present energy sources in that it may be perceived by some countries to offer another route to security--through acquisition of nuclear weapons. The evaluation of the impact of nuclear power on national welfare is very complicated as a result. As Figure 4.4 shows, an individual government may judge nuclear power's potential benefits for military security sufficient to overcome its relative shortcomings in resource security, safety, and public acceptance as compared to other sources. Both sorts of benefits accrue to individual countries, while the deficits--increased chance for nuclear war--would seem to occur internationally. Thus, there is a potential "tragedy of the commons," where the benefits accrue individually but the deficits accrue collectively.[133]

These have been the basis of arguments about nuclear power since the 1950s, as discussed here. Since the 1964 Chinese nuclear explosion, a number of international and unilateral efforts have been made to avoid the tragedy of the proliferation commons by imposing outside restraints on nonweapons countries embarking on nuclear power programs. Here there is space only to discuss

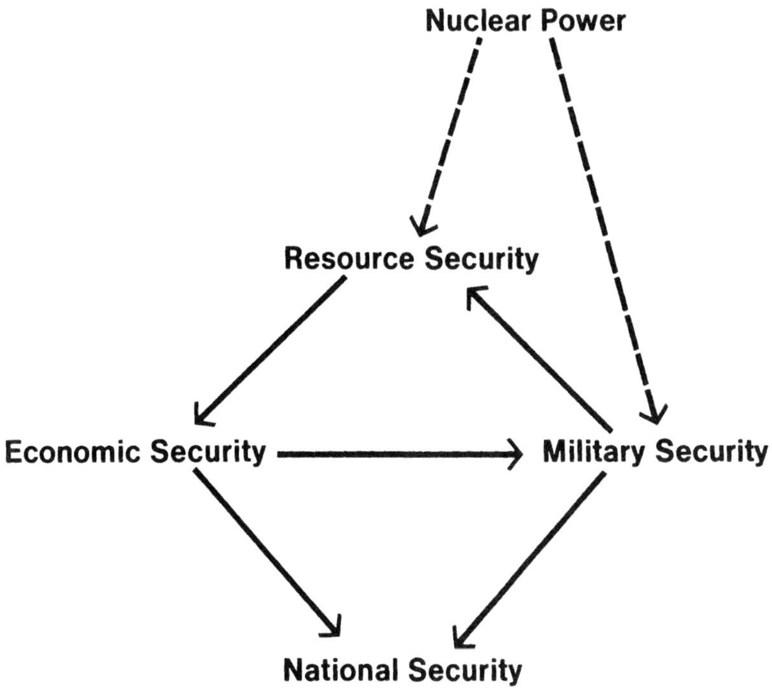

Figure 4.4. National security is dependent on a number of factors, some of which are shown in this figure. Nuclear power provides electricity, an important resource, but brings supply uncertainties that can lead to unwanted shifts in resource security. Nuclear power capability may also bring access to nuclear weapons to some countries. This can increase the apparent military security of these countries at the expense of international security. (Source: Kirk R. Smith, "Comments on U.S. Nonproliferation Policy after INFCE," in Stuart Harris and Keichi Oshima [eds.] Australia and Japan: Nuclear Energy Issues in the Pacific [Canberra and Tokyo: Australia-Japan Economic Relations Research Project, 1980]: 109-112.)

---

briefly the major efforts and how they have affected nuclear power in Asia.

Nonproliferation Treaty (NPT, 1968)

Twelve years after the original 99 countries signed the Nonproliferation Treaty (NPT), there were 114 signatory countries, of which 72 had NPT safeguards agreements in place. These 72 included many Asian countries, as shown in Table 4.10. Important exceptions are noted in the last column. India and Pakistan have neither signed nor placed all their facilities under IAEA safeguards (that is, opened them up to international inspection).

TABLE 4.10
Nonproliferation Treaty and IAEA Safeguards Agreements in Asia (1981)

Country	NPT Signatory	IAEA Safeguards	Non-NPT Little Activity	Non-Safeguarded Activities
Australia	X	X		
Bangladesh	X	–[a]		
Bhutan			X	
Burma			X	
Cambodia	X			
China				X[d]
India				X
Indonesia	X	X		
Japan	X	X		
Laos	X			
Malaysia	X	X		
Maldives	X	X		
Nepal	X	X		
New Zealand	X	X		
North Korea		X[c]		
Pakistan				x
Papua New Guinea			X	
Philippines	X	X		
Singapore	X	X		
South Korea	X	X		
Sri Lanka	X	–[a]		
Taiwan		X[b]		
Thailand	X	X		
Vietnam			X	

Source: IAEA Bulletin 22, 3/4 (1980): insert, and 23, 1 (1981): 32.

Table 4.10 (cont'd)

[a]Awaiting entry into force.

[b]Taiwan ratified the NPT but was removed by vote of the other members in 1972. All nuclear activities are currently covered IAEA safeguards under bilateral agreements with supplier countries.

[c]Non-NPT IAEA safeguards cover all nuclear activities.

[d]Weapons state.

China is already a nuclear weapons state and, though not a signatory to NPT, is considered unlikely to breach the NPT by transferring significant amounts of nuclear technology to other countries without safeguards agreements. China has also indicated interest in joining the IAEA. Japan, even under pressure from the United States, did not ratify the treaty until the mid-1970s.[134]

One of the weaknesses of the NPT is that it does not restrict transfer of nuclear technology to countries that have not signed, but only requires that the particular facilities involved be placed under IAEA safeguards. Thus, India has some facilities under IAEA safeguards (the Tarapur BWRs, for example) and some that are not, including research facilities, reprocessing plants, and its indigenous HWR power program. An intermediate type of arrangement applied to the heavy water research reactor built in 1955 with Canadian assistance. Fuel of Canadian origin was safeguarded, but not fuel of Indian origin. Because the heavy water necessary for operation and obtained from the United States was also not safeguarded, separating the plutonium from fuel of Indian origin for use in India's 1974 nuclear explosion did not strictly violate any agreements, although both the United States and Canada claimed that it violated the "peaceful uses" pledge of India.

India and Pakistan are among a small but vocal and influential group of developing countries who have refused to sign the Nonproliferation Treaty. They feel that the NPT is a one-way street. In return for asking for a guarantee not to develop nuclear weapons, the weapons countries would neither offer them guarantees against nuclear aggression nor reduce the number of their own weapons. In addition, the supplier countries were not bound to treat NPT countries favorably in sales or other transfers of nuclear technology.[135]

By some measures, the NPT might seem to have been quite successful, for it has brought together nearly all the countries of the world to agree on a system of international controls. The Indian explosion, however, greatly shook global confidence in the nonproliferation system, even though the NPT itself had not been violated. An immediate result of this event was that in late 1974 France reiterated its pledge to require IAEA safeguards on transfers of French nuclear technology, even though France was not a party to the NPT. At hand in Asia were negotiations to build reprocessing plants in Korea and Pakistan. In both cases, apparently partly as a result of U.S. pressure, the deals were eventually canceled. The large plant planned in Pakistan was widely believed not to be justifiable on grounds of economics or energy security since the Pakistani nuclear power program was so small and slowly developing.

The second NPT review conference in 1980 (held every five years) failed to address effectively the issues of concern to nonnuclear-weapons countries--assurance of access to peaceful nuclear technology and the failure of U.S.-Soviet arms negotiations--as well as those of concern to weapons countries--third-country transfer of sensitive technologies, for example.[136] The Philippines was one of the most outspoken about the need for supply assurances.[137] To some observers, the failure of the NPT Review Conference indicates that global confidence in the treaty has so weakened that one or two more nuclear tests by presently nonweapons states might lead to the complete collapse of NPT.[138] Asia contains one of the principal candidates for conducting such a test--Pakistan.[139]

The Israeli raid on Iraq's small reactor in 1981 is also likely to have a further negative effect on global confidence in NPT and IAEA safeguards. It will emphasize how insufficient many observers believe IAEA safeguards to be for actually assuring that facilities will not be turned to weapons generation. The French-built reactor came under IAEA safeguards but was nevertheless considered a serious enough threat to provoke military actions that will result in considerable political costs to Israel.

London Suppliers' Club

The Indian explosion of 1974 led directly to a meeting of major nuclear supplier countries in London in 1975 convened by the United States. This group published a set of guidelines in 1978 restricting the transfer of sensitive technologies--enrichment, reprocessing, and heavy water production. Although nominally directed at non-NPT countries such as India and Pakistan, these guidelines result in increased restrictions to NPT signatories as well, although most, including Japan, have submitted to them. Many developing countries view them as an attempt to maintain commercial control of the nuclear technology and as a violation of that part of the NPT that promised unrestricted transfer of nuclear technology to countries with safeguarded nuclear programs.[140]

Unilateral Nonproliferation Initiatives

In 1976, the Ford administration initiated a major study, to be called the Nonproliferation Alternative Systems Assessment Program (NASAP). Restructured in 1977 to conform to President Carter's nuclear power policy and published in 1980, this study provided the technical, economic, and political analyses to support the U.S. position against immediate reprocessing and for delay of the breeder.[141] This policy was first fully articulated by President Carter in April 1977 and was set into practice by Congress through the Nuclear Nonproliferation Act of 1978 (NNPA).

Although couched in terms of supply assurances, the consequence of the NNPA has been unilaterally to impose safeguards conditions beyond those of the IAEA, particularly on transfers of low-enriched fuel to other countries. Prior Congressional approval was required for reprocessing of fuel processed in U.S. facilities or used in U.S.-built reactors. These conditions complicated life for many countries, such as Japan, even though they subscribed to the NPT and submitted to full-scope safeguards under the IAEA. The NNPA also stipulated that fuel will not be transferred to nonweapons countries that have not accepted full-scope safeguards. In Asia, this later provision directly affects India, which, as discussed earlier, has been denied uranium shipments guaranteed originally under the joint agreement signed in the early 1960s. It could also affect Pakistan, the other major non-NPT nonweapons state in Asia.

Japan, while expressing irritation about what some view as an infringement on national sovereignty, has accepted the provisions of NNPA to the extent of agreeing to apply for the "M-10" permits required for reprocessing of fuel of U.S. origin. There has been considerable concern, however, that the U.S. position is an abrogation of the basic tenets under which nuclear power has been developed for a generation.[142] In this view, reprocessing and the breeder are logical and necessary steps to secure economic and reliable energy supply in the long term. The U.S. position, on the other hand, as reflected by NASAP and NNPA, is that proliferation is a much more serious threat than previously thought (largely because of the Indian explosion); that the slower growth of the nuclear power industry, lower uranium costs, and higher capital charges make reprocessing and the breeder economically marginal; and that fuel cycles that include reprocessing are significantly more proliferation prone.[143]

Complications attributable to the new U.S. policy have also contributed to delays in the Philippine reactor and led to Taiwan's insistence that permits be obtained in advance of bids by U.S. vendors. As discussed earlier, these complications stem mainly from those parts of the policy directed toward environmental and safety concerns.

The supplier country most deeply affected by the Indian explosion was, understandably, Canada, which had supplied the reactor used to make the plutonium. Late in 1974, Ottawa announced a number of new, more stringent conditions to be applied to all nuclear transfer contracts present and future including prior approval of reprocessing and enrichment of Canadian fuel. These conditions were tightened in 1976 to preclude any Canadian cooperation with nonweapons countries without full-scope safeguards. As a result of the failure of bilateral negotiations over these new conditions, nuclear relations between Canada and both India and Pakistan ceased entirely in 1976. In the case of Pakistan, Canadian pressure was also aimed at stopping the construction of the planned reprocessing plant. The cutoff of relations has been inconvenient and expensive for both Asian nations. India, for example, had to seek heavy water supplies from the Soviet Union under safeguards agreements nearly as strict as those proposed by

Canada.[144] In the same year, on the other hand, Canada and South Korea were able to reach the accord necessary to begin construction of the CANDU at Wolsung-Kun. Stated South Korean plans at the time (later abandoned) to build a reprocessing plant were apparently not a major stumbling block.[145]

The other major Asia-Pacific nuclear supplier, Australia, also announced a unilateral nonproliferation policy in the years immediately following the Indian test. In 1977, Prime Minister Fraser revealed general conditions that would have to be met if reprocessing of Australian uranium would be permitted. Although less detailed, these were perhaps more stringent than the U.S. and Canadian conditions and much more restrictive than NPT and IAEA safeguards. In addition, in justifying the resumption of Australian uranium exports, he argued that this would be a way for Australia to delay the global need for reprocessing and breeders as well as to have a more active voice in global nuclear fuel-cycle discussions. Although, as in the United States, there has been criticism that nonproliferation actions have not always backed the official words, it is probable that Australian efforts to obtain bilateral safeguard agreements with uranium buyers delayed growth in the uranium export industry--an indicator of significant political will.

From 1972-1980, Australia was apparently able to obtain only a single contract for uranium sales, one with South Korea in which deliveries will begin in 1983. The soft uranium market and safeguards requirements both played roles in this difficulty. As mentioned previously, shifts in government policies in 1980 about prior consent seemed to be leading toward more sales in the early 1980s.

Australia's resolve on nonproliferation objectives did not extend to support of the U.S. initiative to study the possibility of an international spent-fuel storage site in the Pacific. Indeed, Australia supported an objection to this idea passed by the South Pacific Forum in 1979. Apparently behind this stand is Australia's severe case of nuclear allergy in regard to nuclear waste and the government's worry that even mere discussions of waste disposal so close to home would reignite domestic opposition to uranium exports.[146]

INFCE

It has been pointed out that unilateral action to control proliferation by exerting constraint on nuclear fuel cannot be expected to be effective in the long term.[147] A more complete international system of safeguards is needed. Recognizing that the present system was in danger of collapsing completely and that the London Suppliers' Club and unilateral actions were received with considerable skepticism by many countries, President Carter proposed the International Nuclear Fuel Cycle Evaluation (INFCE) in 1977.[148] Although it was designed only to provide an international forum, he undoubtedly hoped that the study would also buttress the U.S. position on the relationship of reprocessing and breeders with proliferation. Taken up with the understanding that it was to direct itself to technical and economic issues only and was not a forum for negotiation, the study officially began in 1978 and concluded in 1980 with 66 countries involved. INFCE, through the IAEA, published nine volumes analyzing the fuel cycle in detail.[149]

It is generally agreed that INFCE resulted in a significant increase in information about nuclear power and the proliferation potential of alternative fuel cycle arrangements. INFCE, however, did not identify any technical means to prevent proliferation, and its working groups felt uneasy in establishing the relative proliferation potential of different fuel cycles. It identified the need to link safeguards with supply assurances, although not strongly enough to satisfy some developing countries, including India.[150] It failed to make recommendations concrete enough to enable supplier countries such as Canada to restructure their nonproliferation policies without further international efforts. It failed to provide guidelines definite enough to distinguish among reprocessing options and thus did not lead to an obvious conclusion of the Japan-U.S. negotiations on the Tokai Mura reprocessing facility, which had been waiting upon INFCE's outcome. Although this is perhaps an unrealistic expectation, given INFCE's mandate, a number of decisions that had been delayed until its conclusion are once again recognized to be capable of resolution only in political forums and not to be much easier to make even with INFCE's expanded database.[151]

The United States found little support in INFCE for its official nonproliferation stance but little directly negative evidence, either. The U.S. ambassador-at-large and Presidential representative on nonproliferation has interpreted INFCE to support large-scale enrichment and reprocessing; to affirm the marginal economics of plutonium recycle in LWRs; and to confirm the viability of spent-fuel storage. The INFCE reports are written in such a way, however, that quite different conclusions can and have been drawn. U.S. officials have also pointed out that some of the basic premises used by INFCE, such as those about nuclear power growth, were unrealistic.[152]

One recommendation of INFCE that has apparently been accepted by the three Asia-Pacific suppliers is the need to study more completely the creation of an International Plutonium Storage (IPS) facility. Some U.S. officials believe that credible international control over separated plutonium might allow the United States to convert the present prior consent requirement of NNPA for reprocessing to the status of a mere formality.[153]

CONCLUSION

Although worlds apart in their environmental and safety implications in most people's minds, nuclear power and some of the renewable energy systems that might come into importance for power production in the next decades share two important handicaps--capital intensiveness and technological uncertainty. These handicaps tend to prevent these systems from becoming as economically attractive during rapid changes in the oil market as the price of oil alone would indicate. Depending on national monetary policies, some observers believe that there can be a relationship between oil price rises and high interest rates. If so there may be a positive feedback mechanism (vicious circle) operating as follows: Very approximately, in a present value calculation of the type used to make financial comparisons of alternative investments, a rise of interest rates can effectively cancel the relative advantage that capital-intensive energy systems might gain over fuel-intensive systems from an increase in the escalation rate of fuel prices. Thus, a rise in interest rates from 15 to 18 percent for nuclear plants might effectively counterbalance the competitive advantage they might have enjoyed from a rise in the rate of coal price escalation from 2 to 4 percent. There are many factors to be considered in such calculations but, in the short term, capital-intensive electric power plants may be less attractive investments than might be expected.

The potentially huge financial impact of nuclear accidents as evidenced by Three-Mile Island, the uncertainty about waste disposal procedures, and the apparently uncontrollable impact on nuclear power development of further deterioration in the international nonproliferation system make nuclear power investments seem much riskier than they otherwise might appear. In terms of financial calculus, this effectively adds a hefty risk premium to an already large interest rate and further handicaps such a capital-intensive system.

Thus, even though many observers believe that oil will escalate at several percent per year in real terms (see Chapter 2), the volatility of the oil market may to some extent create an economic environment protective of itself. In the past, nuclear power investments have been partly insulated from high interest rates by operation of the import-export banks in the supplier countries, which have offered substantial interest discounts. There seems, however, to be a growing belief that such subsidies and other forms of aggressive salesmanship may not serve the[154] long-term interests of either supplier or customer.

This economic environment seems likely to persist for some time and consequently favor alternatives to oil that are less capital intensive. Foremost among these in Asia is coal. Coal, too, has many problems, and Chapter 6 addresses how environmental problems in Asia may affect its development. In addition, the considerable financing necessary for development of coal-handling infrastructure may be difficult to find compared to financing for nuclear plants.

Besides its effect on economic comparisons, security of supply is a critical issue in itself (see Chapter 7). Nuclear power clearly has a very different profile of energy security than petroleum--completely different technical characteristics and actors are involved. Although most observers seem to believe that uranium is preferable to oil in this regard, it is unclear how well nuclear power compares to other alternatives such as coal.[155]

Ease of stockpiling is its principal positive security characteristic at present. The fuel efficiency of breeders can potentially make countries with very little uranium independent, but only with elaborate, expensive, and potentially dangerous indigenous fuel cycle facilities--an option open to only a few countries in Asia in the next generation. Although the uranium market apparently was briefly influenced by a cartel in the 1970s, the cartel was easily broken and there are few prospects for its reemergence.[156] For nuclear power, the chief negative impact on energy security, however, is, as it has been since the opening of the Atomic Age, the connection with proliferation. As discussed earlier, there are signs of collapse in the international nonproliferation arrangements. Although this is accompanied by a growing awareness that there are a number of routes to proliferation other than nuclear power programs, the Israeli raid emphasizes more than any previous event how serious the connection is perceived to be. Thus, it seems that the energy security offered by nuclear power will continue to be doubtful. Although countries can continue to alleviate this doubt somewhat by stockpiling, the growth of nuclear power programs will remain hampered for some time as a result of these concerns.

The considerations discussed in this chapter are likely to lead to lower growth rates in nuclear power expansion among the East Asian trio, which now have the most aggressive programs in the world. Nuclear power will, however, remain an extremely important means to help reduce oil dependency, although these three countries may well choose a different mix of nuclear, coal, and LNG for this purpose than has been planned in the recent past. Indeed, official indications in 1981 are that nuclear projections are being revised downwards, partly, of course, because of lowered demand for electricity. Nuclear programs in the Philippines, India, and Pakistan are likely to continue to move slowly. The other Asian developing countries may wish to obtain one reactor for the technical experience and apparent energy insurance it provides, but probably few will embark on major nuclear development programs in this century.

The major exception to this trend may be the only country in the region with a complete nuclear fuel cycle already developed through its nuclear weapons program--China. Since before Marco Polo, China has operated largely as a world in itself, although with increasing ties to the outside world. In recent years, China has made decisions regarding energy that have not always been consistent with international trends--sometimes with considerable success. Nuclear power development may fall into this category in the next decades. The next chapter examines the Chinese energy system in some detail and the ways it may interact internationally.

NOTES

1. William H. Overholt, ed., <u>Asia's Nuclear Future</u> (Boulder, Colo.: Westview Press, 1977).

2. Bertrand Goldschmidt and Myron B. Kratzer, *Peaceful Nuclear Relations: A Study of the Creation and the Erosion of Confidence* (New York: International Consultative Group on Nuclear Energy, Rockefeller Foundation, 1978).

3. Munir A. Khan, *Nuclear Energy and International Cooperation: A Third World Perception of the Erosion of Confidence* (New York: International Consultative Group on Nuclear Energy, Rockefeller Foundation, 1978).

4. U.S., Department of Energy, *Federal Support for Nuclear Power: Reactor Design and the Fuel Cycle* (DOE/EIA-0201/13), 1981.

5. Bruce C. Netschert and Sam H. Schurr, *Atomic Energy Applications with Reference to Underdeveloped Countries* (Baltimore, Md.: Johns Hopkins Press for Resources for the Future, 1957).

6. Keichi Oshima and Mason Willrich, eds., *Future U.S.-Japanese Nuclear Energy Relations* (New York: Rockefeller Foundation, 1979).

7. Taiwan was removed from the IAEA in 1971 but still submits to IAEA safeguards under NPT.

8. *Nucleonics Week* [hereafter *NW*], 10 April 1980, p. 3.

9. Kim Woodard, *The International Energy Relations of China* (Stanford, Ca.: Stanford University Press, 1980).

10. Thailand Load Forecast Working Group, *Load Forecast for Thailand Electric System: Interim Report, Power Tariff Study* (Bangkok, August 1980).

11. Joseph R. Egan and Shem Arungu-Olende, "Nuclear Power for the Third World?" *Technology Review* 82, no. 6 (1980): 46-55.

12. *Nuclear Engineering International* [hereafter *NEI*], November 1980, p. 4.

13. M. R. Srinivasan, "Power Development and Indian Design Capability," *Nuclear India* 18, nos. 2, 3 (1979): 3.

14. *Business India*, 16 February 1981.

15. C. Taylor, C. White, and M. Gellerson, *India: Economic Issues in the Power Sector* (Washington, D.C.: World Bank, 1979).

16. *Nuclear News* [hereafter *NN*], August 1980, p. 72.

17. *NW*, 18 December 1980, p. 6.

18. *NW*, 30 October 1980, p. 4.

19. *NW*, 2 October 1980, p. 7.

20. *NEI*, December 1980, p. 10.

21. *NW*, 24 January 1980, p. 9.

22. Shirin Tahir-Khali, "Pakistan's Nuclear Option and U.S. Policy," *Orbis* (January 1978).

23. *NW*, 4 September 1980, p. 11.

24. Zalmay Khalilzad, *Nuclear Power and Economic Development, Seven Cases: Brazil, India, Iran, Pakistan, the Philippines, South Korea, and Turkey*, prepared for the U.S. Arms Control and Disarmament Agency (Los Angeles: Pan Heuristics, 1978).

25. Ibid., p. 13.

26. Agha Shahi, "International Relations Review," *Pakistan Affairs* 32, no. 20 (1979): 2-4.

27. *Nuclear Power Planning Studies for Bangladesh* (Vienna: International Atomic Energy Agency, 1975).

28. *NEI*, January 1981, p. 3.

29. Oshima and Willrich, *Future U.S.-Japanese Energy Relations*, p. 84.
30. Ibid., p. 87.
31. *NW*, 25 September 1980, p. 8.
32. Richard Lester, *Nuclear Power Plant Lead-Times* (New York: International Consultative Group on Nuclear Energy, Rockefeller Foundation, 1978).
33. *NW*, 13 March 1980, p. 10.
34. Tatsuro Omura, "The Japanese Approach to Nuclear Technologies" (Paper presented at Future of Nuclear Power Conference, Honolulu, October 1979).
35. *NEI*, July 1980, p. 7.
36. *NW*, 13 March 1980, p. 15.
37. M. Toyoto and N. Itoh, "Improving the Performance of Nuclear Power Plants," *NEI*, December 1979, pp. 62-65.
38. Richard Masters, "Toward Japanese Technology," *NEI*, August 1979, p. 15.
39. Omura, "Japanese Approach to Nuclear Technologies," p. 4.
40. Hiroshi Murata, "Meeting the Need for Advanced Reactors and Fusion," *NEI*, December 1979, pp. 68-70.
41. *NW*, 30 October 1980, p. 5.
42. *NEI*, November 1980, p. 7.
43. *NEI*, August 1980, p. 9.
44. John Marcom, "Japan Has High Hopes for Reactor Exports," *Asian Wall Street Journal*, 13 March 1981.
45. *NW*, 16 October 1980, p. 3.
46. *NW*, 25 September 1980, p. 6.
47. *NN*, June 1981, p. 184.
48. *Energy Daily*, 13 December 1979.
49. *NN*, June 1981, p. 91.
50. Richard Suttmeier, *The Nuclear Power Option: Technological Promise and Social Limitations* (Berkeley, Ca.: Institute of East Asian Studies, University of California). In press.
51. Tokuo Suita, "The New Safety Commission Gets Down to Work," *NEI*, December 1979, pp. 60-62.
52. Geoffrey Greenhalgh and Walter Patterson, *Impact Abroad of the Accident at the Three-Mile Island Nuclear Power Plant: March-September 1979* (Washington, D.C.: Congressional Research Service, January 1980).
53. *NW*, 6 September 1979, p. 8; *NW*, 7 August 1980, p. 10.
54. Roger Gale, *Tokyo Electric Power Company: Its Role in Shaping Japan's Coal and LNG Policy* (Washington, D.C.: Wilson Center, 1981).
55. *NW*, 13 November 1980, p. 10.
56. Richard Suttmeier, "Japanese Reactions to U.S. Nuclear Policy: Domestic Origins of an International Negotiating Position," *Orbis* (Fall 1978): 651-680.
57. Suttmeier, *Nuclear Power Option*, pp. 20-22.
58. *NN*, September 1980, p. 46.
59. Toyoaki Ikuta, "The Energy Issue and the Role of Nuclear Power in Japan," in *Australia and Japan: Nuclear Energy Issues in the Pacific*, eds. S. Harris and K. Oshima (Canberra: Australian National University, 1980).

60. Taiwan Power Company, *Taipower and Its Development* (Taipei, May 1980).

61. David Chu, "A Conceptual Idea for Establishing Nuclear Power Industry in Taiwan," *Energy Quarterly* 11, no. 1 (January 1981): 3-7.

62. *NW*, 19 February 1981, p. 2.

63. John L. Moore, "How the EXIM Bank Helps U.S. Nuclear Exporters," *NEI*, June 1981, pp. 27-29.

64. N. D. Markettos, "Ambitious Nuclear Programme Planned for Taiwan," *NEI*, March 1980, pp. 13-14.

65. *NN*, March 1981, p. 21.

66. Khalilzad, *Nuclear Power and Economic Development*, p. 17.

67. J. H. Cha, "Nuclear Power Program in Korea" (Paper presented at the Workshop on the Future of Large-Scale Energy Systems, East-West Center, Honolulu, 1979); and Michael Parrott, "Korea: An Expanding Economy Trying to Gain Energy Independence," *NEI*, February 1980, pp. 13-14.

68. Arthur D. Little, Inc., *U.S./Republic of Korea Energy Assessment: Electric Sector Evaluation*, draft report prepared for Argonne National Laboratory, Argonne, Ill., 1980.

69. Ibid., section 4.1.

70. United Nations and World Bank, *Considerations Affecting Korea's Energy Objectives, Policies and Strategy: Draft Final Report*, vol. 1 (New York: United Nations and World Bank, June 1979).

71. Arthur D. Little, Inc., *U.S./Republic of Korea Energy Assessment*, section 8.1.

72. United Nations and World Bank, *Considerations Affecting Korea's Energy Objectives, Policies and Strategy*, p. 25.

73. S. Levy, *Review of Safety Aspects of Nuclear Power Program in Republic of Korea* (Washington, D.C.: World Bank, June 1980).

74. Ibid., section III, p. 5.

75. *NW*, 8 May 1980, pp. 8-9.

76. Librado D. Ibe, "Nuclear Energy Utilization--the Philippines Approach" (Paper presented at the Conference on the Survival of Humankind--the Philippine Experiment, Manila, 1977), pp. 77-86.

77. Gary Makasiar, "Philippine Five-Year Energy Program 1981-1985," in *Proceedings of the 1980 Asia-Pacific Energy Consultative Group* and in *Energy* 6 (1981).

78. World Bank, East Asia and Pacific Regional Office, Projects Department, *Philippines Energy Sector Survey*, vols. 1 and 2, draft report (3199-PH), 10 November 1980.

79. Walden Bello, Peter Hayes, and Lyuba Zarsky, "500-Mile Island: The Philippine Nuclear Reactor Deal," *Pacific Research* 10, no. 1 (1979).

80. *Philippine Energy Sector Survey*, annex 4.

81. M. T. Malloy, "Philippines Rules Nuclear Plant Isn't Safe," *Asian Wall Street Journal*, 15 November 1979.

82. *NN*, November 1980, p. 96; Emilia Tagaza, "Nuclear Go-Ahead in the Philippines," *Petroleum News*, April 1980, p. 10.

83. Levy, *Review of Safety Aspects*, p. 2.

84. *NW*, 15 May 1980, p. 7.

85. NW, 27 March 1980, p. 4.
86. NEI, May 1980, p. 5.
87. Budi Sudarsono, "The Need for Nuclear Energy in Indonesia," Indonesian Quarterly 8, no. 3 (July 1980): 62-70.
88. Guy Pauker, Energy in ASEAN, draft report prepared for Rand Corporation, Santa Monica, Ca., 1979.
89. OECD Nuclear Energy Agency and the International Atomic Energy Agency, World Uranium Potential (Paris: Organization for Economic Co-operation and Development, 1978).
90. Thomas Neff and Henry Jacoby, The International Uranium Market, MIT-EL-80-014 (Cambridge, Mass.: Energy Laboratory, Massachusetts Institute of Technology, 1980).
91. Justice R. W. Fox, G. G. Kelleher, and C. B. Kerr, Ranger Uranium Environmental Inquiry, reports I and II (Canberra: Australian Government, 1977).
92. See note 90.
93. NN, October 1980, p. 23; NEI, March 1981, p. 9.
94. Nuclear Exchange Corporation, Monthly Report on the Uranium Market, no. 153 (Menlo Park, Ca.: Nuclear Exchange Corporation, May 1981).
95. Neff and Jacoby, International Uranium Market, chapter 3.
96. Ibid., pp. 5-10.
97. Richard Suttmeier, "United States-Japanese Nuclear Relations: Implications for the Pacific Region," in Harris and Oshima, eds., Australia and Japan: Nuclear Energy Issues (Canberra and Tokyo: Australia-Japan Economic Relations Research Project, 1980), pp. 178-197.
98. Neff and Jacoby, International Uranium Market, pp. 5-21.
99. P. Bauder, A. Horncastle, G. Lurf, and M. Stephany, "Competing for the Non-USA Enrichment Markets," NEI, October 1980, pp. 37-41.
100. Ibid., p. 40.
101. Nuclear Fuel, 27 April 1981, p. 4.
102. See note 45.
103. NEI, May 1980, p. 46.
104. NEI, June 1980, p. 5.
105. The Hindu, 25 February 1981, p. 2.
106. Thomas Neff and Henry Jacoby, "Non-Proliferation Strategy in a Changing Nuclear Fuel Market," Foreign Affairs 57, no. 5 (Summer 1979): 1123-1143.
107. International Nuclear Fuel Cycle Evaluation (INFCE), INFCE Summary Volume (Vienna: International Atomic Energy Agency, 1980).
108. NEI, June 1980, p. 9.
109. NEI, March 1981, p. 6.
110. NEI, June 1979, p. 27.
111. S. Mohan, B. Nagaraj, S. Seshan, and P. K. Rohatji, "Nuclear Power in the Indian Context," in International Energy Studies, ed. P. K. Pachauri (New Delhi: Wiley, 1980).
112. NEI, November 1980, p. 5.
113. Harris and Oshima, eds., Australia and Japan: Nuclear Issues in the Pacific.
114. See note 81.

115. Atoms in Japan, March 1980, pp. 18-21; Radioactive Waste Management Center, Low-Level Radioactive Waste Dumping in the Pacific (Tokyo: Nuclear Safety Bureau, Science and Technology Agency, 1980).

116. World Environment Report, August 25, 1980, p. 1; Honolulu Star-Bulletin, 20 May 1981, p. C-4.

117. Osamu Okada, Japanese Fisherpeople's Fight Against Nuclear Power Plants, monograph no. 5 (Tokyo: Pacific-Asia Resources Center, 1980).

118. Kirk R. Smith and Mark J. Valencia, "Whither Asia's Nuclear Waste?" Perspectives 1, no. 3 (Summer 1980): 11-16.

119. M. Kravanagh, "Ships and Nuclear Fuel Transport," BNFL News, March 1979, pp. 4-5.

120. NN, September 1980, p. 143B.

121. NW, 4 October 1979, p. 1.

122. NEI, July-August 1980, p. 20.

123. Atoms in Japan, April 1980, pp. 4-8.

124. NN, February 1981, p. 81.

125. NW, 18 December 1980, p. 10.

126. NW, 5 June 1980, p. 11.

127. NW, 9 October 1980, p. 12.

128. H. N. Sethna, "India's Atomic Energy Programme - Past and Present," IAEA Bulletin 21, no. 5 (1979): 2-11.

129. See Srinivasan, "Power Development and Indian Design Capability," and Ravindra Tomar, "Indian Nuclear Power Program: Myths and Mirages," Asian Survey 20, no. 5 (May 1980): 517-531.

130. John E. Gray, M. B. Kratzer, K. E. Leslie, H. W. Paige, and S. B. Shantzis, International Cooperation on Breeder Reactors (New York: Rockefeller Foundation, 1978).

131. Khalilzad, Nuclear Power and Economic Development, pp. 105-107; and Tomar, "Indian Nuclear Power Program."

132. Ryukichi Imai, "What Is Nuclear Nonproliferation?" in Harris and Oshima, eds., Australia and Japan: Nuclear Energy Issues in the Pacific, pp. 150-174.

133. Kirk R. Smith, "Comments on U.S. Nonproliferation Policy after INFCE," in Harris and Oshima, eds., Australia and Japan: Nuclear Energy Issues in the Pacific, pp. 109-112.

134. B. Goldschmidt, "The Negotiation of the Nonproliferation Treaty (NPT)," IAEA Bulletin 22, nos. 3 and 4 (August 1980): 73-80.

135. See note 3.

136. Frank Barnaby, "The NPT Review Conference--Much Talk, Few Results," Bulletin of the Atomic Scientists 36, no. 9 (November 1980): 7-8.

137. NN, November 1980, p. 140.

138. William Epstein, "On Second Review of Nonproliferation Treaty," Bulletin of the Atomic Scientists 37, no. 5 (May 1981): 57-60.

139. Onkar Marwah, "India and Pakistan: Nuclear Rivals in South Asia," International Organization 35, no. 1 (Winter 1981): 165-179.

140. See note 3.

141. Nonproliferation Alternative Systems Assessment Program (NASAP), Nuclear Proliferation and Civilian Nuclear Power: Program Summary, vol. 1 (Washington, D.C.: U.S. Department of Energy, June 1980).

142. See note 132.

143. Ryukichi Imai and Henry Rowen, Nuclear Energy and Nuclear Proliferation: Japanese and American Views (Boulder, Colo.: Westview Press, 1980).

144. Tomar, "Indian Nuclear Power Program."

145. D. W. Campbell, The International Nonproliferation Regime: A Historical Overview (Ottawa, Canada: Joint Parliamentary Committee on Nuclear Energy, 1979).

146. Marten Indyk, "Safeguarding Nuclear Energy in the Pacific: The Role of Australia," in Harris and Oshima, eds., Australia and Japan: Nuclear Energy Issues in the Pacific, pp. 113-145.

147. See note 106.

148. Pierre Lellouche, "International Nuclear Politics," Foreign Affairs 58, no. 2 (Winter 1979/80): 336-350.

149. See note 107.

150. NEI, April 1980, p. 22.

151. See note 148.

152. Victor Gilinsky, "Nuclear Reactors and Nuclear Bombs," News Releases U.S. Nuclear Regulatory Commission 6, no. 43 (December 1980): 4-6.

153. NN, April 1980, p. 60.

154. Bijan Mossavar-Rahmani, "Nuclear Bazaar," OPEC Review 3, no. 4, and 4, no. 1 (Winter 1979/Spring 1980): 128-143.

155. For an attempt to derive a quantitative vulnerability measure see B. Chow et al., "Workplan for Study of Nuclear Fuel Cycles and Energy Security: Criteria, Calculational Methods, and Data," draft report prepared for Pan Heuristics, Los Angeles, March 1980, p. 3.

156. Nuclear Fuel, 27 April 1981, p. 2.

5
China in Asia's Energy Development

Kim Woodard

INTRODUCTION

No book on energy development in the Asia-Pacific region would be complete without at least a passing glance at the energy prospects of the largest and most populous country of the region--the People's Republic of China. Along with Japan, whose industrial prowess, level of technology, and standard of living are the highest in the region, China dominates regional energy statistics by virtue of its sheer size, magnitude of potential energy resources, and progress achieved to date by its burgeoning commercial energy industries. In aggregate terms, the People's Republic already possesses the third largest energy system in the world, despite a lagging indigenous technological infrastructure and low per capita income standards.[1] The gap between China's energy system and those of the United States and the Soviet Union is still quite large, but Beijing may be able to close that gap or greatly narrow it over the next two decades, assuming that ambitious Chinese plans for modernization proceed without serious disruption. Indeed, modernization of China's personal living standards would require significant increases in per capita energy consumption under virtually any social design, thereby imposing a staggering production burden on the country's indigenous energy resources.

Whatever the prospects that China may in fact achieve its development goals between now and the end of the century, the development of the Chinese energy system is increasingly intertwined with energy development and the flow of energy commodities throughout the entire Asia-Pacific region. Just a short decade ago a pariah in the international community, the People's Republic of China, is now accepted with all the honors and responsibilities incumbent upon a major power. Recognized and courted by Japan, Western Europe, Australia, and the United States, Beijing is considered an important strategic asset for the industrial democracies. China's international trade continues to expand at a dizzying pace, reaching $39.2 billion in 1980.[2] In 1980, U.S.-China trade doubled to $4.8 billion and exceeded bilateral trade between the United States and the Soviet Union. About 90 percent of China's foreign trade is conducted with noncommunist countries. Roughly 70 percent of China's exports go to countries in the Asia-Pacific region and Beijing obtains 50 percent of its imports from the region. These figures illustrate the rapid integration of the Chinese economy with the world economy as a whole and with Asia and the Pacific in particular (see Table 5.1).

Crude petroleum, coal, and refined petroleum products are among the leading commodities in China's exports to Asia and the Pacific. Energy technology, particularly in the form of energy plants and equipment from Japan, ranks among the leading items in China's imports. The Chinese energy system is thus increasingly drawn into direct contact with energy development in other Asian countries through the normal expansion of trade. In addition to energy-related trade, however, the integration of China's energy system with regional energy development reflects general economic and political imperatives. Despite 30 years of relatively rapid industrialization, China is still largely an agricultural and rural country, characterized by great urban-rural income differentials, poor communication and transportation among various subregions, and high population density and labor intensity in the rural areas. Industrial output grew by 13.5 percent in 1978, 8.5 percent in 1979, and 8.7 percent in 1980, whereas agricultural output (by value) grew by 8 to 9 percent during 1978 and 1979, and then by 2.7 percent in 1980.[3] The relative emphasis on industrial development is reflected in the allocation of commercial energy fuels. In 1978, 65 percent of commercial energy production was allocated to industry, and only 4.8 percent to agriculture.[4] Even if one includes noncommercial fuels, the "modern sectors" of the Chinese economy, which include industry, transportation, and the cities and employ roughly 20 percent of the Chinese work force, consume fully half of all the various forms of energy produced in the People's Republic. Despite the fact that three quarters of China's population still lives and works in the traditional sectors of the economy, the engine of growth remains in the rapid expansion of the modern sectors. This pattern of economic development has close parallels in a number of other Third World Asian countries, increasing the long-range prospects for successful integration between the Chinese energy system and the energy systems of other countries in the region. I will return to this theme in more detail later.

Another background condition that may affect the direction of energy relations among China and the other countries of the Asia-Pacific region is what could be termed the "plateau phase" in the world oil market. As documented by a number of sources, the world is presently reaching a maximum level in the annual level of trade in crude petroleum and refined petroleum products.[5] This ceiling in the volume of the petroleum trade is due to a number of factors, including basic resource constraints, a sharp decline in petroleum demand in the industrial countries under the pressure of price rises, steady increases in domestic energy demand in the oil-exporting countries, the perception that limiting oil exports simultaneously raises the price of crude oil and saves oil for future export, and chronic political instability in the oil-rich

TABLE 5.1
China's Energy Commodity Exports to Japan, 1977-1980[a]

Year	Coal Earnings (US$ million)	Crude Petroleum Exports: Volume (000's b/d)	Crude Petroleum Exports: Price (US$)	Crude Petroleum Exports: Earnings (US$ million)	Petroleum Products Earnings (US$ million)	Total Energy Earnings (US$ million)	Other Exports (US$ million)	Energy Share of Export Earnings (%)
1977	19.1	135	12.60	629.1	9.5	657.7	1,028.0	44.3%
Growth rate	60.9%	11.2%		14.3%	82.1%	17.4%	12.5%	
1978	35.1	151	13.20	726.1	21.6	782.8	1,164.7	40.2%
Growth rate	62.1%	-2.7%		26.9%	157.7%	35.9%	36.1%	
1979	65.3	147	17.70	950.5	104.6	1,120.4	1,670.3	40.2%
Growth rate	58.1%	6.6%		73.4%	109.1%	76.5%	17.7%	
1980	116.7	157	34.50	1,980.0	311.4	2,408.1	1,994.1	54.7%
1979 (fourth quarter)	20.5	138		281.9	50.8	353.2	442.3	44.4%
Growth rate	44.5%	29.8%		62.4%	39.2%	63.9%	16.4%	
1980 (fourth quarter)	32.0	186	33.975	562.1	75.2	669.3	520.9	56.2%
Growth rate	n.a.	-11.4%		-1.5%	n.a.	n.a.	n.a.	
1981 (first quarter)	n.a.	166[e]	37.00[e]	560[e]	n.a.	n.a.	n.a.	n.a.

Source: Official Japanese trade returns as reported in U.S. Central Intelligence Agency, China: International Trade Quarterly Review, Fourth Quarter 1980 (Washington, D.C.: National Foreign Assessment Center, May 1981), pp. 22, 30.

[a] All figures are in current U.S. dollars and have not been adjusted for inflation. First quarter 1981 estimates are my own, based on recent press reports.

n.a. = data not available.

countries bordering the Persian Gulf. According to recent statistics, world crude petroleum production declined by about 5 percent in 1980 from 1979 levels.[6] Meanwhile, the price of crude petroleum rose from $14 per barrel to $35 per barrel in less than two years, before stabilizing, at least for the moment (early 1981). This jump in prices induced an even greater decline in consumption than in petroleum production, yielding a net surplus in world crude petroleum supplies. Whatever the short term trends, many indicators point in the direction of an extended plateau in the world oil market, or perhaps even to a gentle decline in the amount of oil in trade each year.

The plateau phase in the world oil market can be expected to have an immediate and serious impact on the energy ecomomies of the developing countries of Asia and the Pacific as well as on economies elsewhere in the world. Despite massive balance of payments pressures generated by sharp increases in the price of imported oil, many Asian countries continued to expand their oil imports during the 1970s. This was not the result of accident or careless planning; in terms of economic growth, East and Southeast Asia are among the most dynamic regions of the world. An arc of countries stretching from Korea in the north and east to Singapore in the south and west grew at real average annual rates (GNP) of 3 to 6 percent per year over the last decade. This general economic growth was spurred by rapid expansion in the modern sectors of the economy--industry, transportation, and the cities. These are precisely the sectors that rely on commercial energy and, for many of the countries of the region, on petroleum imports. If the modern sectors of Asian economies are to continue to expand at 5 to 10 percent per year, there will be a persistent rise throughout the region in demand for commercial energy commodities, particularly petroleum. The link between economic development and access to increasing amounts of commercial energy commodities is direct and structural. Even with sharp price increases, energy conservation programs, and attention to decentralized rural energy development, the countries of the region must find new sources of commercial energy in the decade ahead, or face stagnation in the modern sectors and a prolonged hiatus in economic growth. The plateau phase in the world oil market spells even higher oil import bills and greater foreign exchange pressures for Asian developing countries in the short term. In the longer term, it may constrain the amount of imported oil available to Asian economies at any price.

As the following discussion will show, the People's Republic of China faces its own intractable constraints on energy development and on its petroleum exports. Beijing cannot be expected to come charging to the rescue of other Asian developing countries, given these very real limits. But the strength of the link between energy problems and economic development provides an area of commonality between the People's Republic and the other developing countries of the region. Depending on the success of oil exploration along the continental shelf, China might be in a position to carry some of the burden of oil exports to the region, but within a relatively low ceiling. Beijing may also be able to offer some elements of its own energy development experience, both through trade in energy production machinery and through direct technical cooperation. But the areas of exchange must be carefully selected to ensure a proper fit between Chinese technology and the needs of other countries. Indeed, several of the ASEAN countries may have access to energy production technologies that would be of interest to the Chinese in their own development. In general, I would suggest that the pressures generated by the plateau phase of the world oil market, if carefully managed, could yield an additional incentive for integration between the Chinese energy system and the energy development of the region as a whole.

Finally, we should be aware of powerful political incentives for greater energy cooperation between China and the other countries of the region. During the 1970s the Chinese government carefully manipulated its oil exports to Asian countries to achieve political objectives. In the case of Japan, the Philippines, and Thailand, the flag followed petroleum exports as each of these countries first established agreements with the People's Republic calling for specified levels of petroleum trade, and then moved in stages toward normal commercial and political relations. More recently, in the face of what the Chinese perceive as a strategic threat from the entente between the Soviet Union and Vietnam, Beijing is pressing for closer security cooperation with other Asian countries, seeking regional coalitions that would resist the onslaught of Soviet power in the western Pacific and Vietnamese expansion in Indochina.[7] Energy cooperation is a convenient glue that helps to bind China's new coalition strategy together. Both in the triangular relationship with Japan and the United States and in Beijing's relations with the ASEAN countries, energy trade and technical cooperation provide a framework for greater cooperation in facing regional political security concerns.

It is against this general background of economic and political factors that we should assess China's role in the energy development of the Asia-Pacific region. In the sections that follow, I summarize basic patterns in China's domestic energy development, discuss the link between offshore oil exploration and China's energy commodity exports, and examine Beijing's approach to the key problem of energy technology transfer. I return finally to a more general economic and political framework in projecting probable patterns of future energy development and cooperation in the region.

THE POLITICS OF ENERGY DEVELOPMENT

The overall pattern of China's domestic energy development cannot be understood in isolation without reference to the broader structure of the economy and the evolution of the political system of the People's Republic. The Chinese Communist Party came to power in 1949 after a century of national humiliation at the hands of the West, a revolution of some 40 years in duration, 15 years of occupation and national degradation at the hands of the

Japanese, and four years of barbarous civil war with the Kuomintang. In 1949, the country was absolutely prostrate, having suffered the loss of millions to war, revolution, and starvation and the near total destruction of an embryonic industrial system. Total national income (GNP in 1979 U.S. dollars) stood at $64 billion,[8] providing an average per capita income just in excess of $100. Actual conditions were far worse than reflected in this average figure, since the money economy had broken down completely in the rural areas and vast areas of the countryside were surviving on subsistence agriculture and local barter. Under these conditions, Mao Zedong and the Communist Party imposed a draconian revolutionary dictatorship, redistributed and then collectivized the land, and turned to the Soviet Union for trade, technical assistance, and a model of industrial development. More people perished in purges and famines, but from 1949 until 1957, the country began to recover and the economy achieved an average annual growth rate of 8.6 percent per year. Primary energy production (commercial) grew at an average annual rate of nearly 18 percent during the same period.

By the late 1950s, however, deep political fractures had begun to mar the appearance of unity. The new regime yielded not one but two models of economic management. Faced with an acute shortage of technical and managerial personnel, the Communist Party was forced to incorporate growing numbers of bureaucrats in government and technocrats in industry who gave lip service to the Party but felt a deeper loyalty to national independence and the imperatives of economic modernization than to the history and cultism of the revolution. This group was reinforced by intellectual elements within the Communist Party who had fled the coastal cities to Yenan during the Japanese occupation. The Russians themselves encouraged the growth of this nationalist-modernist cadre by simultaneously stressing technical training in their assistance programs and trampling on Chinese national pride at every turn.[9]

The development of the nationalist-modernist group during the 1950s produced what the Chinese refer to as "the struggle between the two lines," in government, in industry, and indeed within the Communist Party itself. The modernizers were arrayed against Party ideologues, many of whom had risked their lives for the revolution and had attained power in the inner circle around Mao. The ideologues stressed iron political control, the purity of Marxism-Leninism as a standard of absolute egalitarianism, and the training of a generation of "revolutionary successors." Mao Zedong himself attempted to stand above the two groups and to use their struggle to consolidate his own power. But for at least two decades (1958-1978), the struggle between the two lines yielded a series of convulsions in political, economic, and foreign policy that included the Great Leap Forward, the Sino-Soviet split, the Cultural Revolution, and the purge of the "Gang of Four." These political convulsions were interspersed with short periods of relative calm, but the overall impact on economic development was disastrous. In 1961, the gross national product stood at the same figure as it had in 1955. The Cultural Revolution caused a decline of at least 4 to 5 percent in the GNP in a single year (1967) and resulted in a second period of zero growth between 1966 and 1969. The energy industries suffered wild fluctuations in management, capitalization, labor efficiency, and output, but still managed to grow at an average annual rate of 8 to 9 percent between 1957 and 1977, about half the growth rate between 1949 and 1957.

Following the excesses of the Cultural Revolution, which the current leadership claims caused thousands of needless deaths and total disruption of the economy, Mao Zedong's grip on the Chinese polity began to weaken and the modernizers slowly gained the upper hand. Zhou En-lai led an opening to the West that started with Kissinger's secret visit to Beijing in July 1971 and dominated Chinese politics until 1976. Lin Biao, Zhou's principal opponent, heir-apparent to Mao, titular head of the Chinese military, and a leading ideologue, was killed in a plane crash in Mongolia while trying to escape to the Soviet Union after a failed coup attempt in September 1971, just three months after the Kissinger visit.[10] His death sealed prospects for a new turn toward the Soviet Union and helped legitimize Zhou's opening to the West. With the passing of the revolutionary leadership in 1976 (Zhou En-lai, Ju De, and Mao Zedong died in the space of a single year), the modernizers, led by thrice-purged Deng Xiaoping, arrested and suppressed the second-level leadership (Gang of Four) of the ideologue faction. In the four years since 1976, the modernizers have systematically consolidated their power in the Communist Party, the government, industry, and foreign policy. By the end of 1980, the leadership of the Gang of Four was on trial for their lives and Hua Kuofeng, Mao's second heir-designate, had been quietly pushed out of power and the position he held as chairman of the Party eliminated.

This brief political history is essential to an understanding of current trends in the development of China's domestic energy system. The People's Republic of China is still very much a centrally planned economy, within which the basic objectives and the detailed structure of the economy are determined by conditions in the political system. China has a "political economy" in the truest sense of the term. The current modernization program has to be understood not only as a blueprint for development, but also as a means of legitimizing political power for the generation of Chinese leaders that has struggled for that power, often at the risk of imprisonment or death, since the early 1950s. This legitimizing function lends the modernization program an urgency that it might not otherwise have had. It also deeply politicizes the development of China's energy industries. Energy is now perceived in the People's Republic as a key factor in industrial development, and despite the lip service paid to rural and agricultural development, industry is seen as the key sector pushing the economy toward its ambitious modernization goals. Given the powerful political imperatives behind the modernization program, it is indeed difficult to overstate the ambitiousness of its objectives. The Chinese leadership is aiming at nothing less than the total transformation of agriculture, industry, science and technology, and defense (the "Four Modernizations"). This transformation, to be achieved by the year 2000, is supposed to bring with it a quadrupling of Chinese living standards within just two decades.[11] The current leadership would like to achieve an average per capita income of $1,000 by the end of the century, a level reached by Japan in 1966. Achievement of anything like this objective will require a huge expansion in the economy and the Chinese energy system. In 1979, each dollar of per capita GNP was supported by the consumption of energy (including noncommercial energy) of about 1.8 kilograms coal equivalent (kgce). Even assuming a substantial increase in energy production and end-use efficiencies, the People's Republic would have to surpass the Soviet Union in primary energy output. (The Soviet Union is currently the world's largest producer of crude petroleum and the second largest producer of coal and natural gas.) This would imply tripling China's current commercial energy output, or a sustained growth rate in China's energy industries of 5 percent per year over 20 years, plus a massive investment program to replace antiquated energy-inefficient machinery in industry and other sectors of the economy.

How realistic is this energy development objective? After all, the Chinese managed to expand their energy industries at 17 to 18 percent per year in the 1950s and at 8 to 9 percent in the 1960s and 1970s. The implied 5 percent per year growth in primary energy production between now and the end of the century might seem well within reach. Indeed, China's energy planners believe that it is "realistic." The modernizers are mobilizing an energy development program that is designed to yield the required pattern of sustained growth over the next two decades. The principal elements of that program include: (1) the institution of rational management procedures and realistic national plans, (2) a massive infusion of foreign energy technology, (3) a sustained energy conservation program, (4) balanced exploitation of indigenous energy resources, and (5) joint offshore oil and gas development ventures coupled closely with the careful expansion of energy commodity exports.

If the goals of China's energy planners are ambitious and the programs extensive, it is also the case that Beijing faces some very serious constraints on the pace of commercial energy development. After three decades of rapid expansion, total primary energy production (commercial) grew by less than 3 percent per year in 1979 and 1980. Crude petroleum production, which had led the other energy industries in growth, slipped to near stagnation. Critical constraints on further growth affect virtually all energy industries and some of the constraints appear intractable. The modernizers currently at China's political helm cannot afford more than a few years of "readjustment" in the basic pattern of energy growth without losing the underpinnings of their own legitimacy. And so it is that we turn to the resource, production, and other economic constraints that affect China's basic energy industries.

ENERGY RESOURCES

The size and diversity of China's energy resource base is important to the energy development of the Asia-Pacific region as a whole. This is true from several perspectives. If China's fossil fuel resources are extensive and can be exploited at reasonable cost, then the region may look to the People's Republic as a source of badly needed energy commodity imports to fuel the next stages in industrial and economic growth. If, on the other hand, China's conventional energy resources are small relative to its own energy consumption and prospective demand, then the other countries of the region should not bank on energy trade with the People's Republic, regardless of conditions in the international market. The discovery of vast energy resources in the country would also accelerate the pace of industrial development within China, providing an ever-expanding base for the growth of Chinese military and economic power within the region. If China is an "energy giant" by the end of the century, then it will be a political and commercial giant as well. The governments of neighboring Asian countries have mixed feelings about the prospect of China as a third superpower, given its communist government, large population, and unknown geopolitical ambitions.

In 1980, we know a great deal more about many aspects of Chinese energy development than we did just a few years ago. As the modernizers in Beijing have become increasingly confident in the exercise of their new political power, they have loosened controls on the dissemination of basic energy data. Relatively detailed production and financial data were released by the State Statistical Bureau in June 1979 and again in April 1980.[12] Visitors are given extensive access to construction and production data for individual enterprises--coal mines, oil fields, and power plants. But the Chinese government is still very reluctant to make detailed reserve and resource data available to the outside world. Aggregate estimates of coal reserves and hydraulic potential have recently been published. But detailed reserve and resource data for the critical oil and gas industries is still considered "classified" for national security reasons. Our various estimates of crude petroleum and natural gas resources are based on speculative projections by foreign geologists, compiled from whatever geological data the Chinese have published on the structure of individual fields.[13] This situation is compounded by the fact that the Chinese, by their own admission, have explored only about 20 percent of the total onland area of potentially petrolific sedimentary basins.[14] Under these circumstances, it is best to provide a reasonable range of possible oil and gas resources rather than to attempt a detailed assessment.

Coal

There can be little doubt that China's coal resources are among the largest in the world. Vast coal deposits have been discovered in numerous basins that lie along a great arc stretching from Jilin Province in the northeast to Yunnan province in the south. Official Chinese estimates place "verified reserves" at the staggering figure of 600 billion tons.[15] This figure includes at least 30 billion metric tons that have been measured in deposits that are thought to be readily shaft mined, or enough to allow production at current levels for 50 years.[16] Great areas of Shanxi Province in the northcentral part of the country are underlain by coal deposits, and the province is reported to have about one third of China's total coal reserves, or 200 billion tons.[17] Fully half of the province's reserves are comprised of "high-grade coking coal," and most deposits are reported to be "easily accessible" at a depth of 300 to 400 meters (but too deep for surface mining). The great coalfields in northcentral China may be among the largest in the world and are comparable to the vast bituminous coal deposits in the northwestern United States (Wyoming and Montana). Current Chinese plans call for extensive development of the Shanxi coal deposits over the next two decades and include the construction of very large pit-head thermal power stations. Beyond these immense verified reserves of coal, recoverable coal resources were estimated by Chinese sources at 1.5 trillion tons in the 1950s; published foreign estimates in recent years are as high as 5 trillion tons.[18] Clearly, there will be no resource constraint on the development of the Chinese coal industry for hundreds of years. Coal resources are more than sufficient to supply domestic Chinese needs for the foreseeable future and, providing that key production constraints can be overcome, the People's Republic could become an important exporter of coal to the Asia-Pacific region and indeed to the world market. But the constraints on coal development and export are still very serious, as will be seen later in this chapter when we examine the growth of China's coal industry.

Oil and Gas

If the few published statistics on China's coal resources leave little doubt that the Chinese coal industry has enormous potential, the very absence of statistics on crude petroleum and natural gas resources leaves considerable uncertainty regarding the future of the oil and gas industries in China. There was a period of euphoria following Beijing's opening to the West in the mid-1970s when foreign "estimates" of China's oil and gas potential soared and the People's Republic was occasionally compared to the Persian Gulf as a potential oil-producing region.[19] The Chinese government, which was negotiating long-term trade agreements with Japan, Thailand, the Philippines, and other Asian countries at the time, did nothing to deflate this speculation and occasionally spurred it onward with subtle "hints" from high officials that China would soon be in a position to export large volumes of crude petroleum. But by the late 1970s, as foreign experts took a closer look at individual oil fields and as the pressure of domestic Chinese demand for oil and gas became increasingly evident in press reports, the euphoria disappeared, to be replaced by a sober evaluation of the serious constraints on rapid expansion of the Chinese petroleum industry.[20] In 1980, because of a hiatus in the growth of crude petroleum output with virtually stable onland output over three years (1979-1981), the Chinese government has been forced to renege on commitments to Japan and has stabilized crude oil exports to that country at 8.3 million tons (mt) per year (166,000 b/d). This has produced a wave of depression in foreign "estimates" of China's oil and gas resources that matches earlier euphoria.

The simple truth is that we have very little evidence regarding crude oil and natural gas reserves and resources, either on land or in offshore sedimentary basins that have been identified along China's continental shelf. We do know the identity and location of most of China's onland oil and gas discoveries. The major producing "fields" (actually clusters of fields in defined producing zones) are all located in the northeast part of the country. Daqing (Great Celebration) oil field, long the backbone of the Chinese oil industry, is located in the far northeast, close to the Soviet border, and produces roughly half of China's crude petroleum output (50 mt per year, or 1 million b/d). Official Chinese sources admit that Daqing is well into the "peak" or "plateau" phase of development, although small and medium fields continue to be discovered around its periphery.[21] Cumulative production at Daqing at year-end 1979 was 490 mt, indicating ultimate recovery of about 1 billion tons.

Exploration of the oil fields of the northeast is greatly complicated by their geology. Daqing and the onland fields that border the Bohai Gulf (Shengli and Takang oil fields) are comprised of relatively small individual fields that are located in stratigraphic traps in lacustrine (lake-formed) sedimentary basins, along the sides of large graben structures. The pay-zones are also heavily faulted by the high level of tectonic activity in the region. This combination of factors means that fields are relatively small in volume, hard to see on seismic profiles, and difficult to hit with the drill. Per-well yields are low, averaging around 200 b/d, a figure that would not be considered economic in many areas of the world. The crude produced varies in quality but tends to be waxy and have a high residual content, complicating refining and export. The Chinese have had better luck at the fourth major northeast field--Renqiu. Located a short distance southwest of Beijing, it consists of 40 individual fields on the eastern and western fringes of a graben that covers an area of 200 sq km. Renqiu has fewer and larger structures than the other northeast fields, and per-well productivity averages over 4,000 b/d, with the record well yielding some 21,900 b/d a major producer by world standards.[22] Renqiu oil field produced more than 10 mt of crude petroleum in 1979, or about 10 percent of total crude petroleum output in that year.

In the last few years, as the limits on the northeast fields became evident, Beijing has shifted the main thrust of its exploration activity to the south and west. New fields have been discovered in Jiangsu Province and in the Pearl River delta near Guangzhou (Canton). Aside from the offshore basins, the largest unexplored sedimentary basins in the People's Republic lie in the far western part of the country. Oil and gas development in the

Caidam Basin has been accelerated, and a new oil field has recently been discovered in the southwestern part of Qinghai Province.[23] The Chinese have barely begun to explore the oil potential of the vast Tarim Basin in Xinjiang Province, which has sediments covering some 560,000 sq km. At year-end 1979, there were three producing wells in the Tarim Basin, with daily yields of "hundreds of tons [thousands of barrels] of crude oil and nearly a million cubic meters of natural gas."[24] The Chinese are also enlisting the aid of foreign oil exploration companies for their projects in the far west. In July 1980, an American company (Geosource) signed a contract for a $34.2 million geophysical survey of the Caidam Basin.[25] These and other signs point toward greater emphasis on the definition of whatever oil and gas potential may exist in the great western deserts over the next few years. But it is still far too early to speculate on the success or failure of those efforts.

Foreign experts have become increasingly cautious in recent years in estimating and projecting China's probable oil and gas resources. One of the last major studies, an analysis published by A. A. Meyerhoff and J. O. Willums in 1977, projected total crude petroleum recovery on land and offshore at 9.5 billion tons (70 billion barrels).[26] Meyerhoff based his projection of onland recoverable crude oil (5.4 billion tons or 39.5 billion barrels) on a detailed compilation and structural analysis of known individual producing fields. Willums's even more speculative projection of offshore resources (4.1 billion tons, or 30.0 billion barrels) is based on computerized analogue analysis using volumetric data on offshore sediments that resulted from an early seismic survey conducted in 1969 by the Committee for Coordination of Joint Prospecting for Mineral Resources in Asian Offshore Areas (CCOP). The Meyerhoff-Willums projections can by no means be considered final, but they do provide us with a general benchmark with which to compare China's likely crude petroleum resources to other parts of the world. If this benchmark is taken as locating the correct order of magnitude, then China's endowment of crude petroleum resources would be about one quarter the original crude oil resource endowment of the United States (cumulative production plus projected recoverable resources). If the Meyerhoff-Willums projections were considered conservative (they appear optimistic in 1980), one could establish a reasonable projected range of Chinese crude petroleum resources at 10 to 20 billion tons, or 70 to 140 billion barrels. This range would be one quarter to one half the original recoverable U.S. crude oil resources. This general range was confirmed by Chinese officials as resembling their own projections during the visit by U.S. Secretary of Energy James R. Schlesinger to China in November 1978. But having explored only 20 percent of their onland basins and drilled only a few exploratory wells offshore, the Chinese themselves are in no position to make more than order-of-magnitude estimates.

If aggregate estimates of crude oil resources are hazardous, foreign estimates of current oil and gas reserves or of natural gas resources are little more than guesses. If I were forced to guess at year-end 1980, I would place current crude petroleum reserves (proven plus probable) in the general range of 1 to 2 billion tons (7.5 to 15 billion barrels). This guess is simply twice the likely remaining recoverable resources of Daqing oil field. Since Daqing is currently about halfway through its production cycle and produces half of the total figure for onland output, one could reasonably guess that total onland reserves are twice to four times cumulative production at Daqing. This is less than half of my year-end 1978 guess and would indicate that China's current reserves-to-production ratio may be as low as 10:1, a factor that would go a long way toward explaining the current hiatus in growth of the oil industry.[27]

Natural gas reserves and resource estimates are even more speculative than those for crude petroleum. In 1977, Meyerhoff and Willums projected recoverable natural gas resources at 5.7 billion cubic meters (200 trillion cubic feet) onland and 2.9 billion cubic meters (100 tcf) offshore.[28] Sichuan Province has a number of producing gas fields and, according to CIA estimates, produces about half of national output.[29] A recent American visitor to Sichuan reported reserves of 225 billion cubic meters (8 tcf).[30] Since much of the discovered gas reserves in China is either in the form of associated gas at known fields or "locked in" at remote fields such as those in Caidam that lack pipeline transport, I would place total natural gas reserves (proven plus probable) in the range of 0.5 to 1.0 trillion cubic meters (17.5 to 35 tcf). This is not a large natural gas endowment for a country of China's size and state of industrialization. But official Chinese natural gas production figures are far lower than previously estimated by foreign experts, indicating that the development of the gas industry has suffered in comparison with the oil industry.[31] If this is the case, then serious exploration for natural gas may still be an open option.

In 1980, the contrast between China's vast coal resources and the relatively modest oil and gas resources discovered in onland exploration has produced some pessimism regarding China's overall energy position during the coming decade. With growth rates in excess of 20 percent per year during the 1960s and early 1970s, the oil and gas industries had been thought of as the cutting edge of China's energy development, and the industries most likely to provide substantial foreign exchange from exports. If this pessimism is tonic for the euphoria of the mid-1970s, there is still some room for hope that China will resume growth, perhaps at a more modest pace, in oil and gas output during the 1980s. Offshore development, in joint ventures with foreign oil companies, is tied directly to future oil export prospects and may provide a badly needed boost to the industry in the late 1980s (see later discussion). Eighty percent of the area of China's onland sedimentary basins remains to be surveyed and drilled. As the Vice-Minister of Geology put the situation in a recent interview:

> Some people abroad are of the opinion that the reserves of China's coal, oil and other fuel minerals have already been determined in the main, and that the development of its coal and oil industries has reached the "peak." This is not so. Our energy sources still have large untapped potentials.[32]

Hydropower

Moving beyond fossil energy resources, China has what may be the largest hydropower resources in the world, more than commensurate with its land area of 9.6 million sq km. Most of China is ribbed by mountain chains, limiting agricultural land but greatly increasing hydraulic potential. Current official estimates place the country's theoretical hydraulic potential at 680 gigawatts, of which 370 GW can be exploited.[33] If China's hydraulic potential were fully developed, they could run an electric power system nearly the size of that in the United States on hydropower alone. At the present time, less than 5 percent (16 GW) of China's hydraulic potential has been developed, and recent plans call for an acceleration of dam construction. Beijing is capable of planning and constructing hydropower installations using domestic technology and equipment in the 2 to 3 GW range.[34] But the government is planning a project for the Yangtze Gorges that would be an order of magnitude larger at 25 GW, which, if completed, would be the largest hydropower station in the world. The Chinese are seeking American technical assistance in the design of the Three Gorges Dam, but there is considerable skepticism among American experts regarding the economics and feasibility of the project.[35] In total, eleven large hydropower stations were under construction at year-end 1980, with a designed capacity of 10.9 GW.

At the other end of the scale, encouraged by various government incentives some 90,000 small-scale hydropower stations have been constructed in the Chinese countryside, with a total capacity of 6.33 GW, or 39 percent of total installed hydro capacity.[36] Three fourths of China's 2,100 counties have small hydro stations, and one third of the counties in the country depend on small-scale hydro as the principal source of electric power. The stations, averaging just 70 kW in capacity, are highly subject to variable local runoff rates, yielding a low utilization ratio of just 1,880 kWh per kilowatt. This has led to some increase in the average designed size of the stations. In 1980, about 2,500 small stations were under construction, with an average installed capacity of about 200 kW.

This quick review of China's conventional energy resource base suggests that the People's Republic does indeed have enormous potential for further growth in its commercial energy industries. Its coal and hydropower resources are among the very largest in the world and will provide for the needs of centuries to come. But China's conventional energy resource base may not be as well balanced as those of the United States and the Soviet Union. The petroleum industry is already facing serious resource constraints on further growth in onland production. The basic resource constraint on the oil and gas industries may be eased by exploration in the far west or in the offshore theaters, but the new oil discovered by these efforts may not be available until late in the 1980s. Meanwhile, China will continue to depend on oil fields in the eastern half of the country, the largest of which have already reached production maturity. It is still highly uncertain whether the present hiatus in the growth of the petroleum industry will merely be a "bump" on the general growth curve, or whether new production capacity will simply replace old capacity, creating a prolonged production plateau at just over 100 million tons per year (2 million b/d). Even with China's gigantic coal and hydropower resources, a plateau in crude petroleum production at current levels would be very damaging to the development of the modern sectors of the Chinese economy. In addition, crude petroleum is now the most valuable single export commodity available to Beijing. A prolonged plateau in crude petroleum output would permit domestic oil demand to overtake supply, eliminating even the marginal oil exports that China has been able to muster so far. The critical position of the petroleum industry in China's energy development and the close link between offshore oil development and oil exports warrants separate attention later.

ALTERNATIVE ENERGY RESOURCES

Beyond China's vast and partially explored conventional energy resources lies a completely unknown array of resources for future energy production using technologies that are at or beyond the threshhold of the modern energy sciences. Alternative energy resources that are known to exist in China and to have considerable potential for the distant future include uranium, biomass, geothermal resources, wind, solar resources, tidal resources, and others. China currently maintains at least small laboratory-scale programs for the development of alternative energy technologies for most of the new technologies that are now being explored in the industrial democracies, including such sophisticated technologies as magnetohydrodynamic electric power production and controlled thermonuclear fusion. There are certain technologies, such as OTEC (ocean thermal energy conversion) for which the People's Republic has not developed even an experimental program. But Beijing prefers to keep low-cost experiments going in most other areas in order to train competent scientists and keep abreast of developments in other countries.

Biogas

The development of biogas generation in the Chinese countryside is the most advanced such program in the world, at least in terms of the diffusion and active use of the technology. Encouraged by a government-sponsored program, rural villages have installed some 7 million biogas digesters fueled by human, animal, and agricultural wastes.[37] Most of these digesters are small, serving a single family and providing fuel for household purposes. But there are also some 635 larger digesters fueling power generators for rural agriculture and industry. The Chinese experiment with biogas is well regarded abroad, and a recent conference on the subject organized under the joint sponsorship of the Chinese government and the United Nations Industrial Development Organization attracted participants from 21 developing countries. There is some evidence, however, that biogas is replacing the use of traditional fuels in the countryside without supplementing total rural energy availability. Chinese experts estimate 1979 rural energy production from traditional fuels (firewood, rice stalks, animal dung, and so on) at about 290 mtce.[38] This figure is probably quite stable, or expands only slowly with the growth of the population. The introduction of biogas, at least for household consumption, replaces less convenient traditional fuels, conserving forest and soil resources and improving sanitation, but does not add significantly to total rural energy output or consumption. One foreign expert recently placed total biogas production from China's 7 million digesters at about 2.5 billion cubic meters per year, or the equivalent of just 2.2 million tons of standard coal, which would be less than 1 percent of total rural energy consumption from traditional fuels.[39] Biogas may make life easier for Chinese peasant households, but it does little to ease the overall pressure of consumption on current production. There are some indications that the spread of biogas generators in the Chinese countryside may be slowing down, perhaps constrained by locational, feedstock, and climatic conditions. Even if the use of biogas were to grow by another order of magnitude, it would provide the annual equivalent of just 25 million tons of coal, or the output of a single major mining complex, and would add only 25 kgce to average per capita energy consumption, assuming no compensating decline in the use of traditional fuels.

Nuclear-Power

At the other end of the technology scale from biogas, the development of nuclear power in China remains an open question. The Chinese government has had an active military nuclear program since the mid-1950s. Uranium has been discovered in a number of provinces, including Xinjiang, Inner Mongolia, Qinghai, several provinces in the northeast, and the southern provinces of Guangdong and Guangxi, where there are least three active uranium mines. Uranium output may have been about 1,000 to 2,000 metric tons per year of U_3O_8 in the mid-1970s; at year-end 1978, based on comparisons to other uranium-producing countries with similar levels of output, I placed the order of magnitude for Chinese recoverable uranium resources at $30 per kg at 100,000 metric tons.[40] But it must be understood that these figures are purely speculative and that the size of uranium discoveries in China is a closely held secret (see Table 4.9 in Chapter 4).

If information on China's uranium resources is highly speculative, projections of future nuclear power development are hazardous indeed. Beijing has long been interested in the commercial nuclear option.[41] In late 1978, China contracted with Framatome for the purchase of two 900-MW pressurized water reactors (PWR). But the preliminary deal with the French conglomerate had been signed at the height of the expansionist euphoria that gripped Chinese economic planning in 1978. During the subsequent retrenchment, nuclear power came under attack within the planning hierarchy as too expensive and unnecessary and the deal was canceled in May 1979.[42] During 1980, however, there was a revival of interest in nuclear power, which has led some analysts to speculate that there is an active nuclear "lobby" in the planning hierarchy. The first meeting of the Chinese Nuclear Congress was held in Fabruary 1980, and an agreement of cooperation was signed between the Chinese and American Nuclear Societies in October 1980.[43] In early 1981, an information exchange agreement was signed between General Atomic Company of San Diego and the Institute of Nuclear Energy and Technology of Beijing for cooperation on the development of a high-temperature gas-cooled reactor (HTGR).[44] The Shanghai Design Institute is rumored to be working on a design for a domestically produced 300-MW PWR to be constructed near Shanghai.[45] In addition, there are reports that China is working on a 125-MW prototype heavy water reactor (HWR).[46] In late 1980, following a meeting between Chinese Premier Zhao Ziyang and French President Giscard d'Estaing, a panel of 100 Chinese scientists recommended the construction of six 900-MW nuclear power stations--two each in Guangdong, east China (probably Shanghai) and Liaoning Province.[47] Indeed, as little as 2,000 MW and as much as 15,000 MW are considered possible by the president of the Chinese Nuclear Society, formerly the deputy director of the Dubna Nuclear Research Institute near Moscow.[48] The panel pointed out that Guangdong Province can meet only 61 percent of current electric power demand and that the province lost about $5 billion in industrial output during 1979 because of power shortages. With this carefully orchestrated rationale in place, Beijing announced "agreement in principle" to revive the 1978 reactor deal with France. Some announcements indicated that the financing of, and production from, the Guangdong project would be shared with Hong Kong.[49] In spite of these signs, however, senior Chinese officials still voice some doubt that major foreign contracts will be awarded soon.[50]

There is some concern outside China regarding the safety and proliferation consequences of any active program of commercial nuclear development in China. Chinese nuclear scientists are known to be interested in safety and waste disposal problems encountered in the nuclear reactor programs of the industrial democracies. They followed the Three-Mile Island accident with obvious interest, and within the Second Ministry of Machine Building an office of nuclear regulation is apparently being established.[51] There has been a tendency in the People's Republic to adopt a "double standard" regarding safety issues, dismissing safety problems in the industrial countries as irrelevant to the Chinese context. This results in a rather bullish approach to the safety issue:

> Nuclear specialists attending the symposium noted that at present there are more than 200 reactors operating in nuclear power stations in the world. Their safety, reliability, and economic value have all been proved in practice. In comparison with coal and oil, nuclear power causes the least amount of air pollution, and is thus the "cleanest" energy source. At present many countries in the world are developing nuclear energy in a big way.[52]

There may also be proliferation consequences for an active nuclear power program in China, and these consequences must be considered carefully before Western countries agree to provide the People's Republic with advanced fuel cycle technology. There is no need to worry about direct proliferation of weapons technology to China, since that country already possesses a gaseous diffusion plant, at least two plutonium reactors and plutonium separation facilities, and a moderate stockpile of both fission and fusion warheads. China also has several experimental reactors, including a relatively large (125 MWth) unit which went into operation in late 1980 near Chengdu, Sichuan.[53] The country has a trained community of nuclear scientists and several atomic energy research institutes, facilities that have been carefully shielded from the vagaries of recurrent political upheavals. Nonetheless, it is also the case that Beijing is signator to just one international nuclear control agreement (the Treaty of Tlatelolco, which establishes a Latin American nuclear weapons free zone), and in the past has eschewed the Nonproliferation Treaty and the SALT agreements as superpower machinations. There have been occasional rumors that both Pakistan and Libya have sought Chinese assistance for their covert nuclear weapons development programs.[54] The Chinese, who recognize the threat that proliferation might pose for their own security, have adopted a policy of unilateral restraint and have refused these blandishments. But as of 1980, the political system in Beijing is not known for its stability and there remains a serious risk that at some future date China might unilaterally adopt a less restrained attitude toward the spread of sensitive nuclear technologies and equipment. This raises the spectre of "secondary proliferation" if the Western countries export sensitive nuclear technologies to China with no safeguards attached. Indeed, the development of nuclear power in China might provide some measure of incentive for the Chinese to join the rest of the international community in adopting safeguards on nuclear technology in international trade. There were, in fact, indications in early 1981 that the Chinese might consider joining the International Atomic Energy Agency or adopting certain basic safeguards.[55]

Other Alternative Energy Resources

Knowledge regarding China's other alternative energy resources must be considered embryonic. Geologically, China lies at the intersection of three major tectonic plates and there is a high level of seismic activity in most parts of the country.

Exploration has reportedly yielded some 2,500 potential geothermal sites; by year-end 1977, there were eight very small geothermal generators in operation.[56] But a great deal more exploration and technical development must be undertaken before Beijing could consider geothermal power a major option in the design of its electric power grid. Since the early 1970s, China has been conducting small-scale solar energy experiments, operating scattered solar units for heating buildings and bathhouses, particularly in Tibet and Xinjiang, where firewood is scarce.[57] But it cannot be said that China has a major solar energy program underway and the press has not reported any experiments with central station solar power generation. Rural planning agencies may be interested in wind energy conversion systems for small-scale power generation in areas that lack hydraulic potential. There is one two-way tidal power station in operation in Zhejiang Province with a capacity of 3 MW , and the country has four other small "semi" tidal stations on river-fed estuaries that operate on ebb tides only.[58] At the upper end of the technology scale, the People's Republic has been experimenting with coal-fired magnetohydrodynamic power generators and with both laser and magnetic fusion.[59] But these experiments are on a limited scale and face severe funding and personnel constraints. They are intended principally for scientific training and to keep the country posted on larger experiments conducted abroad.

Once again, we should expect that Beijing will concentrate heavily on the development of China's vast conventional energy resources over the next twenty years, laying particular emphasis on coal and hydropower. Small-scale hydro stations and biogas digesters will provide an important supplement for traditional fuels in the countryside, making life more convenient and less back-breaking for hundreds of millions of Chinese peasant families. Nuclear power stations will probably be constructed near certain large cities that have concentrated and rapidly expanding demand for electric power, and where large-scale hydropower resources or pit-head coal-fired thermal stations are not an option. Large-scale capital investment by the government will continue to be concentrated on the coal, petroleum, and electric power industries and will result in the construction of even larger producing units as the Chinese follow the classic path toward efficiency of scale and industrial growth. To the extent that China is following the "soft energy path" in the countryside (one never sees the phrase used in Chinese commentary), it is only as an expedient to relieve pressure on future resources and provide some visible short-term progress in living standards. The main thrust of government programs is toward conventional commercial energy development that stresses supplyside economics. But in 1980, a few dramatic publications showed that Beijing, becoming aware of some very serious constraints that face further expansion of the conventional energy industries, has begun considering energy conservation as a cost-effective way of relieving chronic energy supply shortages.

GROWTH CONSTRAINTS AND ENERGY CONSERVATION

In August 1980, the theoretical journal of the Chinese Communist Party, Hongqi (Red Flag), published a general article on the state of China's energy system.[60] The opening sections of the article paint a rather stark picture of inefficiency and waste in the management of all four of the major energy industries. China's overall energy end-use efficiency (primary energy production divided by final energy use) is placed at just 30 percent, which is compared unfavorably to end-use efficiency of 51 percent in the United States, 40 percent in England, and 44 percent in Japan. According to this Chinese analysis, about 70 percent of primary energy production is wasted through inefficient production, wasteful transportation, and utilization of antiquated, energy-inefficient equipment, particularly in industry. The end-use efficiency of Chinese industrial energy consumption is rated at 39 percent, compared to 70 to 80 percent in the industrial countries. End-use efficiency for household consumption is rated at a meager 20 percent, compared to 70 to 80 percent in the industrial countries.

Although Western experts might find fault with certain parts of the statistical analysis and comparisons in this article, technical visitors to energy installations in the People's Republic tend to support the view that there are many areas of inefficency and technical backwardness in each of the major energy industries. One foreign expert believes that there are "deep structural deficiencies" in the Chinese energy system that lead directly to low production, conversion, transportation, and utilization efficiencies.[61] The coal industry, for example, is characterized by low levels of mechanization, poor labor productivity, and poor conversion ratios in the use of coal to produce electric power and steel. The oil industry suffers a lack of refinery capacity and consequent waste of the lighter petroleum fractions by the direct burning of crude petroleum to fire boilers. The electric power industry, taxed by chronic excess demand from industry, tends toward very high load factors, poor maintenance, and frequent interruption of supply.

The general inefficiency of the Chinese energy system causes serious local bottlenecks and constant pressure on available supplies. This situation has been greatly exacerbated during the past two years because of a decline in the growth rate for primary energy output and for the output of the coal, petroleum, and electric power. The annual growth rate for total primary energy production fell to 2.7 percent in 1979, an incremental gain that was nearly canceled by a decline of 1.7 percent in commercial energy output in 1980 (see Table 5.2). This compares to average annual growth rates in primary energy production of 10.1 percent in the 1970-1975 period and 9.1 percent in the 1975-1978 period. Electric power output, a key constraint on industrial growth, increased by 6.4 percent in 1980, down from an average growth rate of 9.5 percent per year from 1970 until 1979.[62] This decline in the growth of the primary energy industries is important when set against the background of very high growth rates in the industrial

TABLE 5.2
China's Primary Energy Production, 1970-1980

Year	Raw Coal (million tons)	Crude Petroleum (million tons)	Natural Gas (billion cubic meters)	Electric Power[a] (billion kWh)	Total Primary Energy (mtce)[b]
1970	327.4	30.0	3.84	107.0	279.9
Growth rate	7.6%/year	18.9%/year	17.5%/year	11.2%/year	10.1%/year
1975	478.0	77.0	9.19	187.0	463.6
Growth rate	8.6%/year	10.1%/year	13.4%/year	10.5%/year	9.1%/year
1978	618.0	104.1	13.73	256.6	609.3
Growth rate	2.7%	2.0%	5.5%	9.4%	2.7%
1979	635.0	106.2	14.51	282.0	625.8
Growth rate	-2.4%	-0.2%	-1.7%	6.4%	-1.7%
1980	620.0	105.95	14.27	300.6	615.5

Source: Figures for 1970 through 1977 are my own, derived primarily from official Chinese government sources. Figures for 1978 through 1980 have been released by the State Statistical Bureau of China in a series of annual statistical reports. The annual statistical report for 1980 has just been published in *Beijing Review* 24, 19 (May 11, 1981):24, and includes marginal upward adjustments over earlier official data for 1980.

[a] Figures for electric power output include both hydropower and thermal power; figures for total primary energy output include only hydropower. Hydropower output in 1980 was about 55 million kWh.

[b] mtce = million metric tons coal equivalent.

sector, which consumes 65 percent of commercial energy output.[63] The industrial sector expanded at an average rate of about 8.9 percent per year from 1970 through year-end 1980.[64] For the first time in Chinese economic history, industrial growth is outpacing growth in primary energy production, and by a considerable margin. Previously, the People's Republic has often experienced shortages in commercial energy commodities, particularly electricity, but this is the first time that there has been a radical imbalance between the growth rates for industry and for the energy sector. The result has been new attention to the problem of energy conservation, particularly in industry.

Chinese planners openly admit to declining output rates in the energy industries, attributing slower growth to the exigencies of the "readjustment" in the national economy, which is now being stretched from the original three years (1979-1981) to five years or more. But it is also obvious that energy shortages could pose a serious threat to the readjustment process itself unless shortages are carefully managed through better allocation of available energy supplies. This threat of further economic dislocation comes at a critical time for the "modernizers" who now hold power and would be held politically responsible for a breakdown in their ambitious programs. The leadership has responded with a nationwide program of energy conservation, seeking to turn the greatest weakness of the Chinese energy system--its relative inefficiency--into a short-term asset. In August 1980, Vice-Premier Yu Qiuli, who heads the newly created State Energy Commission, announced that energy conservation and the development of new production would henceforth be equal priorities.[65] By September, the government claimed that significant conservation measures had already been undertaken during 1979 as part of the readjustment program, yielding such significant results as a decline in energy consumption per ton of steel produced from 2.51 mtce (1978) to 2.28 mtce (1979), a 10 percent efficiency improvement in a single year.[66] Energy savings in the industrial and transportation sectors were reported to be 18 million tons coal equivalent (mtce) in 1979, about 20 mtce in 1980, and are targeted for 24 mtce in 1981.[67] Though it is not clear what these figures mean since the base from which they are calculated is not specified, it is nonetheless true that industry continued to expand rapidly during 1979 and 1980 even as the growth in energy output slowed. In November 1980, further steps were taken to accelerate the energy conservation program, as the State Council budgeted about $2 billion for energy conservation in 1981 and the People's Bank of China announced that loans would be available to selected industrial enterprises (particularly metallurgical, cotton textile, and dyeing industries) for energy-saving equipment.[68]

Given present inefficiencies in the Chinese energy system, a conservation program may yield significant short-term gains that would tend to offset declining growth rates in the primary energy industries for the next few years. The Chinese government, however, cannot expect too much from its conservation program under conditions of rapid industrial and economic growth, even if that program is completely successful. If, for example, conservation measures moved the People's Republic from 30 percent end-use efficiency to 40 percent efficiency by 1985, that would be equivalent to producing an additional 64 mtce of commercial energy supplies, or about a 2 percent annual equivalent add-on to the primary energy production growth rate. At least an additional 3 percent annual increase in primary energy output would be necessary to keep pace with the planned 5 percent increase in effective energy supplies required to fuel the next stages in Beijing's modernization program. If the industrial growth rate continues at 8 to 10 percent per year, industrial energy requirements could only be met by continued curtailment of energy supplies to other sectors of the economy and by very careful allocation of key energy commodities, particularly electric power. In addition, after the most obvious waste is eliminated, further improvements in energy efficiency will require costly replacement and renovation of existing industrial plant and equipment. The energy conservation budget for 1981 will constitute about 5 percent of total national government allocations for capital construction, a staggering figure considering other capital requirements for the energy sector alone.

Beyond the immediate problems of the readjustment period (1979-1981) lies considerable uncertainty regarding the long-range prospects for resumption of high growth rates in the principal Chinese energy industries. To a large extent, this slowing simply reflects the maturation of the Chinese energy system. Slower growth rates are typical of large industrial and energy systems as new increments to production become increasingly difficult (the "scale effect"). In simple terms, the easy feats have already been performed and the new tricks are both technically intricate and organizationally sophisticated. Newly discovered oil fields tend to be deeper, smaller, and more remote, making them more difficult and expensive to develop. Transportation systems, such as railroads and transmission lines, get overburdened with new production. The cumbersome traditional bureaucracy can no longer respond by simply adding a new ministry. Planning requirements mushroom and must be replaced by efficient lower-level organizations. Inefficient enterprises may no longer be sustained by the government without the loss of capital needed in more critical units. To its credit, the Chinese government has recognized many of these problems that follow along with the maturation of the energy industries. But the constraints on further growth are ever more difficult and diffuse. Key constraints affecting virtually all energy enterprises in China lie in the areas of resources, capital, organization, technology, manpower, transportation, and, above all, time itself. The average lead time for construction of a new coal mine or a new hydroelectric station in China, from the time ground is broken to the time the first production begins, is now on the order of five to seven years. Current output is already pressing hard on capacity, and new increments in output can only be delivered by the costly and time-consuming avenue of new additions to capacity. The combination of long lead times and pressure on current capacity puts a high premium on coherent long-range planning and ultimately calls for a high degree of political stability over the next twenty years. This is precisely the stated program of the "modernizers" as they take the reins of government from the revolutionary generation. But it remains to be seen whether the Chinese political system has yet evolved a sufficient degree of flexibility and responsiveness to avoid the recurrent upheavals of the past and permit implementation of the modernization program. Politics, it seems, is as important to energy development in China as it is in the Middle East.

ENERGY COMMODITY EXPORTS AND OFFSHORE DEVELOPMENT

China's oil exports are now among the most visible aspects of emerging energy relations with the countries of Asia. Trade in crude petroleum and refined petroleum products with Japan, Hong Kong, Thailand, the Philippines, and North Korea benefits Beijing in a number of ways. Oil exports provide the largest single source of foreign exchange to pay for imported plants, equipment, and technology for the modernization program. The complex negotiations carried on with these countries to establish price levels and the volume of oil exports provide a convenient venue for the development of strong political relations and visible evidence of Chinese friendship. Most important for the next 20 years, oil exports to the region provide a direct and organic connection between China's modernization program and economic development throughout Asia. The ability to satisfy domestic energy needs with a margin to spare for export demonstrates China's new role as an economic power within the region, thereby conferring status and legitimacy on the "modernizers" in their struggle to consolidate domestic political power in Beijing.

The history of China's energy commodity trade reaches back to the establishment of the People's Republic of China and even a bit beyond. Before World War II, China imported limited quantities of refined petroleum products, particularly kerosene, for use in coastal cities. During the war, Japan developed coal mines in Manchuria and shipped the coal off to fuel Japanese war industries. This general pattern continued during the 1950s. Beijing exported coal and very small quantities of coke, at first to Japan and then to the Soviet Union, Hong kong, and East Pakistan. Coal exports continued throughout the 1960s and 1970s, peaking at 1.7 mt in 1967, falling to about a half million tons per year in the early 1970s, and then growing again to 3 million tons in 1978 and perhaps 5 million tons in 1979.[69] The foreign exchange that was generated by coal exports was used to pay part of the import bill for refined petroleum products. Refined petroleum imports from the Soviet Union were introduced immediately after the end of the revolution in 1949 and simply replaced the supply previously provided by Western oil companies. China imported a growing quantity of gasoline, kerosene, and diesel fuel from the Soviet Union during the 1950s. This trade peaked in 1961, just after the Sino-Soviet split, at about 2.9 mt. Refined oil imports then declined rapidly during the early 1960s, causing serious disruption in transportation and contributing to a prolonged hiatus in economic growth. The loss of Soviet refined products was only partially offset by rapid development of the Daqing oil field and spot imports from Eastern Europe and the Middle East.

The People's Republic has long been a marginal exporter of petroleum by-products, particularly mineral jelly and wax. Beijing also provided limited quantities of crude oil and refined products to its communist allies, North Vietnam and North Korea, beginning in about 1964. But it was not until 1973 that China made its debut as a net petroleum exporter. In January of that year, Premier Zhou En-lai made an offer to export 0.2 mt of Daqing crude during a visit by the Minister of International Trade and Industry, Yashuhiro Nakasone. Led by big production increases at Daqing, Chinese oil output jumped by 24 percent in 1973 and 18 percent in 1974. This led to a sudden squeeze on refinery capacity and left crude petroleum available for export. Crude oil exports to Japan jumped from 1 mt in 1973 to 3.9 mt in 1974 and 7.8 mt in 1975. By 1975, there was enough export crude available to begin exports to Thailand, the Philippines, and Romania. Significant refined products exports to Thailand and Hong Kong were initiated at about the same time. Total petroleum exports peaked quickly at 10.4 mt (0.2 million b/d) in 1975. But the growth rate for crude petroleum output dropped to 12 percent per year in 1975 and 1976 and then to 7 percent in 1977. This drop from the astronomical growth rates of the early 1970s was seized on by the ideologues, who pointed to tightening domestic oil supplies (oil consumption was rising even faster) and attacked oil exports as an invention of "capitalist roaders" (the modernizers). The ensuing political battle, which took place in the winter and spring of 1976, led to a sudden drop in crude oil exports to Japan. Total petroleum exports stabilized in 1976 at about 10 mt and then followed a pattern of slow growth to nearly 15 mt in 1980.[70]

It is important to keep China's energy commodity trade in perspective. At the height of its dependency on the Soviet Union for refined products in the 1950s, China was importing roughly 50 percent of the country's domestic consumption of critical petroleum products. But it is also the case that on an energy equivalent basis, net energy imports have never exceeded 3 percent of commercial energy consumption. The other 97 percent was produced domestically. In 1980, the People's Republic exported about 4 percent of primary energy production. It is true that net oil exports now account for about 14 percent of annual crude oil output. The total export of crude petroleum and refined products earned China about $4 billion in foreign exchange between 1973 and 1978, a source of revenue that matched almost exactly the outlays for foreign energy plants and equipment during the same period.[71] By 1979, oil revenues had jumped to about $2 billion, or about 15 percent of total export earnings.[72] As significant as these figures may be in funding the modernization program, it should be kept firmly in mind that the production and consumption sides of the energy balance are held in close balance as a deliberate matter of policy. Beijing may be willing to undertake energy conservation programs to maintain oil exports, but it is highly unlikely that any Chinese leadership could export oil to the point that industrial or economic growth would be constrained. Domestic consumption needs still take priority over exports.

The basic reluctance of the Chinese leadership to export oil beyond the point where domestic growth would begin to suffer is vividly illustrated by recent developments. Under the terms of a trade agreement signed with Japan in 1978, crude oil exports were to rise to 15 million tons by 1982. During the readjustment in 1979, this agreement was renegotiated, and the 15 mt goal stretched out until 1985. But when crude petroleum output fell to a growth rate of about 2.0 percent per year in 1979 and then declined by 0.3 percent in 1980, the Chinese government reneged on the stretched-out version of the trade agreement as well. In September 1980, Chinese officials indicated that exports to Japan in 1981 and 1982 would be limited to 8.3 mt (0.17 million b/d) for each year.[73] This is a little more than half the volume called for in the first trade agreement. Chinese officials make no secret of the fact that domestic demand is pressing hard on production and that it will take priority over export commitments. This does not, incidentally, mean that export revenues from the oil trade have stopped growing. Chinese crude oil prices follow the international market very closely. Daqing crude was selling at $34 per barrel in mid-year 1980, up from $12 per barrel just a year and a half before. Oil revenues jumped by about one third in 1979 and increased by an even larger margin in 1980. In a reversal of classic supply-demand imperatives, rapid increases in the market price of oil may soften the incentive for higher exports and harden insistence on meeting domestic requirements first.

Following this line of argument one step further, it is clear that the combination of the resource constraint on growth of the oil and gas industries and the consumption constraint on oil exports defines a set of boundaries for China's future energy relations with the Asia-Pacific region. Just as in several other Asian countries, rapid growth in the modern sectors of the Chinese economy is generating high consumption pressure on available energy supplies. Given careful planning, the Chinese government can hope to shift some of this rising demand toward coal and to slow the demand curve itself through vigorous conservation programs. But petroleum remains critical to the further growth of industry, transportation, and the cities. Steady growth in these sectors, in turn, is critical to the economic and political strategy of the modernizers in Beijing. The flattening of the crude petroleum production curve at just over 100 mt per year (2 million b/d) could not come at a worse time for the modernization program. If the limits on onland production persist, the "readjustment" in the Chinese economy will be indefinitely extended, with lower industrial growth rates and consequent failure to achieve the ambitious goals of the modernization program. Oil exports will be held to present levels and may be forced into a premature decline by rising domestic consumption requirements. The result would be to constrain oil revenues, crimp the import of advanced foreign technology, and slow the development of energy relations with Japan and Southeast Asia. Coal exports are growing rapidly and could expand to 15 to 20 mt by the late 1980s. But coal is more difficult than oil to extract and transport, brings less return, and faces stiff competition on the international market from other suppliers. In 1979, coal exports to Japan generated just $65 million in earnings, about 6 percent of earnings from petroleum exports.[74]

In light of these trends, China's offshore oil and gas exploration program takes on a new urgency. If that program is successful in discovering and producing crude petroleum from the enormous sedimentary basins that cover much of China's continental shelf, the growth of the petroleum industry would resume, demand pressure on available oil production would ease, and crude petroleum exports would expand, generating vast new revenues for China's modernization program. What is now an impending plateau in oil output would be reduced to a mere fluctuation in the growth curve. There is a direct link between the exploitation of offshore oil and the prospects for oil exports. In the absence of very large discoveries in the far western basins (Xinjiang and Qinghai), new onland production will be absorbed by higher domestic demand or will simply replace declining capacity at the older fields of the northeast. Onland production will mostly be reserved, therefore, for domestic consumption. That leaves exports heavily dependent on the success of the offshore program. Chinese planners themselves, aware of the direct link between offshore development and oil exports, have begun to establish a framework of joint venture agreements with foreign oil companies that will facilitate the export of much of the crude petroleum discovered and produced along the continental shelf.

The first efforts to determine the hydrocarbon potential of the continental shelf were begun, oddly enough, by foreign seismic surveys that were conducted without the acquiescence of the Chinese government. In 1966, the United Nations Economic Commission for Asia and the Far East established an organization to help survey the enormous expanses of continental shelf that extend from the shorelines of East Asia, Southeast Asia, and the Indonesian Archipelago. The initial intent of the Committee for Coordination of Joint Prospecting for Mineral Resources in Asian Offshore Areas (CCOP) was to provide bathymetric maps, weather data, general magnetic and seismic profiles, and other scientific data on the continental shelves of the region. But interest rapidly shifted in the direction of preliminary assessments of hydrocarbon potential in the vast sedimentary basins that were discovered on many shelf areas. In 1968 and 1969, with encouragement from Taiwan, a survey ship provided by the U.S. Navy, and multinational teams drawn from interested countries, CCOP conducted a series of preliminary geomagnetic and seismic surveys along the Chinese continental shelf. Sediments discovered along the shelf were of sufficient thickness and volume to prompt a wave of speculation regarding the presence of hydrocarbon potential.[75] This led a group of countries that lie around the perimeter of the shelf, including Japan, South Korea, Taiwan, South Vietnam, and the Philippines, to initiate their own offshore exploration programs. The first step in exploration in each case was to lay claim to exaggerated areas of the shelf surrounding each country. The second step was to invite foreign oil companies, usually American, to explore in "concession zones" granted without reference to conflicting claims by other countries. In the final step, the foreign companies were encouraged to drill any areas within their concessions that might seem promising. The assumption was that a discovery by oil company _x_ in a concession granted by country _y_ would reinforce that country's claim to the shelf. The companies went along in the hopes of gaining a foothold with very low initial investments. Unfortunately, the scramble that ensued undid itself, as the risks of drilling in contested areas of the shelf became obvious and as the cost of planting the flag with the drilling bit mounted. Of 110 offshore wildcats drilled in concession zones around China's perimeter between 1972 and 1977, 22 were listed as oil or gas "shows" but only two or three (in the Nido complex off Palawan in the Philippines) were commercially developed, and then only with substantial government prodding. By 1977, the foreign oil companies had departed or were relinquishing concessions in zones claimed by South Korea, Taiwan, Cambodia, and South Vietnam. The concession zones themselves reverted to the status of jurisdictional claims and an angry war of words set in that is with us today. Only in one area, the Korea-Japan Joint Development Zone off the southern tip of Korea, did the contestants reach a creative solution to the jurisdictional dispute. And that solution was made without the concurrence of China, which also has claims to parts of the Zone.[76]

The Chinese themselves were not unaffected by the oil fever generated by the CCOP surveys. The Chinese government and press took on all of the other contestants and added disputes with North Korea and North Vietnam, erstwhile allies. The Chinese jettisoned the median line criterion, which favors Japan in the East China Sea, and began insisting on the line of "natural prolongation" (greatest depth) as the principle for division of the shelf. The dispute with Japan in the East China Sea is particularly sensitive for Beijing because of the Taiwan factor. China would reject any settlement that might prejudice its historical claim to sovereignty over the island of Taiwan. China claims the entire area of the South China sea, based on historical records and traditional fishing patterns. This raises the prospect of an interminable verbal dispute with Vietnam and the Philippines over ownership of the shoals, islets, and reefs that dot the South China Sea. The Chinese dispute with Vietnam was greatly complicated and intensified by Hanoi's occupation of Cambodia and the subsequent "punitive" Chinese attack on Vietnam in early 1979.

The jurisdictional dispute is interesting in itself as a historical artifact and a weathervane for relations in the region. But it is also an important deterrent to exploration of the continental shelf. The thickest sediments along the entire shelf, in the "Taiwan-Sinzi Folded Zone" lie precisely astride the area of greatest contention between China and Japan in the East China Sea. Several of the first commercial-level oil discoveries to occur along the shelf in the South China Sea are to the northeast of Hainan Island in the Tonkin Gulf, very close to waters claimed by Vietnam and backed up by Soviet-assisted offshore surveys. It is unlikely that China will be able to move the exploration effort into contested zones in the absence of an agreement, or at least a tacit standoff by the other claimants. This important political factor confines the People's Republic (and its joint-venture partners) to near-shore areas and lengthens the likely time horizons for exploration of the shelf as a whole.

The beginnings of China's own offshore exploration program in the early 1970s are shrouded in the haze that surrounded all reporting on China at the time. Two major oil fields, Dagang and Shengli, were discovered on land along the margins of the Bohai Gulf in the mid-1960s. The area is characterized by flat salt marshes and an ambiguous shoreline. By 1969, it is thought that the Dagang oil field had been extended a short distance into adjacent marine areas, at first by the construction of causeways and fixed platforms extending out from shore. From a geological perspective, the first offshore production was simply an extension of onland drilling as Chinese crews followed the edge of the Dagang graben across the shoreline.

The early CCOP surveys stimulated interest within the People's Republic, as well as elsewhere in the region, in the hydrocarbon potential of the Chinese continental shelf. Chinese marine geologists had conducted a bathymetric study of the shelf as early as 1959. But Beijing was poorly prepared for the seismic, magnetic, and other research entailed in mapping offshore sedimentary basins. It was not until 1974 that China was able to launch broad mineral surveys of the continental shelf. During that year, Beijing commissioned a French company to do a seismic survey of the Bohai Gulf, purchased a French survey ship, and began to train Chinese crews. Between 1974 and 1978, the State Geological Bureau conducted surveys of the shelf in the Yellow Sea, the East China Sea, and the South China Sea.[77] These were broad-gauge surveys, roughly comparable to the CCOP work conducted in 1968 and 1969, to map the area, thickness, and volume of the principal sedimentary basins.

Meanwhile, China had also begun to develop what later became a domestically owned fleet of offshore drilling rigs. Two drilling barges were built in the late 1960s, probably for use in the extension of the Dagang oil field. The first Chinese jack-up rig was constructed at a Dalian shipyard in 1972. This was supplemented by the purchase of a secondhand Japanese jack-up from Mitsubishi in 1973. A Shanghai shipyard built a jury-rigged catamaran drillship in 1974. But by 1976, the limits on China's domestic rig-building capabilities were becoming evident and Beijing began ordering more sophisticated jack-up rigs from Singapore shipyards, loaded with American exploration equipment packages. Six rigs were ordered from Singapore between 1976 and 1980 at an average cost of about $30 million each.[78] These were supplemented by two additional jack-up rigs built in Japan and a secondhand Norwegian semisubmersible with a water-depth capability of 1,000 feet.

China had considerable difficulty absorbing the level of technology required to operate and maintain its new fleet of rigs. There were repeated problems in the early operation of the Singapore rigs, and American crews had to be flown in to demonstrate the use and care of equipment that was unfamiliar to Chinese crews. Chinese-built rigs proved to be less operable than the foreign rigs and were simply taken out of service after initial trial runs. On November 25, 1979, Chinese inexperience and bureaucratic blundering resulted in the capsizing of the first Japanese rig ("Bohai 2") in a heavy storm and the loss of 72 lives.[79] This brought a shake-up in the industry that reached the highest levels in the Ministry of Petroleum. All these difficulties reflect the early state of China's technological capabilities in the offshore field. In the longer term, they will probably be seen as early wrinkles in an otherwise ambitious program. But technical problems that plague the offshore drilling program have greatly slowed the pace of that program, demonstrating the need for cooperation with foreign companies in joint ventures for offshore exploration.

No precise data are publicly available on the extent or success of early Chinese drilling activity. That activity has tended to concentrate heavily on relatively well-known areas in the Bohai Gulf and on near-shore drilling in the South China Sea. Several oil and gas shows have been reported in the Pearl River Delta area and northeast of Hainan island in the Tonkin Gulf.[80] But the precise well yield data needed to assess the commercial viability of these discoveries is treated as proprietary information by the Chinese. Some of the results have been shared with foreign oil companies to stimulate interest in joint-venture agreements.

Beijing began to open the door to foreign participation in offshore development during the spring and summer of 1978. American, Japanese, and European oil companies were invited to give a series of technical seminars in China and discussions ensued that explored various contract options for joint-venture or service agreements. As in most cases when China is seeking foreign technology, early discussion resulted only in vague promises, while Chinese engineers and officials probed and established a range of possibilities without making firm commitments. The discussions were accelerated by the normalization of relations with the United States in December 1979, the prospect of a formal U.S.-China trade agreement, and the establishment of a Chinese legal framework for joint ventures with foreign companies. In June 1979, final agreement was reached for a first phase of intensive seismic surveys in nine zones of about 25,000 sq km each. Six of the zones were designated in the South China Sea, one in the Bohai Gulf, and two in the Yellow Sea.[81] Control of each zone was placed in the hands of a major oil company or consortium, which then subcontracted parts of the surveys to various operators. The companies bore the cost of the surveys (about $10 to 20 million per zone) with no formal guarantees of future rights, but with an understanding that they would be given first refusal for future joint-exploration agreements on promising structures discovered in their zones. In return, the companies were to hand over processed data to the Chinese. The original deadline for completion of the surveys and delivery of the data was year-end 1980, but this has recently been extended to mid-1981.[82]

During the two years (June 1979 to June 1981) allotted for completion of the seismic surveys, Beijing is setting up the framework for negotiation of joint-venture exploration and production contracts in the subsequent phases of offshore development. Once again, the Chinese are taking a relatively cautious and protracted approach that reflects the realities of their inexperience. Two experimental joint-venture contracts have been signed with French and Japanese companies to "test the water" before entering negotiations with American companies for large-scale exploration projects. In December 1979, a branch of the China Petroleum Corporation signed a contract with subsidiaries of the Japan National Oil Corporation (JNOC) for exploration of several promising blocks in the southern half of the Bohai Gulf. The importance of this contract is that is establishes the so-called "Bohai formula" for the sharing of risks and production.[83] Under the terms of the formula, the Japanese agreed to split total development costs with the Chinese in a 49:51 ratio that preserves nominal Chinese control of the enterprise. JNOC must first match previous investment by the Chinese side in their blocks (estimated at $210 to 220 million), and then split subsequent development costs according to the 49:51 formula. Of a projected 630 million barrels in production over the next fifteen years, 15 percent would be considered "cost oil," the China Petroleum Corporation would receive 42.5 percent, and JNOC would purchase an equivalent amount.[84] JNOC hopes to receive 65,000 b/d (3.25 mt per year) at peak production. China will impose a 5 percent tax on Japanese cost and profit oil and would, of course, be free to export some portion of the oil it receives from the project. If China chose to export part of its share to Japan, that could increase total crude oil shipments to Japan by several million tons. By itself, the Bohai contract would neither relieve Japan's oil dependency on the Middle East nor relieve China's balance of trade headaches. But the "Bohai formula" sets an important precedent for cooperation with Japan and with oil companies from other countries in prospecting and developing China's subsea hydrocarbon wealth. During 1980, similar contracts were signed with two French companies: with Elf-Aquitaine for exploration in a Bohai tract north of the JNOC blocks, and with Total for exploration in the Tonkin Gulf, indicating the Beijing plans to move forward under the Bohai formula. Bids will be requested in July 1981 for exploration in the zones being surveyed by American and British oil companies.[85]

One of the most interesting problems faced by the new joint ventures is the actual division of labor. Beijing would naturally like to pay as great a proportion of its investment share as possible in the form of Chinese equipment and services. The Chinese have allocated at least seven jack-up rigs, including a Chinese-built rig, for work in the Japanese and French joint ventures. The Chinese rig, "Bohai 3," has reportedly managed to drill just a single well since its construction in 1979 and has been inactive because of broken and corroded pumps.[86] But this has evidently not deterred the Chinese, who are now building five additional jack-up rigs in Dalian shipyards and a semisubmersible in Shanghai.[87] Beijing apparently plans to construct the frames for these rigs and then provide them with imported American equipment. In addition, China may attempt to build four more semisubmersibles of advanced design under licence with a Norwegian firm (Aker).[88] Aside from the rigs, China can expect to provide the joint ventures with shore services from oil bases built up along the Chinese coastline. Training of Chinese crews and their participation in exploration efforts will also be a feature of the joint ventures. It is clear that the new contracts do not constitute "concession agreements" and that the People's Republic intends to continue pursuing maximum "self-reliance" even in the offshore theaters, if under a new label.

It is still too early to project the outcome of China's offshore development program with any degree of certainty. Rumors filtering back through the oil companies conducting the seismic surveys indicate general optimism regarding hydrocarbon potential in the South China Sea portions of the shelf, but considerable caution regarding the institutional arrangements implied by the structure of the joint-venture agreements. A careful study by the National Council for U.S.-China Trade suggested that American companies may be able to begin exploratory drilling by August 1981.[89] Production should not be expected before 1984, at least in the American zones. This estimate should be considered optimistic. In the North Sea, where there was agreement on division according to the median line, tremendous capital and technical resources, and an existing framework of economic cooperation provided by the European Economic Community, it took four years following seismic surveys and the sale of blocks to achieve the first production and ten additional years to define 3 billion tons in reserves and to reach production of 1 million b/d (50 mt per year).[90] One should hardly expect the Chinese to beat this schedule.

This brings us back to the subject of China's oil export potential. At year-end 1978, based on computerized projections of the Chinese energy balance that allowed for both the resource constraint and the consumption constraint on exports, I projected crude petroleum exports at 25 to 40 mt per year (0.5 to 0.8 million b/d) for the late 1980s.[91] That is hardly another Saudi Arabia. But even that projection was made just before the sudden drop in growth rates for oil output from onland fields. Because of the current flattening of the growth curve for onland production, total petroleum exports (including refined products) should be expected to level off at about 15 million tons per year (0.3 million b/d) in the early 1980s. If, as expected, the joint ventures in the Bohai Gulf begin to come on line by 1985, exports may resume a modest increase in subsequent years. If all goes well in exploration of the great western basins and in the joint ventures with American companies, China may be able to muster the lower end of my projected range of oil exports (i.e., 0.5 million b/d) by 1990.

This projection does not augur well for a greatly expanded Chinese role as a supplier of the oil imports so desperately needed by the Asia-Pacific region. (It certainly does not suggest more than marginal Chinese oil exports to the market for low-sulfur, high residual crudes on the West Coast of the United States.) Back home in Beijing, the modernizers will be hard pressed, even in the event of substantial offshore discoveries, to meet the energy demands of the burgeoning Chinese economy. Emphasis on coal and hydropower development will help from the supply side. New conservation and industrial efficiency programs will help from the demand side. But it should still be expected that high growth rates in the modern sectors of the Chinese economy will generate additional consumption pressure on available oil supplies. In this way, China is not very different from Indonesia, Malaysia, Nigeria, or Venezuela. There are, indeed, great areas of commonality between the People's Republic and the energy problems of other developing countries in East and Southeast Asia. It may be that direct cooperation in meeting common problems generated by the energy crisis and the impending plateau phase in the world oil market will be at least as important as oil exports in building China's future energy relations with the Asia-Pacific region.

CHINA AND ENERGY DEVELOPMENT IN THE ASIA-PACIFIC REGION

Our review of the general features of China's domestic energy development and energy commodity trade reveals a picture of stark contrasts. On one hand, there is little doubt that the People's Republic of China already possesses one of the largest commercial energy systems in the world and has a vast resource base for further expansion of its energy industries. On the other hand, China is sharply constrained in its energy development by the sheer size of its population, by the finite limits on key oil and gas resources, by the current state of technology, by competing pressures on available capital resources, and by the binding delays imposed by the natural evolution of Chinese political and economic organization. My various projections of China's energy balance, constructed with the aid of computer modeling and allowing for a wide range of resource and growth assumptions, indicate that under almost any set of assumptions, China will join the United States and the Soviet Union as an energy and economic superpower by the year 2000.[92] I believe that only a major international conflict or sporadic domestic political upheavals could alter this projection. Even if the ambitious programs of the modernizers in Beijing are fully implemented, per capita energy consumption and living standards will remain modest by world standards. The pressure of domestic consumption requirements will limit energy commodity exports and will prevent China from becoming a major oil exporting country, unless offshore crude petroleum discoveries are larger than expected by a significant order of magnitude. China is now and will increasingly be an energy giant by world standards, but a hungry giant, bound by thousands of constraining threads to its own past, and struggling with enormous effort for each step taken into a brighter future.

The awakening of the modern Chinese giant is regarded with both empathy and apprehension among neighboring countries in the Asia-Pacific region. Empathy with the Chinese experience and with China's ambitious modernization programs is very strong throughout Asia and is frequently reflected in the enthusiastic reactions of Asian visitors to the People's Republic. The region as a whole is on the move, seeking higher living standards, greater regional cooperation, and a firm voice in international councils. But there is also apprehension. In terms of population and land mass, China is as large as the rest of East and Southeast Asia combined. The Chinese government is still avowedly communist and inclined to support the forces of violent revolution in neighboring countries. China's domestic politics has been marred by convulsive changes in direction and difficulties in transferring power to new leaders. Here again there are echoes among China's neighbors, for whom the shadow of political instability lurks in the wings of economic development programs. Even Japan watches China's emergence into a position of regional power with evident wariness.

As one looks down the corridor of the next decade, two key determinants of China's future energy relations with the region as a whole are visible--<u>technology</u> and <u>security</u>. The acquisition of modern technology for the development of China's energy industries, and the sharing of China's own energy technology with other developing countries of the region, will determine both the success of the modernization program and the scope of energy cooperation with neighboring countries. Sharing of technology among Asian countries is especially important in facing the implications of the impending plateau phase in the world oil market. Every country in the region must adapt its commercial energy system to the existing indigenous energy resource base and must explore a wide range of alternative future technologies to replace dependence on imported petroleum. These goals cannot be accomplished in the absence of significant sharing and cooperation in the development of energy technologies.

The period since the breakthrough in Sino-American relations in 1971 has been one of astonishing change in Chinese attitudes toward the acquisition of foreign energy technology. Beginning in 1972, immediately after the first Nixon visit and concurrent with the achievement of diplomatic relations with Japan, Beijing shifted its energy technology acquisition programs toward Japan and the West. The Chinese government began signing import contracts for a growing volume of trade in modern energy plants and equipment, hoping to acquire the technology as part of the deal. Between 1972 and 1978, China imported about $4 billion in energy production plants and equipment. These new orders were balanced almost precisely by export earnings in the crude petroleum trade. A wide range of modern equipment and complete plants was purchased, including components and production facilities for the coal industry, the electric power industry, petroleum exploration and development, refineries and petrochemical plants, offshore oil platforms and seismic survey equipment, and a huge array of support equipment for pipelines, ports, shipping, and other facilities. Japan and Western Europe split about 85 percent of this trade evenly, leaving American companies with 10 to 15 percent of the orders by value.[93] The program was greatly accelerated between 1976 and 1978 as the modernizers rose to power and issued a stream of orders for multibillion-dollar development projects.

The important characteristic of the period between 1972 and 1978 from the perspective of technology acquisition is that Beijing consciously abandoned the extreme interpretation of technological self-reliance ballyhooed during the Cultural Revolution and shifted the basic prototypes for its energy industries to Western models. The energy plants and equipment ordered from Japan, Western Europe, and the United States were thought to carry advanced technology to them, as if the crates landing on Chinese docks contained technological and scientific arcana as well as steel components. China also sent numerous technical teams abroad for training in the use of the new equipment, opened exchange relations with universities in the United States and elsewhere, and revitalized the Chinese intellectual community, which had been crushed by twenty years of abuse and the excesses of the Cultural Revolution. But before 1979, the Chinese government and its energy-planning ministries believed that the primary route for technology acquisition was the direct purchase of advanced plants and equipment from abroad.

The early bubble of enthusiasm for foreign machinery burst in the opening months of 1979. Citing the sweeping "readjustment" in its modernization program, Beijing canceled or delayed billions of dollars in preliminary contracts signed just months before. A number of factors helped to prick the bubble. Prominent among these were the limits on foreign exchange reserves imposed by a continuing imbalance of trade with the industrial countries; the low ceiling on earnings from crude oil exports; difficulty absorbing new plants and equipment efficiently into the existing industrial infrastructure; and, most important, growing awareness that advanced technology could not be imported in crates of equipment but could be acquired more directly and cheaply by direct technology purchase contracts. During 1979 and 1980, Chinese planners shifted sharply in favor of direct technology purchase as the principal strategy for technology acquisition:

> In the case of complete sets of large equipment imported in the early 1970s, it is estimated that every U.S. dollar spent on imported equipment needed, on an average, an investment of about four yuan [$2.50] in ancillary equipment at home. If foreign funds are used on a scale exceeding the ability to provide investment for ancillary equipment, it will hold up the progress of construction and cause serious waste . . .
>
> Most of the complete sets of large equipment we imported in 1978 consume much energy and this has exacerbated the

> shortage of fuel and power. We must never do this again in the future . . . In the past we suffered from importing relatively more equipment but too little technology.[94]

This shift in favor of direct technology purchase is already evident in the pattern of Sino-American trade. According to the U.S. Department of Commerce, the two largest deals during 1980 both involved the direct purchase of advanced American energy technology (Westinghouse steam turbine technology and Hughes Tool drill bit technology).[95]

Direct technology purchase is an important step toward integration of the massive Chinese economy with the world economy and with the economies of the industrial democracies. To implement this step, the Chinese leadership is creating a basic infrastructure of domestic law and organization to interface with the market economy countries. The promulgation of a joint-venture law, the Chinese acceptance of international patent conventions, a new Chinese tax law for foreign investment, banking arrangements that facilitate the flow of long-term credits, the creation of special economic zones, membership in the World bank, and the new contracts for joint venture offshore oil exploration are among measures designed to facilitate the flow of advanced foreign technology directly into the mainstream of Chinese industry, agriculture, and commerce.

On the other side of the ledger, what would China bring to technology cooperation with the other countries of the Asia-Pacific region? It is, perhaps, precisely because of the difficulties of energy development in the People's Republic and the Chinese adjustment to these difficulties that Beijing may have something to offer in the way of advice and technical assistance for the energy development of other countries of the region. Take the Chinese emphasis on intermediate technology production facilities in the electric power industry as an example. Chinese machine-building factories, operating under what in the West would be considered impossible equipment, supply, and maintenance standards, produce the great bulk of generating equipment used in Chinese power plants. Power generators are mass-produced in many locations according to standard specifications. The generators produced are relatively small, simple, and easy to install and maintain. They may not be as efficient as advanced American generators, but they are far cheaper and can be produced in indigenous factories. The same description would apply to Chinese factories that produce basic petroleum equipment, refinery components, and mining equipment. Most of the equipment produced is simple, rugged, and would be considered obsolete in Western industry. But it fits Chinese energy development needs admirably, at least until the People's Republic begins to require greater efficiency, high-voltage transmission technology, larger generating units, and comparable advances in other energy industries.

China also has a great deal to offer the other developing countries of the region in the design of rural energy development programs. Chinese experiments with small-scale hydropower and with biogas digesters are already well known in Asia, and formal institutes and international conferences have been established to assist in the sharing of the Chinese experience in rural energy development. It should be possible for agricultural engineers and government planners from other Asian countries to learn directly from Chinese experience with diffuse rural energy production technologies, and that learning should include a sharp awareness of the limits and constraints that affect small-scale energy production units as well as their advantages.

The Chinese "model" may, therefore, have applicability in commercial energy development programs in other Asian countries, despite deficiencies in the model itself. The conditions under which Chinese energy development is taking place are very similar in many respects to conditions throughout Asia and in other parts of the Third World as well. It is true, of course, that China has an unusually large land mass and is richly endowed with natural energy resources. Commercial energy development involves certain elements of scale and geology that give larger countries a natural advantage. But this fact is simply an argument for greater regional cooperation in the exploitation of energy resources and does not eliminate the relevance of the Chinese development experience. I would guess, for example, that the aggregate energy resources of Southeast Asia are roughly comparable to those of China and that these resources could supply regional commercial energy needs if properly developed and shared through trade.

The flow of energy technology, both to China from the advanced industrial countries and perhaps from China to other developing countries of the region, will be an important determinant of the success of China's ambitious modernization program and also of prospects for greater regional energy cooperation. The government in Beijing is making serious efforts to simplify and rationalize its own energy industries so that they may more easily accept a heavy infusion of advanced foreign technology. The Chinese are also streamlining and extending their five- and ten-year plans to maximize the effectiveness of capital allocation and to smooth and integrate growth curves in the various energy industries.[96] All these programs hinge on the direct acquisition of foreign energy technology and its proper absorption into the structure of the Chinese economy. Technology acquisition will therefore be a key determinant of energy policy during the 1980s.

A second major determinant of China's energy relations with the Asia-Pacific region during the decade ahead is the problem of security relations. The decade of the 1970s witnessed a complete reversal of what had previously been thought of as akin to basic laws of gravity in the security relations among Asian countries. The United States was involved in a brutal and futile war against a Chinese ally in Indochina just ten years ago. Japan rested comfortably under the American nuclear umbrella, maintaining minimal "self-defense" forces and focusing the national endeavor on trade. The countries of Southeast Asia, torn by domestic insurgencies and a schizophrenic attitude toward the war in Vietnam, sought to gain distance from American military failures while hoping for some continuing protection from what remained of American military power in the region.

By 1980, security relations in the region had shifted radically. At one pole stood China, Japan, and the United States, drawn together by a common and rising concern with the expansion of Soviet naval, military, and political power in the region. At the other pole stood a new entente between the Soviet Union and a unified Vietnam, formalized by a treaty of friendship and cooperation that calls for consultation in the event of a Chinese threat to either and legitimizes the Vietnamese invasion of Cambodia and Hanoi's hegemony in Indochina. The market economy countries of Southeast Asia, faced with American withdrawal from Vietnam, continuing instability and border wars in Indochina, a flood of Indochinese boat refugees, Soviet use of the naval base at Cam Ranh Bay, and an attack by Vietnamese troops across the Cambodian border into Thailand, extended the functions of the Association of Southeast Asian Nations to include regular Foreign Ministers' meetings on regional security concerns. All these trends may be summarized under what I have called the emergence of "tacit regional security coalitions" in Asia and the Pacific.[97] These coalitions are rapidly replacing unilateral American military power throughout the region.

From Beijing's perspective, regional security relations hinge on defense against what is perceived as an expanding military threat from the Soviet Union. Deng Xiaoping, the leader of China's modernizers, during his much-publicized visit to the United States in January 1979, repeatedly voiced concern with growing Soviet naval power in the western Pacific, Soviet influence in Vietnam, and the continuing buildup of both conventional and nuclear Soviet forces along China's northern border.[98] Since the time of that visit, high-level official exchanges among China, Japan, and the United States have increasingly focused on basic elements of a new program of security cooperation among the three countries.[99] China wants the United States to continue its security commitments to the region, even at the cost of further communist gains in such arenas as Korea or Thailand. Both the United States and China are urging Japan to increase defense spending and to turn Japanese technological and capital resources toward defense of the region. For its part, Beijing is modernizing its conventional military forces and continues to develop a strategic nuclear deterrent against Soviet attack.[100]

Although it is very difficult to project the course of security relations within the Asia-Pacific region during the 1980s, it is quite clear that there exists a direct linkage between nearly emerging security coalitions and regional energy cooperation. This "energy-security interface" rests in part on a classic military concern over the protection of extended oil supply lines that reach from the Middle East to Southeast Asia and Japan. But the linkage between regional energy and security concerns is far more extensive than implied by the defense of shipping lanes. Energy cooperation within the China-Japan-U.S. triangle, or among the ASEAN countries, greatly strengthens intergovernmental organizational ties and the perception of common interests. A framework of cooperation constructed to do battle with the energy crisis can easily be adapted to confront external security pressure as well. This "surfacing" of underlying security cooperation was evident in the wake of the Vietnamese border attack on Thailand in June 1980, when ASEAN Foreign Ministers met immediately and issued a stern collective warning to Hanoi.[101]

One of the key problems facing American security policy in the years ahead is just how far to go in the development of the new security coalition with China. Washington has already abandoned the "even-handed" policy of the early 1970s that limited the sale of defense-related technology to China with a set of restrictions developed for the control of trade in advanced technology with the Soviet Union. The U.S. government now casts a blind eye toward the sale of European military hardware to the People's Republic. Restrictions on American companies that wish to sell military hardware to China are being relaxed, and by early 1981 contracts had already been negotiated for certain "defensive" (logistical and support) equipment as well as so-called "dual use technology." The latter category is particularly important for China's energy technology acquisition program. Washington is now permitting the sale of advanced remote sensing technology, very large computers, and other sophisticated equipment that is essential to the modernization of Chinese energy industries but may also be applied to the modernization of Chinese defense industries.[102] Similar equipment and technology is denied the Soviet Union. China now has access to American Landsat data and is purchasing satellite communication gear with important military applications under the "dual use" formula.[103] Once again, these steps in American policy illustrate the important interface between energy and security relations in the emerging coalition with China.

It is still too early to judge the outcome of China's emerging security relationship with other countries of the region. Any substantial growth in Chinese military power should be handled with extreme caution by the other countries of the region, balancing the need for security cooperation with regional restraints on the use of that power. It is almost certain, however, that energy cooperation with the People's Republic will race ahead of security cooperation, preparing the way for a Chinese political role in the region that is commensurate with the country's size and emerging power. To do otherwise would be to risk the emergence of renewed isolation and radicalization of the Chinese regime, untrammeled by cooperative ties to neighboring countries. It is in the interest of peace and security in the region as a whole to proceed toward greatly expanded energy cooperation with the People's Republic of China.

NOTES

1. Kim Woodard, The International Energy Relations of China (Stanford, Ca.: Stanford University Press, 1980), pp. 262-284. Taking both the production and consumption sides of the commercial energy balance into account and allowing for noncommercial rural

use of traditional fuels, China's energy system ranks third behind the United States and the Soviet Union.

2. Imports plus exports f.o.b. in 1980 U.S. dollars. U.S. Central Intelligence Agency, China: International Trade Quarterly Review, Fourth Quarter 1980 (ER CIT 81-003), May 1981, pp. 16, 18.

3. State Statistical Bureau of the PRC, Main Indicators, Development of the National Economy of the People's Republic of China, 1949-1979 (Beijing, 1980), p. 1; State Statistical Bureau of the PRC, "Communique on Fulfillment of China's 1980 National Economic Plan," Beijing Review [hereafter BR] 24, no. 19 (May 11, 1981): 23.

4. Wu Zhonghua, "Resolving the Energy Crisis with Energy Technology," Hongqi [Red Flag, in English] 357, no. 17 (August 1980): 31-43.

5. For example, U.S. Central Intelligence Agency, Office of Economic Research, The World Oil Market in the Years Ahead (ER 79-10327U), August 1979, p. 8; Fereidun Fesharaki, "Global Petroleum Supplies in the 1980s: Prospects and Problems," OPEC Review 4, no. 2 (Summer 1980): 27-49.

6. "Crude Oil Output Said to Fall in 1980," New York Times, 29 December 1980, p. D-5.

7. Kim Woodard, "The Second Transition: America in Asia under Carter," SAIS Review. In press.

8. U.S. Central Intelligence Agency, "China, Major Economic Indicators," unclassified data sheet issued by the National Foreign Assessment Center, Washington, D.C., February 1, 1980.

9. For example, Moscow first promised and then withheld technical assistance for China's nuclear weapons program. See "Statement by the Spokesman of the Chinese Government: A Comment on the Soviet Government's Statement of August 3," BR 6, no. 33 (August 16, 1963): 7-19.

10. Hua Fang, "Lin Biao's Abortive Counter-Revolutionary Coup d'Etat," BR 23, no. 51 (December 22, 1980): 19-28.

11. Yu Youhai, "U.S. $1,000 by the Year 2000," BR 23, no. 43 (October 28, 1980): 16-18. My projection takes into account a wide discrepancy between official Chinese GNP figures ($253 per capita in 1979) and CIA estimates derived from Table 5.1 ($536 per capita in 1979). The discrepancy itself is the result of different methods of translating Chinese GNP into dollars. The Chinese use foreign exchange rates and the CIA uses an "equivalent value" (what it would cost to "buy" the Chinese GNP) formula. When both data series are indexed to a common base year, they match almost exactly. All growth rates cited in this paper for GNP, industrial output value, and agricultural output value are adjusted for inflation.

12. State Statistical Bureau of the PRC, "Communique on Fulfillment of China's 1978 National Economic Plan," BR 22, no. 27 (July 6, 1979): 37-41; "Communique on Fulfillment of China's 1980 Economic Plan," BR 22, no. 27 (July 6, 1979): 12-15.

13. A. A. Meyerhoff and J. O. Willums, "Petroleum Geology and Industry of the People's Republic of China," in United Nations Economic and Social Council for Asia and the Pacific, CCOP Technical Bulletin, vol. 10 (Bangkok: United Nations, December

1976): 103-212; Woodard, *International Energy Relations of China*, ch. 14.

14. Based on an interview with Chinese Minister of Geology Cheng Yuqi, reported in Ji Zhe, "Opening Doors to Underground Wealth," *BR* 23, no. 42 (October 20, 1980): 20-26.

15. Year-end 1979 measured plus inferred (proven plus probable) reserves. "Coal Deposits--Enough to Last Five Centuries," *BR* 23, no. 11 (March 17, 1980): 8.

16. "Stepping Up Coal Production in the 80s," *BR* 23, no. 13 (March 31, 1980): 6.

17. "Shanxi--An Ideal Energy Base," *BR* 23, no. 42 (October 20, 1980): 16-17.

18. Woodard, *International Energy Relations of China*, pp. 380-383; Dori Jones, "The Dawning of Coal's 'Second Golden Age,'" *Chinese Business Review* [hereafter *CBR*] 7, no. 3 (May-June 1980): 38.

19. Selig S. Harrison, *China, Oil, and Asia: Conflict Ahead?* (New York: Columbia University Press, 1977), ch. 3.

20. Randall W. Hardy, *China's Oil Future: A Case of Modest Expectations* (Boulder, Colo.: Westview, 1978).

21. "New Oil Deposits," *BR* 23, no. 21 (May 26, 1980): 6.

22. "Renqiu Oilfield," *BR* 23, no. 18 (May 5, 1980): 7-8; Dori Jones and Stephanie R. Green, "Renqiu's Buried Riches," *CBR* 7, no. 6 (November-December 1980): 20-24.

23. "New Oilfield in Northwest China," *BR* 23, no. 10 (March 10, 1980): 6.

24. "Oil Potential in Tarim," *BR* 23, no. 8 (February 25, 1980): 6-7.

25. Dori Jones, "Enticing Foreigners Inland," *CBR* 7, no. 5 (September-October 1980): 50-52.

26. Meyerhoff and Willums, "Petroleum Geology and Industry of the PRC," p. 103.

27. Woodard, *International Energy Relations of China*, p. 266.

28. See note 26.

29. U.S. Central Intelligence Agency, *Electric Power for China's Modernization: The Hydroelectric Option* (ER 80-10089U), May 1980, p. 23.

30. Stephanie Green, "Sichuan Journal," *CBR* 6, no. 6 (November-December, 1979): 14-18.

31. Compare note 29 (new CIA figures) with old CIA and other estimates in Woodard, *International Energy Relations of China*, pp. 430-434.

32. Quoted in Ji Zhe, "Opening Doors to Underground Wealth," p. 22.

33. "Speeding Up Building Hydropower Stations," *BR* 23, no. 31 (August 4, 1980): 5-6.

34. Xiang Rong, "Building a Dam on the Changjiang River," *BR* 22, no. 12 (March 23, 1979): 17-18; Nicholas H. Ludlow, "Gezhouba on the Yangtze," *CBR* 7, no. 3 (May-June 1980): 11-15.

35. Nicholas H. Ludlow, "Harnessing the Yangtze," *CBR* 7, no. 3 (May-June 1980): 16-21, 23.

36. "More Small Hydropower Stations," BR 23, no. 46 (November 17, 1980): [illegible].

37. "Conference on Methane," BR 23, no. 35 (September 1, 1980): 30.

38. Wu Zhonghua, "Resolving the Energy Crisis with Energy Technology," p. 31.

39. Vaclav Smil, "China's Energy Wonders: A Closer Look," mimeographed paper prepared for U.S. Agency for International Development, Workshop on China's Energy, Washington, D.C., December 18, 1980, p. 9.

40. Woodard, International Energy Relations of China, pp. 264, 268, 353, 398.

41. For a history of China's nuclear development and analysis of commercial nuclear prospects, see Woodard, International Energy Relations of China, pp. 343-357.

42. Kevin Fountain, "On the Back Burner: Nuclear Power in China," CBR 6, no. 6 (November-December 1979): 38-41.

43. Nuclear News [hereafter NN], December 1980, pp. 53-57.

44. NN, February 1981, pp. 56-57.

45. Bernd Knoll, "Revival of China's Nuclear Faction," Petroleum News: Southeast Asia 11 (August 1980): 14-15.

46. See NN, March 1981, p. 34 and Nuclear Engineering International [hereafter NEI], September 1980, p. 3.

47. "Proposal for Nuclear Power Stations," BR 23 (December 8, 1980): 3-4; and Craig Hosner, "A Visit to China," NEI, February 1981, pp. 15-16.

48. Wang Ganchang, "Perspective of Nuclear Energy Development in the PRC," in Proceedings of the Annual Joint Meeting of the American and European Nuclear Societies (Washington D.C.: November 1980).

49. Wall Street Journal, 12 December 1980.

50. NEI December 1980, p. 4-5.

51. NN, December 1980, p. 53-57.

52. "Proposal for Nuclear Power Stations," BR 23 (December 8, 1980): 3-4.

53. See BR (February 23, 1981): 7-8 and Nucleonics Week, 19 February 1981, p. 5-6.

54. For example, "Columnist Asserts Libyan Leader Tried to Buy A Bomb from China," New York Times, 16 April 1979, p. A-9.

55. NN, March 1981, p. 34.

56. Kevin Fountain, "New Energy Sources in China," CBR 6, no. 5 (September-October 1979): 28-33.

57. "Xinjiang Uses Solar Energy," BR 22, no. 8 (February 23, 1979): 30; "China's First Solar-Heated Building," BR 22, no. 30 (July 27, 1979): 30.

58. "Tidal Power Station," BR 23, no. 23 (June 9, 1980): 6-7.

59. For example, "New Progress in Laser Fusion Research," BR 23, no. 5 (February 4, 1980): 30.

60. Wu Zhonghua, "Resolving the Energy Crisis with Energy Technology," pp. 31-33.

61. Vaclav Smil, "Deep Structural Deficiencies," CBR 7, no. 1 (January-February, 1980): 64-65.

62. "Fulfillment of 1980 Targets," BR 24, no. 1 (January 5, 1981): 5.

63. Wu Zhonghua, "Resolving the Energy Crisis with Energy Technology," p. 32.

64. Yao Yilin, "Report on the Arrangements for the National Economic Plans for 1980 and 1981," speech to 3rd Session of 5th National People's Congress, August 30, 1980, BR 23, no. 38 (September 22, 1980): 30-43.

65. "Energy Policy," BR 23, no. 34 (August 25, 1980): 6-7.

66. Yao Yilin, "Report on National Economic Plans," p. 31.

67. "Energy Conservation," BR 23, no. 47 (November 24, 1980): 4-5. "Stress on Energy Conservation," BR 23, no. 47 (November 24, 1980): 4-5.

68. "Bank Loans for Saving Energy," BR 23, no. 45 (November 10, 1980): 7.

69. Detailed historical energy trade data by commodity, country, and year can be found in Woodard, International Energy Relations of China, pp. 495-534.

70. See table in Helen Kauder, "China's Crude Oil Exports Under Major Long-Term Agreements, 1975-85," CBR 7, no. 1 (January-February 1980): 36.

71. Woodard, International Energy Relations of China, pp. 73-74, pp. 544-564.

72. Crude and refined petroleum exports to Japan, about half of the total by volume were valued at $1.06 billion in 1979. See U.S. Central Intelligence Agency, China: International Trade Quarterly Review, First Quarter, 1980 (ER CIT 80-004), September 1980, pp. 11, 18.

73. Dori Jones, "China's Sagging Oil Production," CBR 7, no. 6 (November-December 1980): 24.

74. U.S. Central Intelligence Agency, China: International Trade Quarterly Review, First Quarter, 1980, p. 18.

75. K. O. Emery et al., "Geological Structure and Some Water Characteristics of the East China Sea and the Yellow Sea," CCOP Technical Bulletin 2 (1969): 41.

76. For a recent restatement of this claim, see: "China's Sovereignty Over East China Sea Continental Shelf," BR 23, no. 20 (May 19, 1980): 6-7.

77. For details, see Woodard, International Energy Relations of China, pp. 198-201.

78. For a recent list of China's fleet of offshore rigs, see table in Dori Jones, "China's Offshore Rigs, CBR 7, no. 6 (November-December 1980): 52.

79. "Investigating the Causes of the Oil Rig Accident," BR 23, no. 31 (August 4, 1980): 7; "Oil Rig Incident Sternly Dealt With," BR 23, no. 36 (September 8, 1980): 7-8.

80. In September 1980, a Chinese official reported that 17 wells had been drilled in the South China Sea, yielding 8 oil shows. On August 13, 1979, a wildcat on a Pearl River estuary came in at 411 metric tons per day (3,000 b/d), easily within commercial range. See Dori Jones, "China's Expanding Offshore Oil Fleet," CBR 7, no. 6 (November-December 1980): 49-51; Kevin Fountain, "The Development of China's Offshore Oil," CBR 7, no. 1 (January-February 1980): 22-35.

81. "Disputes in the South China Sea," PNSEA, supplement, 10, no. 7 (October 1979).

82. Kevin Fountain, "Development of China's Offshore Oil," pp. 22-35.

83. PNSEA 11, no. 3 (June 1980): 3.

84. Ibid., p. 3.

85. Yao Yilin, "Report on National Economic Plans," p. 38.

86. Dori Jones, "China's Expanding Offshore Oil Fleet," p. 50.

87. Liu Mei-yun, "For China, New Rigs are Only Half the Answer," PNSEA 11, no. 3 (June 1980): 16.

88. See note 86.

89. See note 82.

90. Robert E. King, "North Sea Joins Ranks of World's Major Oil Regions," World Oil 185, no. 7 (December 1977): 35-46.

91. Woodard, International Energy Relations of China, p. 335.

92. Ibid., chs. 12, 23, and 24.

93. Ibid., pp. 544-564.

94. Liu Lixin, "On the Use of Foreign Funds," BR 23, no. 24 (August 25, 1980): 23-25.

95. David Denny, Colloquium at the School for Advanced International Studies, Johns Hopkins University, Washington, D.C., December 4, 1980.

96. "China's Five and Ten Year Plans, 1981-1990," CBR 7, no. 2 (March-April 1980): 6.

97. See note 7.

98. "Vice-Premier Deng in Washington," BR 22, no. 6 (February 9, 1979): 8-14.

99. "Joint Press Communique on Premier Hua's Visit to Japan," BR 23, no. 23 (June 9, 1980): 8-9; "Vice Premier Deng Visits the United States," BR 23, no. 24 (June 16, 1980): 8-9.

100. "China to Conduct Experiment in Launching Carrier Rocket," BR 23, no. 20 (May 19, 1980): 4; "On-Site Report: To the South Pacific," BR 23, no. 22 (June 2, 1980): 22-23.

101. Derek Davies, "Victory for the Hardline Hawks," Far Eastern Economic Review 109, no. 28 (July 4, 1980): 12-15.

102. Karen Berney, "Dual-Use Technology Sales," CBR 7, no. 4 (July-August 1980): 23-26.

103. "Remote Sensing Techniques in China," BR 23, no. 24 (June 16, 1980): 29.

6 Environmental Aspects of Energy Developments

Toufiq A. Siddiqi

LINKS BETWEEN ENERGY USE AND ENVIRONMENTAL QUALITY

The close links between energy use and the level of economic activity are well known. Less obvious, perhaps, are the close links between energy use and environmental damage. Each of the steps taken to convert energy into useful work for society is accompanied by effects on the environment that can, in many cases, be substantial. Indeed, energy is in some sense the ultimate resource and the ultimate pollutant.[1] The magnitude of energy use can be a useful indicator not only of the level of economic activity, but also of the potential environmental stress.

One school of energy planners regards any adverse environmental effect as a small price to pay for meeting society's goal of economic development. Environmental concerns are viewed by them as "obstacles to progress." Their counterparts on the other side of the fence include environmentalists opposed to all power plants, oil drilling, and the like who recommend a return to the "good old days." In countries where the energy versus environment debate has gone on for many years and the implications of both positions have been discussed in detail, the majority of the public seems to accept an intermediate position, namely, development of energy resources, but in a way that minimizes adverse effects on the environment. This intermediate position recognizes that the positive and negative aspects of energy uses must be weighed against each other.

Those who propose "unshackling the energy industry from environmental regulations," as well as those opposing any new energy activity in any region, are ignoring the fact that neither "energy" nor "wilderness" are goals in themselves. It would be more appropriate to think of them as means towards achieving the goal of "improving the quality of life." Though it is unlikely that there will ever be a measure of the quality of life (QOL) that would be comparable to the economists' gross national product (GNP), some components of QOL seem to be common across countries and cultures. Basic human needs in all societies include food, shelter, and good health. Affection, education, and mobility are highly valued in most societies. Components of the biophysical environment, such as forests and lakes and the variety of species they contain, are also considered important in many societies.

It is clear that energy is required to provide many of the components of the QOL, such as food, shelter, and mobility. At the same time, the use of energy should be undertaken in ways that do not detract substantially from the provision of other components of the QOL, such as human health--both physical and mental. The ecosystems on which the survival of all species depends need to be managed in a way that is sustainable well beyond the usual short time span of interest to most planners. Indeed, maintenance of ecosystems is necessary for society not only for protection of physical and mental health, but also to assure the continuation of the many environmental services upon which society depends. The economic value to society of nutrient recycling by soils, erosion and flood control by forests, and fish raising by estuaries is immense, for example. It would be extremely difficult and expensive to replace these natural services by technology.

In most cases, ecosystems can be managed in a sound manner that will also permit us to meet society's needs for energy and minerals. The simplest analogy is perhaps that of an oil field from which oil can be produced at varying rates. Generally, if production is started at the maximum rate possible, the total amount of oil recovered will be less than if a lower rate of initial production had taken place. A decision maker interested only in a time span of a year or two might make the decision for maximum initial production, whereas another looking at a 10-year period might opt for a lower initial rate and larger total production.

In a number of countries, including the United States, the energy/environment debate has become polarized because of a number of factors, including the single-issue focus of most lobbying groups. The typical scenario goes something like this: The energy companies lobby for permits to drill for oil offshore, or to mine coal in scenic areas. The environmental groups file a lawsuit to prevent this from happening. The lawsuits and countersuits drag on for several years. Because the energy companies have more money and more lawyers, they usually win the right to drill offshore or mine coal in 90 percent of the areas they wanted (the other 10 percent is given up as a "compromise"). Frequently, the companies do not find any oil (as was the case for the offshore U.S. East Coast), or discover that the cost of producing coal has gone up so much during the years of the lawsuits that they cannot compete with producers who started earlier in nonscenic areas. The energy companies and the environmental groups lost a considerable amount of money and public goodwill, and only the lawyers have something to show for it.

Such an approach is not a viable option for countries that need to reduce their dependence on energy imports in a hurry. Such countries might analyze the experience of other nations and conclude that taking into account environmental concerns would only delay their plans for energy development. There is no inherent reason why this should be the case. Rather than having the agencies responsible for energy and those for environmental management adopt adversary positions, all need to be involved from the very beginning of the planning process. One possibility might be to have both energy and environmental units in planning commissions, the prime minister's office, or whichever organization in a country has the responsibility for overall long-term development planning. Another way may be to have an environment unit in the ministry or department of energy. The latter type of arrangement has been adopted in some countries, including the Philippines and the United States.

The Asia-Pacific region, extending from Iran to the United States and from China to New Zealand, is extremely rich in physical, industrial, and cultural diversity. It includes some of the largest users of energy in the world, such as the United States and Japan, as well as nations in Asia with meager energy requirements. Despite this diversity, the countries share a common desire to increase the prosperity of their citizens and consider energy an essential ingredient for achieving it. The more industrialized countries have been faced for some time with the necessities of controlling the undesired "residuals" that accompany the production and use of energy. Other countries following the path to industrialization have also become aware that there are limits to the environmental and social price they are willing to pay for increased energy use.

The countries in the Asia-Pacific region are emphasizing different energy sources, depending on the situation in each country. The environmental implications associated with each energy system are also different. It is not my intention here to list all of them, nor even to discuss most of them; a number of excellent surveys are available to provide such overviews.[2] One valuable way to categorize the environmental impacts associated with energy use is to separate them according to the steps in each fuel cycle, as shown in Table 6.1. In the limited space available here, the discussion will be restricted to some of the significant environmental implications associated with the energy sources that a number of countries in the region are emphasizing for the next two decades--coal, oil, nuclear fission reactors, and some renewable energy sources. This does not mean that the role, for example, of natural gas, geothermal, or wind power is likely to be unimportant--only that their overall environmental impact in the region is likely to be smaller than that of the sources discussed here.

SIGNIFICANT ENVIRONMENTAL IMPLICATIONS OF SOME ENERGY PROGRAMS IN THE ASIA-PACIFIC REGION

Based on the stated energy policies of the various countries in the region, including the emphasis on particular energy sources, Table 6.2 gives an approximate indication of the types of energy versus environment debates most likely to exist or arise in the countries listed.[3] The very rough nature of this table should be emphasized, since it is clear that, for example, the level of concern about radioactive waste disposal in Pakistan, with its

TABLE 6.1
Some Environmental Impacts of Energy Conversion

	Health	Property	Social	Quality of Life	Environmental Services
Exploration					
Oil/gas	Accidents	--	--	Invasion of wilderness	--
Production					
Coal mining	Accidents, black lung	Loss of farmland, subsidence	Use of aboriginal lands	Defaced landscape	Acid drainage
Offshore oil	Accidents	--	--	Oil on beaches	Oil as a biocide
Hydroelectric dam	Dam collapse	Loss of farmland	Displacement of residents	Loss of wild rivers	Fish passage, wildlife breeding grounds
Processing					
Oil refining	Air/disease	Air/crops	--	Smells, visibility	Pollution of estuaries
Shale processing	Air/disease	Water consumption	--	Waste piles	Water pollution
Conversion					
Coal power plant	Air/disease	Air/crops, buildings	--	Noise, visibility	Acid rain, CO_2 particles/climate
Fission reactor	Reactor accident that breached containment would produce all classes of impact				
Transportation					
Oil tanker	Fire	Fire, collision	--	Oil on beaches	Oil as biocide
Electrical transmission	Electrocution	Restriction on land use	--	Unsightly towers	--
Plutonium	Leak/cancer	Land contamination/quarantine	Terrorism nuclear bombs	--	--
Consumption					
Automobile	Air/disease	Air/crops	Suburbanization	Noise, visibility	Paved environment, heat/climate
Waste management					
Radioactive wastes	Leak/mutations	Land use	Terrorism, sabotage	--	Groundwater contamination

Source: Adapted from John P. Holdren, "Energy Resources," in William W. Murdoch (ed.), Environment (Sunderland, Mass.: Sinamer Associates, 1975).

TABLE 6.2
Environmental Factors in Major Energy Programs in the Asia-Pacific Region

Country	Air Pollution from Increased Use of Coal	Disposal of Radio-active Wastes from Nuclear Power Plants	Effects of Offshore Oil and Gas Develop-ment on Coastal Zones	Land-Use Implications of Hydropower Projects	Deforestation Due to Increased Rural Energy Demand
Australia	XX		XX	X	
Bangladesh				X	X
China	XXX	X[a]	X		
Fiji					
India	XXX	X	X	X	XXX
Indonesia	X		XX		XX
Japan	XXX	XXX			
Malaysia			X	X	XX
Nepal				XX	XXX
New Zealand			X	X	
Pakistan		X	X	XXX	XXX
Papua New Guinea				X	
Philippines	X	X[b]	XX	X	X
Singapore			X		
South Korea	XXX	XXX			X
Sri Lanka				XX	
Thailand			X	X	X
United States	XXX	XXX	XXX		

XXX = major concern

XX = medium concern

X = some concern

[a] There are no known commercial power plants in operation on the Chinese mainland, although some are under consideration. There are two operational plants on Taiwan, and four more under construction.

[b] After lengthy delays due to safety concerns, construction of the first nuclear power plant is again proceeding.

one 125-MW operating nuclear power plant, will be quite different from that in Japan, with 23 units in operation and 11 under construction (see Chapter 4). Even for the selected areas, the discussion must necessarily be extremely brief; the references cited can provide the reader with detailed information.

Air Pollution from Increased Use of Coal

It has been estimated that coal will have to provide one half to two thirds of the additional fuel needed by the world during the next 20 years.[4] Much of the coal produced will come from the Asia-Pacific region, and much of it will be used there. The two major coal-based economies in Asia, India and China, will contribute significantly to this growth. Steam coal shipments from Australia and the United States are expected to increase substantially, with some exports possible from China, India, and Indonesia. Japan, for example, may nearly triple its coal imports to over 200 million tons (mt) annually by the year 2000. Estimates of Asia-Pacific coal resources and reserves that can be mined economically with present technology are given in Table 6.3.

Each step in the coal cycle has environmental disturbances associated with it, as shown in Figure 6.1. Thus, countries such as Australia, China, India, and the United States, which are major producers as well as users of coal, are faced with the task of reducing essentially all of these impacts. In most cases, the cost is not large, although the removal of particular pollutants such as sulfur oxides may appear to be too expensive to some countries. Some cost estimates from a recent study of specific environmental measures related to coal mining and transportation are given in Table 6.4, and for coal utilization in Table 6.5.[5] These costs are generally well below present world costs for coal, which are $50 to $70 in 1978 dollars for imports.

The environmental effects associated with the moving and transportation phases of the coal cycle are usually confined within the country producing or consuming the coal, except for the impacts of an accident affecting a coal-transporting ship at sea. This is not always the case for the emissions from coal combustion, which can affect third countries. Oxides of sulfur and nitrogen can be carried hundreds of miles by wind, especially if power plants use tall stacks to disperse the pollutants. The "acid rain" resulting from such emissions can have a significant effect on lake, forest, and agricultural ecosystems, although the extent of the effect is disputed. In 1980, several European countries reached an agreement to attempt to control emissions likely to cause acid rain in neighboring countries, and similar discussions are also underway between the United States and Canada. With the large increase in coal use expected in Asia, a similar agreement in the region would be desirable at an early date, before the majority of siting decisions are made. Of course, in large countries such as China and India, much of the acid rain could be deposited in other regions of the countries themselves.

Emissions of particulates and sulfur oxides from fuel combustion are presently being controlled in Japan and the

TABLE 6.3
Coal Resources and Reserves in the Asia-Pacific Region

Country	Geological Resources	Technically and Economically Recoverable Reserves	Units[a]	Reference
Afghanistan	470	66	mtce	WEC[b]
Australia	>770,000	63,000	mt	Wocol[c]
Bangladesh	1,050	242	mt	WEC
Burma	150	2.3	mtce	WEC
Canada	>247,000	63,000	mt	Wocol
China	1,440,000	99,000	mtce	WEC
India	85,800	22,600	mt	India Department of Coal
Indonesia	6,530	234	mtce	WEC
Iran	385	193	mt	WEC
Japan	20,300	3,200	mt	Wocol
Malaysia	260	n.a.	mtce	WEC
New Zealand	n.a.	3,700	mt	New Zealand Ministry of Energy
North Korea	4,950	534	mtce	WEC
Pakistan	n.a.	477	mt	Pakistan Ministry of Petroleum & Natural Resources
Philippines	1,150	175	mt	Philippines Ministry of Energy
South Korea	1,165	116	mtce	WEC
Thailand	246	103	mt	WEC
United States	3,970,000	250,000	mt	U.S. Office of Technology Assessment (USOTA)
Vietnam	850	150	mtce	WEC
World Total	10,700,000	688,000		

[a] mtce = million metric tons coal equivalent; mt = million metric tons.

Table 6.3 (cont'd)

[b]WEC refers to the World Energy Conference, Survey of Energy Resources 1980, Federal Institute for Geosciences and Natural Resources, Hanover, Federal Republic of Germany, 352 pp. and maps (1980).

[c]Wocol refers to the World Coal Study, 2 vols. (Cambridge, Mass.: Ballinger, 1980).

n.a. = data not available.

United States. Particulate emissions are also being controlled in Australia, China, India, and many of the other countries in the region, at least from new power plants. Australian and Indian coals generally have a low sulfur content, but the Indian coals have a very high ash content. The ambient air quality standards and emission standards for power plants in the region are listed in a later section. Table 6.6 shows typical pollutant emissions from a 900-MW coal power plant operating at 34 percent thermal efficiency.

Carbon Dioxide. Carbon dioxide, which plays an important role in global ecosystems, is not generally considered a pollutant. During the last few years, however, it has become evident that the level of CO_2 in the atmosphere is growing, apparently because of the increasing combustion of fossil fuels.[6] If the use of coal continues to increase, as presently projected, the CO_2 level in the atmosphere could double during the next century.[2] There is considerable debate on the implications of such a buildup of CO_2:

1. Many of the simulations undertaken show that a global warming of 1°-2° C could occur. This conclusion is not universally accepted since, for example, the effect would be quite dependent on the latitude.
2. Such a warming might result in the melting of some of the polar ice cap, with a corresponding rise in the level of the sea and possible eventual submergence of coastal lands.
3. An average temperature increase of 1° C, if accompanied by a decrease of 10 percent in precipitation, could reduce wheat yields in parts of the United States and the Soviet Union by up to 20 percent.[7] It could also cause a global increase in rice production of about 2 percent.[8] Thus, even if the world became somewhat warmer, some countries might benefit, whereas others might be adversely affected.

Given such uncertainties, it seems premature to abandon policies for increased use of coal on the basis of possible effects of carbon dioxide buildup in the atmosphere. At the same time, this

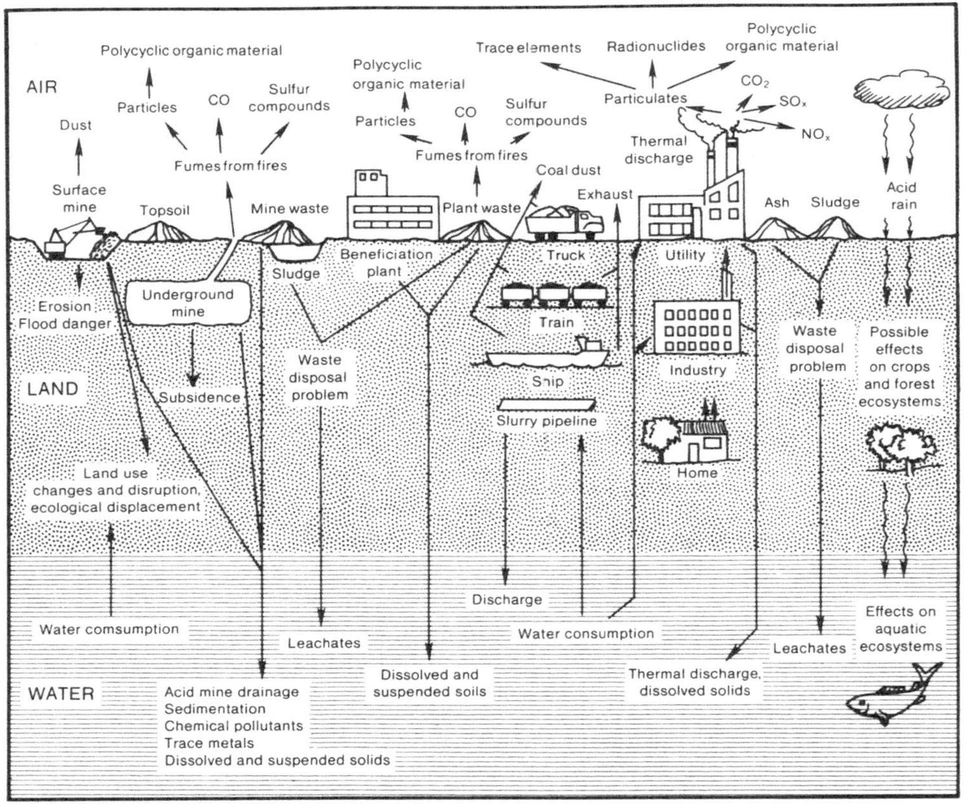

Figure 6.1. Environmental disturbances from coal related activities.

TABLE 6.4
Cost Estimates for Environmental Measures Related to Coal Mining and Transportation
(US$/metric tons of coal, 1978)

	Coal Mining and Cleaning				
	Contour Surface Mining	Area Surface Mining	All Surface Mining	Underground Mines	Comments
1. Reclamation of active mines	3.0-3.2	0.2-1.0		1.1-5.4	Higher for surface mining in steep sloped areas
2. Fee for reclamation of abandoned mines			0.10 (lignite) 0.4 (hard coal)	0.2	U.S. legislation exists; under consideration in other countries
3. Dust control			0.10-0.20		
4. Mine drainage control	0.4-0.5	0.2-0.4		0.07-0.6	1985 technology
5. Occupational health safety requirements				6.4	
6. Coal cleaning, prevention of runoff from storage and wastes			0.10	0.10	Per ton cleaned

	Coal Transportation			
	By Rail	Slurry Pipeline	Harbors	Comments
1. Dust control, prevention of spills, control of runoff	0.05		Unknown	
2. Treatment of slurry water		0.2-0.3		Reduced by evaporating

Source: Based on data from the Organization for Economic Cooperation and Development. See also Organization for Economic Cooperation and Development, Macroeconomic Evaluation of Environmental Programmes (Paris: 1978).

TABLE 6.5
Cost Estimates for Environmental Measures for Electric Utility Coal Utilization, New Sources (US$/metric ton, 1978)

	Australia	Japan	United States
Control of thermal discharges, cooling towers			
Wet	0.42	-	0.8-0.9
Dry	-	-	7.0-11.0
Particulate control			
ESP	0.80	1.6	1.5-2.0
Baghouse	-	5.4	1.6-1.8
SO_x control			
Limestone FGD	-	14.2	7.0-18.0
Regenerative	-	-	17.0
Dry FGD	-	-	9.0-30.0
NO_x control			
Combustion control (where possible)	0.20	0.8	0.2-0.5
Postcombustion control - NO_x selective	-	4.7	6.0-9.0
Postcombustion control - SO_x/NO_x scrubber	-	-	7.5-18.0
Combustion by-products disposal			
Ash - conventional	0.25	3.9	0.3-0.7
Ash - as hazardous material	-	-	5.0-10.0
Ash plus FGD sludge - conventional	-	-	2.5
Wastewater treatment			
Conventional	-	0.9	0.3-2.0
Zero discharge	-	-	2.0-3.0
Noise control, external plant only			
FD fans	-	0.2	0.2-0.5

Source: *World Coal Study*, 2 vols. (Cambridge, Mass.: Ballinger, 1980).

TABLE 6.6
Annual Emissions from a 900-MW Coal-Fired Power Plant[a]

	(tons)
Airborne effluents	
Particulates	1.8×10^3
Sulfur oxides[b]	6.4×10^3
Nitrogen oxides	1.6×10^4
Carbon monoxide	1.2×10^3
Hydrocarbons	230
Liquid effluents	
Organic material	38
Sulphuric acid	48
Chloride	15
Phosphate	24
Boron	190
Suspended solids	290
Solid waste	
Bottom and fly ash	2.1×10^5

Source: Based on United Nations Environment Programme, The Environmental Impacts of Production and Use of Energy, Fossil Fuels, part 1 (Nairobi, Kenya: 1979).

[a]1.75 mtce/year of coal with 2 percent sulfur, 12 percent ash, and 99 percent capture of fly ash.

[b]Flue-gas desulfurization could potentially reduce sulfur oxide emissions by a factor of 10 to 20 with some increase in solid and liquid wastes.

area of research is likely to become increasingly important in the next few decades. Presently, it does not appear to be practical to trap CO_2 emissions during combustion, since the volumes involved are extremely large. Avoiding CO_2 emissions may someday be a powerful argument for increasing the reliance on biomass fuels that are harvested on a sustainable basis. The Asia-Pacific region consumes only about 20 percent of total global fossil fuel production, which approximately indicates its relative contribution to further CO_2 buildup. About 33 percent of global coal consumption occurs in these regions, however--mainly in India and China.[9] As oil declines and coal rises in importance, the region's relative contribution to the CO_2 problem may rise.

Disposal of Radioactive Wastes from Nuclear Power Plants

The topic of radioactive waste disposal is of growing importance in th Asia-Pacific region but will only be discussed briefly here, as it has been examined in detail in Chapter 4.

The outlook for the continued growth of nuclear power in the United States has dimmed considerably since the Three-Mile Island (TMI) accident. Recent estimates of the cost of cleaning up the damaged power reactor exceed $1 billion, plus more than $500 million to bring it back into operation. In a desperate attempt to get assistance from the federal government, the utility company that owned TMI has sued the U.S. Nuclear Regulatory Commission for $4 billion, accusing it of inadequate supervision! Whatever the outcome of the suit, one result will probably be an increase in the cost of building and operating nuclear power plants. The cost of "decommissioning" nuclear power plants and of disposing of their radioactive wastes will also increase.

Although no new nuclear power plants have been ordered in the United States since the TMI accident, some of the other countries in the Asia-Pacific region feel that they do not have the other alternatives available to the U.S. Based on their nuclear power programs, the largest amount of radioactive wastes in the region during the next two decades are likely to be generated in Japan, South Korea, and Taiwan, China (see Chapter 4, Table 4.2). None of these is considered to have good sites for the disposal of radioactive wastes. The prospect of such wastes being disposed of at sea, or transported over thousands of miles to whichever countries accept them for burial, is of great concern to many countries in the region, particularly the island nations of the Pacific. These nations feel that they have not derived any benefits from nuclear power but rather have already paid the costs--in terms of radioactive fallout from weapons tests in the past--and may have to bear the costs of possible accidents from movement of nuclear wastes in the future.

Over the next two decades or so, the only substantial choices for electricity generation on a large scale are coal and nuclear fission. Air pollution, mining damage, and CO_2 buildup accompany the coal cycle; radioactive wastes, risk of reactor accidents, and nuclear weapons are associated with nuclear fission. It is possible that some countries may choose between coal and nuclear primarily on the basis of environmental acceptability rather than resource availability.

Effects of Offshore Oil and Gas Developments on Coastal Zones

The spectacular blowout of an oil well or spillage from an oil tanker often makes the news headlines, but the lasting impact of offshore oil and gas operations is usually on the adjacent coastal lands, and is a more or less continuing process. Norway and Scotland provide dramatic examples of coastal areas that have been significantly changed in less than a decade because of oil and gas exploration and production in the North Sea.[10] Most of the countries in the Asia-Pacific region have some offshore exploration or production going on at present. Some of the major offshore programs include: (1) Bass Strait and Northwest Shelf (Australia), (2) Gulf of Beibu (Tonkin), Bohai, and Pearl River (China), (3) Bombay High and Bay of Bengal (India), (4) Brunei, (5) Java, Sumatra, and Kalimantan (Indonesia), (6) Trengganu, Sabah, and Sarawak (Malaysia), (7) west coast of North Island (New Zealand), (8) Palawan (Philippines), (9) Gulf of Thailand, (10) Alaska and California (U.S.A.).[11] These locations are shown in Figure 6.2, together with the locations of some of the other major energy programs discussed in this chapter.

Fishing and tourism are the industries most frequently affected by offshore oil exploration, production, and shipment.[12] It is thus desirable to identify such areas before exploration begins and design additional safety measures to protect the especially sensitive areas. A number of countries have initiated programs to identify and designate such important coastal areas as coral reefs and mangroves, as well as known fishing grounds and present or planned tourism developments.[13] Several are attempting to draft a set of environmental guidelines that could be used as an initial point for discussions between companies undertaking offshore oil and gas exploration, production, and shipping, and government organizations responsible for environmental management offshore and in the coastal zones.[14]

The oil produced offshore is brought to land by pipelines or tankers. Pipeline operations have a good safety record, and it is worth keeping in mind that most of the oil moved by tankers is oil produced <u>onshore</u> and then moved to distant markets. Most of the oil pollution in coastal areas originates on land and is associated with the refining and utilization of oil. Highly aggregated data for the sources of oil in the oceans are given in Table 6.7.

The development and maintenance of loading and unloading facilities can have a larger impact on ecosystems than the shipment of oil from the well by tankers or pipeline. The effects of these facilities on the coastal zone are of particular interest because the major fisheries of the world are located there. The most vulnerable areas are estuaries, where a variety of interactions take place as freshwater mixes with seawater, providing an especially productive ecosystem. Estuaries also serve as nursing grounds for many marine species.

Oil entering the marine ecosystem forms a "slick" that can rapidly move away from the original source. The subsequent behavior of the slick varies greatly, depending on the type of oil, the temperature of the water, and meteorological conditions. The hydrocarbons may eventually (1) evaporate into the atmosphere, (2) disperse and dissolve in the water, (3) be incorporated into bottom sediments, or (4) be oxidized via chemical and biotic pathways.[15]

There is a great deal of uncertainty regarding the effects of oil on marine ecosystems. The generally accepted view is that the water-soluble components of hydrocarbons, such as the aromatics, are the most toxic. Growth inhibition, increased mortality, and other ill effects have been observed at low hydrocarbon

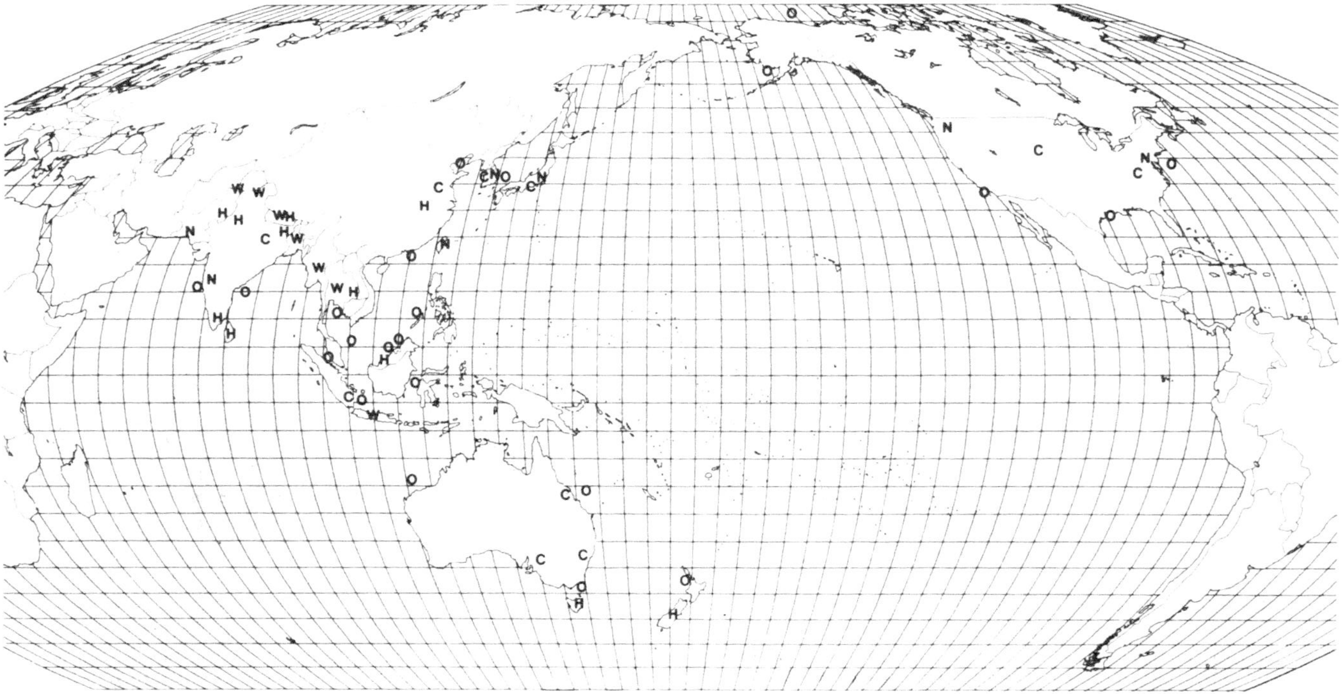

C = Coal   H = Hydropower   N = Nuclear   O = Offshore Oil   W = Wood for fuel

Figure 6.2. Some major energy programs in the Asia-Pacific region, with potentially significant environmental implications.

TABLE 6.7
Sources of Oil in the Oceans

Source	Estimated Contribution (mb/year)	Percentage of Total
Production and transport	16.0	35
Rivers and urban runoff	14.0	31
Atmospheric fallout	4.5	10
National seepage	4.5	10
Direct discharge into oceans:	6.5	14
Coastal refineries		
Municipal waste		
Industrial waste		
Total	45.5	100

Source: U.S. Department of Energy and U.S. Environmental Protection Agency, *Energy/Environment Fact Book* (Cincinnati, Ohio: 1978).

concentration in a number of important marine organisms, including oysters, phytoplankton, and fish eggs.[16]

Oil tanker accidents such as the *Torrey Canyon* near the United Kingdom, the *Argo Merchant* near New England, the *Showa Maru* in the Malacca Straits, and the *Amoco Cadiz* off the coast of France have been studied in detail.[17] The cost of cleaning up after an oil spill can be extremely high. For example, it was estimated that cleanup costs average US$1,000 per barrel of oil spilled.[18] In view of the possible expense involved and the high cost of the cleanup equipment, it may be useful for groups of countries such as ASEAN to have centralized cleanup facilities available to different nations when needed.

Oil refineries are one of the many industries located on coastal zones close to where offshore oil is being landed. These refineries can have a significant effect on the quality of the environment in coastal areas. The annual emissions from a refinery using 25 million barrels of oil a year are given in Table 6.8.

New refineries are being planned in many parts of Asia, including China, India, and Indonesia. Present capacity and those under construction total about 15.3 million b/d in Asia and Oceania (not including the Persian Gulf). The siting of refineries and related industry presents a challenge, as well as an opportunity, for integrated land use and coastal zone planning. Some of the major issues are discussed in the sections on China and India.

TABLE 6.8
Emissions from an Oil Refinery Processing 25 Million Barrels per Year[a]

	(tons)
Gaseous effluents	
Sulfur oxides	21,000
Organic compounds	23,000
Nitrogen oxides	18,000
Carbon monoxide	4,300
Ammonia	2,230
Liquid effluents (1.4 x 10^8 tons waste water) containing:	
Chlorides	24,000
Grease	600
Ammonia nitrogen	600
Phosphate	3
Suspended solids	2,000
Dissolved solids	100,000
Trace metals (Cr, Pb, Zn, Cu)	22

Source: United Nations Environment Programme, The Environmental Impacts of Production and Use of Energy, Fossil Fuels, part 1 (Nairobi, Kenya: 1979).

[a]Equivalent to approximately 80,000 barrels per working (stream) day. This energy flow is roughly equivalent to three times that of the 900-MW coal power plant in Table 6.6.

Land-Use Implications of Hydropower Programs

Renewable energy sources are not entirely benign in terms of their impact on the environment, as shown in Table 6.11. Among the best-known examples of the environmental effects of renewables are those resulting from the construction of the Aswan High Dam in Egypt. About 100,000 people had to be relocated without adequate planning, and food had to be rushed in by the World Food Programme to avoid famine.[19] Two major historical monuments--Abu Simbel and Philae--had to be dismantled and moved to higher locations. Today, the sediment deposited in the reservoir is not only reducing the useful life of the dam, but it has also had several deleterious effects downstream, including the erosion of the river bank and the delta. The construction of the dam has also been held responsible for a reduction in the fish catch in the Nile Delta and, probably, for an increase in schistosomiasis, a debilitating parasitic disease.

In general, human-made lakes such as those formed by hydropower plants can result in a number of undesirable side effects:

1. Loss of valuable land, possibly including rare forests and historical sites
2. Change in sediments downstream, affecting agriculture and fish production
3. Changes in the hydrological cycle (evaporation, rainfall, and river flow)
4. New stresses on the earth's crust, generating seismic movements
5. Changes in the biological and chemical properties of the water leaving the lake, affecting downstream ecosystems

A number of countries in the Asian region have ambitious development plans for adding to their hydropower capacity. Some of the larger projects scheduled in Australia, China, India, Malaysia, Nepal, New Zealand, Pakistan, and Sri Lanka are indicated by "H" in Figure 6.2. A few of the projects, such as those proposed for Silent Valley (India) and Lake Pedder (Australia), have become controversial because of public opposition. In the former case, there is concern that one of the last remaining portions of primary forests in India will be lost. This is discussed at greater length in the section on India.

<u>Lake Pedder Controversy</u>. The opposition to the hydropower plant at Lake Pedder in Australia was based on the probable loss of a beautiful recreational area.[20] The original Lake Pedder was a small (3 sq km) elevated lake of unusual character in an inaccessible area of southwest Tasmania. What made the lake unique was that it had a beach of pink-tinted white-quartz sand, about 3 km long, with a width reaching 1 km during the summer months. The water in the lake was extremely acid, with a high humus content, and an unusual ecosystem had evolved there--about 18 endemic species of flora and fauna. The lake was also a popular recreational spot.

The Hydro-Electric Commission of Tasmania had been considering different ways to tap the potential of the Gordon River and its tributaries for hydropower. These studies indicated that the least expensive approach would be to direct the Upper Huon and Serpentine to the Gordon River. The Commission realized that this would result in the flooding of Lake Pedder and the loss of its famous beach, but felt that the benefits resulting from the project, such as the creation of a much larger lake and a number of beautiful bays and basins suitable for boating, camping, and the like, would outweigh the loss of the white-quartz beach. The costs of the other alternatives, in monetary terms, were much higher.

Some public concern had become evident in 1967 and was revived around 1971, when it was discovered that the lake itself had 13 animal and 4 plant species of relatively rare type that would be unlikely to survive the flooding of the lake.

Public opposition took the form of a major demonstration at Lake Pedder itself, a publicity campaign, and the formation of a Lake Pedder Action Committee. The Commonwealth government was requested by conservationists to provide funding to Tasmania so that the lake could be saved for the entire nation. Both major political parties in Tasmania, however, stated that they would not reconsider the decision to proceed with the project. The attempts by the conservationists to elect their own candidates in state elections were unsuccessful, as were attempts to stop the project on legal grounds and through the intervention of the Commonwealth government. By the end of 1972, flooding of the lake was underway.

Although Lake Pedder could not be saved in its original form, the episode provided a major impetus for detailed environmental evaluations of future projects. In February 1973, the Canberra government established a Committee of Inquiry to look into the circumstances surrounding the flooding of the lake and to report upon its implications for the future planning of major development projects in Australia. The conclusions and recommendations of the committee have validity well beyond Australia and are consequently summarized here:[21]

1. There is a need to adopt a much broader multi-objective approach to the planning and management of water and land resources. Objectives should include the enhancement of national economic development, the enhancement of regional development, the enhancement of environmental quality, the conservation of natural resources, and the improvement of social welfare.
2. Future development projects involving the management of land-based resources should be undertaken only on the basis of adequate land-use planning.
3. At the national level, there is a clear need for the development and implementation of overall co-ordinated land-use and national resource management policies, possibly through a national Land Conservation and Natural Resources Council. The States should be encouraged to establish co-ordinating land conservation and resource management councils having overall responsibilities for the development and implementation of land-use and national resources policies at State and regional levels.
4. An improved approach to project planning should be adopted. Projects should be characterized by a clear statement of overall objectives.
5. Environmental impact studies can be used as an interim evaluative device. Soundly based multi-objective planning, in which environmental factors are included amongst the planning parameters and treated as an integral part of the overall planning process from its inception, was seen to be the most effective way of ensuring that all relevant factors, environmental and otherwise, are taken properly into account in the overall public interest.

6. Public participation in the decision-making process is desirable whenever major resource development projects are undertaken in what is alleged to be the public interest.

According to the Committee, the Lake Pedder controversy has clearly shown how a withholding of information and an unwillingness to give consideration to the public viewpoint can lead to bitter and unnecessary conflict over an environmental issue.[22] The president of the Australian Conservation Foundation, Prince Philip, of the United Kingdom, remarked:

> In a pioneering society, the principal problems are to survive, to exploit and to develop. The Lake Pedder case marks the end of Australia's pioneering days and it ushers in a new phase of conscious concern by all sections of the community for the long-term future of the natural and human environment.[23]

Such rising public concern led the Commonwealth government to introduce the Environmental Protection (Impact of Proposals) Act in 1974 (discussed later). Similar acts were also passed by state governments.

In several other countries of Asia and the Pacific, environmental assessments are now being undertaken before major hydropower or multiple-purpose hydroprojects are initiated. A good example of this policy is provided by the assessments accompanying the Mahaveli project in Sri Lanka.[24] Given the need for increasing electrical capacity in many countries of the region, hydropower with proper planning may be environmentally more acceptable than coal or nuclear power. In particular, the small-scale hydropower plants ("micro-hydro") presently being built in a number of countries offer an attractive option for the decentralized production of electricity.

Deforestation from Increased Rural Energy Demand

Wood is still the major energy source for over a third of the world's population.[25] Used mainly for cooking food and heating homes or water, it provides about 80 percent of all energy in rural areas of the less-industrialized countries.[26] Estimates of the fuelwood contribution to total energy supply in selected Asian countries are given in Table 6.9.

Growth in population, as well as rising awareness of the need for better nutrition, is likely to increase demand for fuelwood, particularly in rural areas where few alternative sources of energy are available. The present demand for fuelwood in Asia requires the cutting of about 0.8 billion cubic meters of fuelwood, substantially contributing to the increased deforestation in the region.[27] It has been estimated that current forestry practices are causing about 5 million hectares of forest to be lost each year in Asia alone.[28] Even without an increase in demand, the remaining tropical forests in the region could vanish

TABLE 6.9
Fuelwood Consumption in Selected Asian Countries, 1976

Country	Per Capita Commercial Energy Consumption (kgce)	Official Woodfuel Consumption (1,000 cubic meters)	Per Capita Official Woodfuel Consumption (kgce)	Woodfuel as Percent of Total Energy
Burma	49	19,611	273	85
India	218	118,179	83	28
Indonesia	218	111,708	360	62
Malaysia	578	5,613	195	25
Nepal	11	8,700	291	96
Philippines	329	22,960	226	41
South Korea	1,020	7,350	91	8
Thailand	308	16,091	160	34

Source: Eric Hyman, "Wood as an Energy Source for Less Developed Countries," unpublished paper.

in 60 to 80 years unless present practices are altered. Of course, the need for rural energy is not the only cause of deforestation--growing requirements for agricultural land, as well as the need for timber and wood products, also bear major responsibility for the loss of forests. It has been estimated that in 1974, about 47 percent of the 2.5 billion cubic meters of wood cut throughout the world was used as fuel, but this percentage could rise to 90 percent in the tropical countries.[29] In the smaller islands of the Pacific, about 50 to 90 percent of the current deforestation is caused by the demand for fuelwood.[30]

The use of fuelwood is expected to increase over the next two decades in most of the developing countries.[31] Projections for the countries of Asia are given in Table 6.10.

The environmental consequences of the large-scale removal of forest cover can be extremely serious, including soil erosion, silting, and flooding in the rainy season. Although direct cause-and-effect relationships are hard to prove, it is generally felt that the increased flooding in Northern India, Pakistan, and Nepal may be a consequence of the substantial deforestation that has taken place in the foothills of the Himalayas.

Although this section deals only with the deforestation implications of fuelwood use, it should be noted that the actual burning of wood also has a significant impact on the local environment. Emissions per unit of fuel can exceed those from fossil fuel combustion. Rough calculations indicate that the combustion of traditional fuels in developing countries has the potential for continual pollution of all the arable regions of those countries

TABLE 6.10
Present and Projected Consumption of Fuelwood in Asia (mtce)

Country	1976	1980 (estimated)	1985 (projected)	1990 (projected)
Afghanistan	2.1	2.3	2.5	2.7
Bangladesh	5.2	6.5	7.9	9.2
Burma	7.1	8.8	10.4	12.4
China	52.7	56.8	61.1	65.6
India	43.8	46.3	50.6	54.8
Indonesia	41.2	45.8	50.6	55.4
Malaysia	2.1	2.2	2.4	2.5
Nepal	3.2	3.4	3.7	3.3
North Korea	1.7	1.9	2.1	2.3
Pakistan	3.1	3.5	3.9	4.3
Philippines	8.5	9.3	10.5	11.6
South Korea	2.7	2.8	2.6	2.5
Thailand	6.0	6.3	6.9	7.4
Vietnam	6.0	6.7	7.1	7.6
Others	6.4	7.1	7.7	8.4

Source: David Hughart, "Prospects for Traditional and Nonconventional Energy Sources in Developing Countries," World Bank Staff Working Paper no. 346 (Washington, D.C.: World Bank, 1979).

to a height of 0.5 km, with the particulate concentrations reaching the ambient levels recommended by the World Health Organization (40 micrograms per cubic meter).[32] This estimate may be subject to debate, but it shows that firewood is probably an underestimated contributor to regional air pollution. A number of other more hazardous air pollutants, including polycylic organic material from incomplete combustion, are also emitted during biomass combustion in small stoves.[33] The exposure levels and consequent health effects on those responsible for daily cooking, primarily women in rural areas, could also be substantial.[34]

To reduce the pressure on forests from fuelwood gathering, there are at least three possible approaches:

1. Provide another source of energy. The absence of a transportation infrastructure in many rural areas makes it difficult and expensive to supply other fuels. Even without transportation costs, it would require, for example, about 1.25 billion barrels of oil to replace the 0.8 billion cubic meters of fuelwood consumed in Asia alone. At present world prices, the cost of the substitution would exceed $50 billion a year, an impossible undertaking for economies already struggling to cope with oil import bills that have increased by factors of 10 or more in most cases!

In some cases, biogas from human and animal wastes is being used as a substitute fuel, particularly in China and India. Dried cowdung has also been in use as a fuel for a long time in South Asia, although the practice is harmful for agricultural production since it removes many of the nutrients required by the soil. Micro-hydro power, wind, and solar energy may be used in appropriate locations. It appears unlikely, however, that the use of wood as a fuel in the rural areas of Asia can be substantially reduced during the next few decades.

2. The efficiency with which fuelwood is used can be increased. It is well known that the cooking stoves used in many parts of the world are highly inefficient in terms of the fraction of energy converted that does useful work. A number of programs have been initiated in several countries with the objective of designing stoves that would be more energy efficient (that is, use less fuelwood to do the same work) and also lower the emissions of pollutants.

3. If wood fuel cannot easily be replaced, efforts need to be made to grow it on something close to a self-sustaining basis. "Fuelwood plantations" are being experimented with in a number of countries, including India, South Korea, and the Philippines. Several of the projects are being supported by the World Bank and other international assistance agencies.[35] Two rapidly growing species--eucalyptus and leucaena--appear to be good candidates for widespread use. It is too early to assess the success of the experiments, but in most places the results will probably depend more on the social arrangements for managing the fuelwood plantations than on the biophysical characteristics of the ecosystem.

Environmental Effects of Some Other Renewable Energy Sources

The discussion in previous sections has centered on most of the energy sources that, at present, provide the largest proportion of energy used worldwide, and it has been shown that significant environmental implications are associated with each of them. Proponents often claim that these "new" and "renewable" energy technologies not only meet national energy goals but are also environmentally more desirable than conventional forms. This claim appears to be true for only some of the "newer" technologies.

Table 6.11 shows a recent comparison of only the air pollutants from the processes used to make the necessary construction materials and the routine fuel cycle operations for some "renewable" and "traditional" technologies.[36] The renewable sources appear attractive in terms of reducing the environmental costs per unit of energy delivered when compared to oil or coal. On present evidence, the most promising renewable options seem to be: (1) passive solar heating and cooling, (2) increased electricity generation through the addition of generators to certain existing dams, (3) electricity generation by wind turbines, and (4) biogasification of sewage and feedlot manures.

TABLE 6.11
Environmental Impacts of Renewable Energy Sources[a]

	Particulates	Oxides of Sulfur	Oxides of Nitrogen	Carbon Monoxide
Heat				
Solar flat plate collectors	12[b]-4,700[c]	3-60	9-50	1,200-19,000
Solar ponds	1-620	0-110	0-20	?
Wood-burning stoves	29,000	760	1,800	560,000
Chemical fuel				
Fuels from biomass	140-320	2,900-11,000	1,500-14,000	120-9,400
Synfuels from coal	40-260	670-1,600	670-880	350-1,300
Refined oil products	110-1,000	900-15,000	2,300-5,000	290-4,100
Electricity				
Solar thermal electric	760-53,000	25-7,600	1-1,300	10,000-70,000
Solar photovoltaic	180-29,000	70-2,800	10-530	1-28,000
Wind electric conversion	150-5,000	7-240	0.1-30	2,800-29,000
Hydroelectric	560-11,000	3-1,900	0-320	1,100-4,100
Wood electric	11,000-56,000	3,300-5,600	22,000-38,000	5,300-230,000
Refuse-derived fuel electric	2,000-20,000	?	19,000	9,400-17,000
Coal electric	4,000-38,000	10,000-35,000	12,000-24,000	1,400-4,400
LWR-fission electric	200-670	80-180	30-65	970-2,100

Source: John P. Holdren, Gregory Morris, and Irving Mintzer, "Environmental Aspects of Renewable Energy Sources," Annual Review of Energy 5 (1980): 241-291.

[a] Air pollutant emissions from materials acquisition and routine fuel cycle operations (grams of pollutants per mtce = 29.3 gigajoules).

Table 6.11 (cont'd)

[b]Low figures represent controlled emissions during materials production and energy system operation.

[c]High figures represent uncontrolled emissions during materials acquisition and controlled emissions during energy system operation.

From an environmental point of view, the least desirable renewable energy sources appear to be satellite solar power stations, new hydroelectric dams, and intensively managed biomass plantations. As mentioned earlier, wood-burning stoves have very high emissions per unit of energy delivered. It should be noted that emissions from solar thermal and photovoltaic electricity can vary by more than two orders of magnitude--that is, they depend very strongly on the particular type of system selected.

Research and development of renewable energy systems are being carried out in a large number of countries in the region, including Australia, China, India, Japan, and the United States. Renewable sources already play a larger role in the energy systems of China, India, and other less-industrialized countries than in the highly industrialized nations, and it is likely that this situation will not change during the next decade or two.

GOVERNMENT RESPONSES TO ENERGY-ENVIRONMENT CONCERNS IN THE ASIA-PACIFIC REGION

In response to growing concern about the impacts of energy use on environmental quality, governments in the region have begun to develop legislation, assessment techniques, regulatory apparatus, and research capabilities. To give a picture of these activities in the limited space available here, it is useful to consider the environmental assessment requirements and air pollution standards within the region and to look in detail at its two most populous countries.

Environmental Assessment Requirements

During the past decade, a number of countries have formulated policies and legislation requiring some type of environmental analysis of major projects, including, of course, energy facilities. Such practices have also been adopted by some organizations that finance international development programs. The World Bank has been a leader in this field, requiring a formal process of environmental assessment for all major projects through an Office of Environmental and Health Affairs.[37] The Office reviews project proposals for their environmental soundness but does not prepare separate environmental impact statements. The U.S. Agency for International Development (USAID) undertakes an "initial environmental examination" for all projects, and a more detailed "environmental impact assessment" is performed when the probable effects of the project are judged to be significant.[38]

Countries requiring an environmental assessment usually require that these evaluations be performed for major projects that are likely to have a significant impact on the environment. Most of the energy projects of the type discussed here fall into this category and consequently are subject to an environmental assessment. The status of such policies and existing legislation is summarized, with some modifications, in Table 6.12.[39]

Air Pollution Standards

In the context of energy conversion, several countries in the Asia-Pacific region have established ambient air quality standards or standards limiting emissions from particular types of facilities such as power plants, refineries, and the like. Countries have used different approaches to define standards, using various means and lengths of measurement. The ambient air quality standards for the more common pollutants are given in Table 6.13.

Among the Asian countries, Japan has set the most stringent air quality standards, since pollution levels in many areas of the country had started to become hazardous to human health. The rapid implementation of these standards has already led to significant improvements in the levels of most air pollutants throughout the country.[40]

Emission standards should be set so that the levels of the different pollutants in the air do not exceed the ambient levels. Some countries, however, have set emission standards based on the "most practical means," taking economic factors into account, but have not yet established ambient standards. It should be mentioned that, at least in the United States, the Environmental Protection Agency was given the mandate to set ambient air quality standards on the basis of protecting public health and without regard to economic factors. This approach has been criticized by some in industry on the basis that the standards are so strict that the incremental cost of meeting them far exceeds the incremental benefits.

China

In recent years, as discussed in Chapter 5, China has been focusing seriously on its severe energy problems. Included among these problems are the environmental impacts of energy production and use.

Environmental Legislation.[41] During the first decade of its existence, the People's Republic of China promulgated two important laws designed to protect environmental quality. "Sanitary Standards for Designing Industrial Enterprises," including standards for air, water and soil for working and living areas, were issued in 1956,[42] and the "Sanitary Regulations for Drinking Water" in 1959. Environmental management, however, did not become major official concern until the country started making preparations for the Conference on the Human Environment held at Stockholm in 1972. Participation in the conference provided an opportunity to analyze the experience of other nations and draw up plans to avoid repeating some of the adverse consequences of industrialization.

China held its first national conference on environmental protection in 1973. The conference made an important contribution to the development of environmental management in the country, and an initial program for protecting and improving the environment was formulated. "Some Regulations Concerning Environmental Protection and Improvement," issued in 1973, is the first legal document in China to deal with the overall concepts and principles of environmental protection. Several other regulations were announced that year, dealing with nature conservation, protection of endangered species, and industrial discharges.

The Constitution of the People's Republic, adopted in 1978, states in Article 2 that "the State protects the environment and natural resources and prevents and eliminates pollution and other hazards to the public." This marks the first time that environmental protection was included as an official goal in the fundamental law of the land and signaled a break from the period of the Cultural Revolution, during which environmental concerns were largely ignored and at times were considered reactionary.

The Standing Committee of the Fifth National People's Congress approved, in September 1979, the "Environmental Protection Law of the People's Republic of China" for "trial implementation." Wide-ranging in scope, this law gives official status to environmental concerns and the academic disciplines involved.[43] Of special interest to the energy sector, "Management Regulations for Ships of Foreign Countries" were also promulgated. China also joined the International Convention of Civil Responsibilities for Oil Pollution and Damages. Among legislation being prepared are the "Law of Water and Soil Conservation" and the "Law of Prevention and Protection Against Air Pollution."

Environmental Management. The primary responsibility for environmental management at the central government level lies with the Leading Group of Environmental Protection under the State Council. The group is composed of senior officials from more than 20 departments and is headed by a vice-premier. The Office of the Leading Group of Environmental Protection is the corresponding executive agency. The Office has a number of divisions, and a research institute and a general monitoring station are being established.

Environmental protection departments or their equivalents have been established in a number of ministries, including the Ministries of Petroleum, Power, Transportation and Communication, and Forestry. This should assist in the inclusion of environmental considerations in energy planning and implementation.

Environmental protection bureaus have been set up in 22 provinces, autonomous regions, and municipalities directly under the central government, and environmental protection offices

TABLE 6.12
Status of Environmental Assessment Requirements at the Central Government Level in Selected Asia-Pacific Countries

Country	Formal Policy Statement or Legislation	Organizations Responsible for Managing the Environment	Types of Projects Requiring Environmental Assessment
Australia	The Environmental Protection (Impact of Proposals) Act of 1974.	Department of Home Affairs and the Environment.	Commissions of Inquiry for major projects and policies require submissions on environment impacts. State regulations call for environmental impact assessments (EIA) on major projects.
China	Constitution, Chapter I, Article 11, March 1978 Law on Environmental Protection of PRC (for trial use) as adopted in principle by the 11th session of the Standing Committee of the 5th National Peoples Congress, September 1979.	Environmental Protection Office in State Council. Environmental Protection Bureaus in provinces and municipalities.	EIA demand for new construction projects or expansion of existing ones.
India	The Constitution Articles 40 A and 51 A. The draft Five-Year Plan (1978-83) Chapter 7 on Environmental Planning and Coordination.	National Committee on Environmental Planning and Coordination established in 1972 (NCEPC). The Department of Environment established in 1980.	Environmental appraisal is obligatory on the part of the project planner. Guidelines and checklist questionnaires are recommended to federal and state agencies in the following sectors: river valley development, hydroelectric projects, thermal power plants, large-scale industrial projects, mining operations.
Indonesia	Repelita III (1979-83) Policy and Programs for Management of Natural Resources and Environment.	Minister of State for Development Supervision and Environment. Proposed Coordinating Board for Environmental Management.	

Japan	The Basic Law for Environmental Pollution Control of 1967 states that "efforts shall be made to balance pollution control against the need for economic development." The Environmental Impact Assessment Law has not yet been passed by the Diet.	The Environment Agency, established in 1971, is involved in project planning and review of other agencies (e.g., harbors, power sources, urban and rural areas). Central Council for Control of Environmental Pollution (80 members) advises Prime Minister.	Reports from contractors are required on some public works projects.
South Korea	Environmental Protection Law (PL 3078, 1977). Presidential Decree No. 9707.	Office of Environment, in the Ministry of Health and Social Affairs.	All major development projects with prior consultation with the Office of Environment.
Malaysia	Third Malaysia Plan (1976-80), Environmental Quality Act of 1974.	Division of the Environment within the Ministry of Science, Technology and Environment.	Proposed plan would include all major projects. Three stages: initial screening, initial environmental evaluation, EIA. Only on a limited number of projects up to 1985; full implementation in Fifth Plan period 1986-90.
New Zealand	Not available	Commission for the Environment established in 1973 to advise, educate and formulate environmental protection and enhancement procedures.	Impact reports on projects requiring government licenses, approval, or finance (65 reports in 1973-79).
Papua New Guinea	Not available	Department of Conservation and Environment (DCE).	Ad hoc project evaluations on major landscape disruptions (e.g., hydropower, mining) contain environmental considerations.
Philippines	Presidential Decrees Nos. 1121, 1151, 1152 in 1977 establishing the National Environmental Protection Council (NEPC), the Philippine Environmental Policy and the Philippine Environment Code.	The NEPC is chaired by the President, comprises most government department heads, and is within the Human Settlements Commission which is chaired by the First Lady. There is an Environment Unit in the Ministry of Energy.	Environmental impact statements (EIS) required for all major projects.

Table 6.12 (cont'd)

	Formal Policy Statement or Legislation	Organizations Responsible for Managing the Environment	Types of Projects Requiring Environmental Assessment
Sri Lanka	Environmental Protection Act of 1979, Chapter 6, Article 27 (14) in the Constitution.	Central Environmental Authority (CEA) with 3 members, Environmental Council comprising representatives from all relevant government departments, and Environmental Appeal Board.	The Central Environmental Authority makes any needed studies, surveys, reports, etc. on impacts of plans or projects on the environment.
Thailand	The Economic and Social Development Plan, 1977-81, Chapter on Development of Conservation of Critical Economic Resources and the Environment. The National Environmental Quality Act of 1978.	The National Environment Board within the Ministry of Science, Technology and Energy comprises heads of other government departments.	Projects designated by NEB unless exempted by Prime Minister or Assembly.
United States	The National Environmental Policy Act of 1969, Executive Orders 11514, 12114, and amendments.	The Council on Environmental Quality (CEQ) in the Executive Office of the President (a) issues annual status and trends report, (b) establishes guidelines for preparation of environmental impact statements (EIS), advises the President on compliance of federal agencies with NEPA. The Environmental Protection Agency administers pollution control laws and reviews EISs from other agencies.	EISs must be prepared on all "major federal actions significantly affecting the quality of the human environment."

Table 6.12 (cont'd)

<u>Source</u>: Richard Carpenter and William Matthews, "NEPA: Environmental Innocence Abroad," <u>East-West Perspectives</u> 1 (Fall, 1980): 6-11; Qu Geping and Li Jinchan, "Environmental Management in China," paper presented at the Workshop on Environmental Management, East-West Center, Honolulu, 1980; Department of Environment, <u>Role of the Government of India in Environmental Protection</u> (New Delhi: 1981); Department of Science and Technology, <u>Report of the Committee for Recommending Legislative Measures and Administrative Machinery for Ensuring Environmental Protection</u> (New Delhi: 1980).

---

have been set up in the rest of the provinces. Most of the prefectures, cities, and counties also have institutions for environmental protection. Whereas the Central Office of Environmental Protection is responsible for policy formulation, legislation, and the establishment of environmental standards, the work of inspection and supervision is the responsibility of the local environmental organizations at the different levels. The division of responsibilities, however, has not been clear, and efforts are underway to improve this situation.

<u>Scientific Research Institutions</u>. Environmental management requires input from a large number of disciplines, and this need is reflected in China in the variety of institutions contributing to the required research. Much of the basic scientific research is undertaken under the auspices of the Chinese Academy of Sciences; research on environmental economics and legislation is undertaken by institutions under the Chinese Academy of Social Sciences. In addition, various ministries and departments under the State Council, as well as universities, colleges, and provincial institutes contribute to the research effort. A Chinese Society of Environmental Science, established in 1979, is helping to increase research as well as to promote environmental public education.

About 300 environmental monitoring stations have been set up or are in the process of being set up. It is planned to provide complete sets of equipment to more than 60 major monitoring stations in the provinces and key cities. The supply of uniform equipment and the introduction of common measuring techniques will provide a basis for the establishment of a nationwide environmental monitoring network.

<u>Some Energy-Related Environmental Concerns</u>.[44] Coal supplies about 71 percent of the primary commercial energy used in China, oil about 23 percent, and natural gas and hydropower about 3 percent each.[45] Thus, the major environmental concerns related to energy use are those associated with the coal cycle. Surface mining with its accompanying environmental effects is given priority, wherever possible. The rail transportation system presents a major obstacle to an increase in the use of coal, which is already running at full capacity. It is at the combustion end of

TABLE 6.13
Ambient Air Quality Standards in Three Countries Compared with World Health Organization (WHO) Recommendations

Pollutant	Average Time	Japan	South Korea	Philippines	United States		WHO Recommendations
					Primary	Secondary	
Particulate matter	Year				75 μg/m^3	60 μg/m^3	40-60 μg/m^3
	Day	100 μg/m^3		180 μg/m^3	260 μg/m^3	150 μg/m^3	100-150 μg/m^3
	Hour	200 μg/m^3		250 μg/m^3			
Sulfur dioxide	Year		0.05 ppm		80 μg/m^3 (0.03 ppm)		40-60 μg/m^3 (0.02 ppm)
	Day	0.04 ppm		0.14 ppm	365 μg/m^3 (0.14 ppm)		
	Hour	0.1 ppm		0.3 ppm		1,300 μg/m^3 (0.5 ppm)[a]	
Carbon monoxide	Day	10 ppm					
	8 hours	20 ppm		9 ppm	10 mg/m^3 (9 ppm)	Same as primary standard	10 mg/m^3 (9 ppm)
	1 hour			30 ppm	40 mg/m^3 (35 ppm)		40 mg/m^3 (35 ppm)
Nitrogen monoxide	Year				100 μg/m^3 (0.05 ppm)		
	Day	0.04-0.06 ppm					
	Hour			0.1 ppm			190-320 μg/m^3 (0.10-0.17 ppm)

Table 6.13 (cont'd)

Pollutant	Average Time	Japan	South Korea	Philippines	United States		WHO Recommendations
					Primary	Secondary	
Photochemical oxidants	Hour	0.06 ppm		0.06 ppm	240 μg/m^3 (0.12 ppm)[c]	Same as primary standard	100-200 μg/m^3 (0.05-0.1 ppm)
Hydrocarbon	3 hours	0.20-0.31 ppm[b]			160 μg/m^3 (0.24 ppm)	Same as primary standard	
Lead	3 months				1.5 μg/m^3		

Source: Y. Iikura, D. James, T. Jimenea, and T. Siddiqi, "Energy-Related Air Quality Standards in the Asia-Pacific Region," working paper (Honolulu: East-West Environment and Policy Institute, 1981).

[a] Three-hour average.

[b] Given not as standard but as guidelines.

[c] This is a revised standard for ozone. The earlier standard, for photochemical oxidants, had been set at 0.08 ppm.

the cycle, though, that coal use has had the most significant impact. It has been estimated that the burning of coal contributes more than 70 percent of the air pollution in China. As shown in Table 6.14, the levels of a few of the pollutants are extremely high in some Chinese cities, exceeding by a substantial margin the standards recommended by the World Health Organization and the U.S. ambient standards.

Air pollution in the cities in northern China is worse during the winter months because of the increased use of coal for heating. Unfavorable meteorological conditions, such as temperature inversion or absence of wind for prolonged periods, aggravate the situation during some months--for example, Peking in December and Shengyang in January have especially high levels of air pollution.

There are two daily peaks of air pollution during the winter --- from about 6 a.m. to 8 a.m., and from 6 p.m. to 8 p.m., caused by domestic cooking and heating. Frequently, these times coincide with temperature inversions in the air, leading to high concentrations of pollutants.

Although the heating requirements in the southern part of China are smaller, the coals in that region are of poorer quality, with higher sulfur and ash content. Since the railroad transportation system is already overloaded, it is impractical at the moment to move higher quality coals from the north for use in the south.

It has been estimated that the total amount of sulfur dioxide emitted into the air in China exceeds 15 million tons a year. The reduction of air pollution is likely to be a more difficult task in China than in many other industrialized nations, since coal is the major energy source for home heating, cooking, and for small industry. Emissions of particulates and sulfur oxides from large installations such as electric power plants can be reduced substantially at an acceptable cost. Similar reductions from homes and small-scale industries are likely to be expensive, and China would have to switch to centralized steam heating or natural gas to reduce air pollution in the cities. Such a massive change in the country's infrastructure would be expensive and slow.

Even though hydropower supplies only about 3 percent of the total commercial energy used in China, it contributed about 18 percent of the electricity used in the country during 1979.[46] Excluding small units (<500 kw), the total installed hydropower capacity in China is about 16 gigawatts (GW). The small units in remote rural areas have an additional installed capacity of about 4 GW. The estimated potential for hydropower that can be economically tapped is 380 GW, providing substantial potential for increasing electricity generation. Hydropower development will be concentrated in specially favorable locations on selected rivers, such as the upper stream of the Yellow River, the Hongshiu River in Guangxi Province, and the middle and upper reaches of the Yangtze River.

The implementation of these hydropower schemes could have significant environmental effects, such as those discussed earlier. The efforts of the Environment Protection Office and

TABLE 6.14
Air Pollution Levels in Some Cities of China

Pollutants	Pollution Level (daily average in $\mu g/m^3$)[a]	U.S. Primary Standard (in $\mu g/m^3$)
Suspended particulates	240-1,980 (cities in the north) 190-950 (cities in the south)	260 (daily average)
Sulfur dioxide	32-2,190 (cities in the north) 14-780 (cities in the south)	365 (daily average)
Nitrogen dioxide	100-170 (cities in the north) 20-130 (cities in the south)	100 (annual average)
Ozone	60-210 (Lanzhou) 70-80 (Beijing)	240 (hourly average)
Lead	0.10-0.6 (range for all cities)	1.5 (three months average)

Source: Bing Lin Wu and Guan Ju-Gen, "Energy Policy and Air Quality Management in the People's Republic of China," in Proceedings of the Conference on Air Quality Management and Energy Policies, Baroda, India, February 16-25, 1981 (to be published).

[a]Micrograms per cubic meter.

the agencies in the provinces, however, are designed to minimize the adverse effects.

Offshore oil and gas exploration and production are just beginning to gain momentum in China, and the onshore effects of this activity could be substantial in some areas within a decade, unless appropriate planning is initiated soon. There is no indication that the Chinese are rushing to grant exploration and production licenses. This provides adequate time for the inclusion of environmental guidelines for offshore oil and gas production, including the construction of onshore facilities for processing the fuels.

China has been the world leader in the use of small-scale biogas plants in rural areas, with over 6 million such units in operation. To the extent that the residues are returned to the soil, this approach has a beneficial effect on the environment and can operate to reduce the number of pathogens in human waste. A number of other developing countries are examining the Chinese experience in biogas use with great interest, with an eye on possible adoption of some of the approaches.

India

The second most populous country in the world also has a number of major environmental problems. Many of these are linked more to population pressures than to energy use directly. In recent years, however, a number of steps have been taken to assess and start controlling impacts from energy production and use.

Environmental Legislation. The Indian Constitution gives explicit recognition to the need for environmental protection in the section on Directive Principles of State Policy. Article 48 states that "the State shall endeavor to protect and improve the environment and to safeguard the forests and wildlife of the country." Article 51-4 states that "it shall be the duty of every citizen of India to protect and improve the national environment including forests, lakes, rivers, and wildlife, and to have compassion for living creatures."

Legislation dealing with specific aspects of environmental protection, such as soil conservation, forest management, wildlife protection, and public health, has been in existence in India for several decades. The Indian Department of Science and Technology has identified over 200 laws at the central government as well as state government level that have a bearing on environmental protection.[47] The Committee for Recommending Legislative Measures and Administrative Machinery for Ensuring Environmental Protection was appointed by the Indian prime minister in February 1980 to examine the adequacy of the existing administrative, legal, and institutional arrangements for protecting environmental assets. After a survey of the existing legislation, the committee concluded that:

1. Many of the existing laws are updated (in a limited sense) versions of earlier ones. They are primarily

meant to promote development and resource utilization for specific economic benefits without a careful analysis of the potential short-term and long-term deleterious effects on the environment.

2. Many of the existing laws relating to management of environmental resources do not clearly state the social objectives they aim to achieve. In the absence of any explicit policy statement on the objectives to be accomplished, the administrative machinery set up to implement the legislations have on their own interpreted their duties from time to time which have often proved to be not in conformity with the intent and purpose for which the enactment was made in the first place. Moreover, changes in national policies and circumstances have progressively rendered several of these Acts obsolete.
3. Some of the laws in force, particularly with regard to land use and management of environmental resources, appear at times to be accomplishing mutually defeating social objectives. Where such resources are shared by more than one State, legislation enacted in one State may have adverse environmental implications for a neighbouring one.
4. The implementing and monitoring machinery of many of these laws are deficient in the scientific and technical expertise as well as other infrastructural resources required to assess and prevent the possibility of adverse environmental impacts.[48]

The committee also pointed out the need to undertake a comprehensive review of some important Acts of the central government dealing with environmental quality--the Insecticides Act of 1968 and the Water (Prevention and Control of Pollution) Act of 1974, the Indian Forest Act of 1927, the National Forest Policy, and the Wildlife (Protection) Act of 1972.

Environmental Management. As noted earlier, even though legislation to protect the environment existed at both the central and state government levels, it had little success in preventing the continuing deterioration of the environment until the end of the 1960s. The need to integrate environmental considerations from the early stages of planning for economic development was explicitly stated for the first time in India's Fourth Five-Year Plan (1969-1974):

> It is an obligation of each generation to maintain the productive capacity of land, air, water, and wildlife in a manner which leaves its successors some choice in the creation of a healthy environment. The physical environment is a dynamic, complex and interconnected system in which any action in one part affects others. There is also the interdependence of living things and their relationships with land, air and water. Planning for harmonious development recognizes this unity of nature and man. Such planning is

possible only on the basis of a comprehensive appraisal of environmental issues, particularly economic and ecological. There are instances in which timely, specialized advice on environmental aspects could have helped in project design and in averting subsequent adverse effects on the environment, leading to loss of invested resources. It is necessary, therefore, to introduce the environmental aspect into our planning and development. Along with effective conservation and rational use of natural resources, protection and improvement of human environment is vital for national well-being.[49]

A formal mechanism for integrating environmental aspects into development planning was established in February 1972, when Prime Minister Indira Gandhi constituted a National Committee on Environmental Planning and Coordination (NCEPC). The preparatory work undertaken for India's participation at the U.N. Conference on the Human Environment held at Stockholm in 1972 helped to provide the priorities for the work of the NCEPC.[50] The activities undertaken by the NCEPC include:

1. Promotion of environmental and ecological research
2. Environmental appraisal of development projects and evolving guidelines for incorporating environmental considerations in development planning
3. Policies for environmental management, including issues relating to institutional support, legislation, education, and publicity
4. Cooperation in the field of environment with international agencies such as UNEP as well as through bilateral agreements

The NCEPC is composed of representatives from government, universities, other research institutions, and the public. It is now served by a Secretariat in the Department of the Environment.[46]

The NCEPC has encouraged the setting up of high-level Environment Boards in several Indian states and Union Territories. It has played the role of the leading <u>advisory</u> body to the government of India on environmental issues. The actual <u>implementation</u> of measures to protect the environment remains the responsibility of the ministry or department appropriate to the sector. Early in 1981, NCEPC was reconstituted as the National Committee on Environmental Planning (NCEP).

A start has been made towards the protection of forests and wildlife by establishing 19 national parks and 202 wildlife sanctuaries covering an area of 76,000 sq km.[51] The Committee for Recommending Legislative measures concluded that this was not sufficient to cover the ecological diversity of threatened habitats, nor was the management of these areas effective enough to ensure the protection of endangered species. It recommended the establishment of Biosphere Reserves as a matter of high priority.

Boards for the Prevention and Control of Water Pollution have been established at the central as well as state levels. Their investigations have shown that the major source of water pollution in rivers, streams, and coastal waters are municipal wastes from cities and villages. Only eight cities in India have adequate sewers and sewage treatment facilities.[52] In most other cases, municipal and industrial wastewaters are discharged without treatment into rivers, streams, or the ocean. The expenditure required to reduce water pollution in a country the size of India, with a population of over 600 million, is huge, but there is increasing awareness of the need to tackle the problem.

Air quality, which is also a serious concern in many areas, is discussed later.

Scientific Research Institutions. The NCEPC has been served by a Secretariat located within the Department of Science and Technology consisting of a team of scientists from various disciplines. This core group forms the staff of the new Department of Environment and has played an important role in ensuring that some of the best scientific talent in the country is made aware of environmental problems and is given an opportunity to contribute to solving them.

Over the last several years, the research program of the NCEPC has been implemented through the Environmental Research Committee (ERC) and the Indian National Man and Biosphere (MAB) Committee. The ERC has focused on the broad areas of environmental pollution and its impacts. It also encourages demonstration projects to convert research results into practical applications. The MAB Committee supports 14 major project areas related to the ecological impacts of human activities on the biosphere, along the lines formulated by the UNESCO Man and Biosphere Program. A list of projects supported by ERC and MAB is included in the report by the Department of Environment.[53]

Much of the research related to environmental management issues is undertaken by established research organizations and universities. In addition, new institutions have been established to address environmental issues specifically, such as the National Environmental Engineering Research Institute (NEERI) at Nagpur and the School of Environmental Sciences at the Jawaharlal Nehru University, New Delhi.

Some Energy-Related Environmental Concerns. The environmental concerns associated with energy conversion in India are quite similar to those discussed for China, since coal, hydropower, and biomass play important roles in energy supply in both countries. Until the 1960s, "noncommercial" energy, primarily fuelwood, cowdung, and crop residues, supplied more than half of the total energy used in the country; even now, it contributes about 40 percent to the total energy used.[54] Fuelwood, cowdung cakes, dried leaves and twigs, and similar materials are used mainly for cooking, outdoors or indoors. The housewives, who usually do the cooking, are often exposed to high levels of particulates and organics.[55] More efficient stoves could be designed to reduce this pollution, but would require government subsidies, if they are to be used widely. Furthermore, the time span for the social acceptance of new designs in rural areas is extremely long. In urban areas where oil or gas have been made available, their acceptance has been quite good.

The air quality in large urban and industrial areas is frequently unsatisfactory, with energy conversion in power plants, industry, and the transportation sector bearing most of the responsibility. Unlike China, most parts of India do not require home heating, and the direct use of coal for this purpose is limited to the northern part of the country.

Coal is the major source of "commercial" energy in India, with over 100 million tons consumed annually. Indian coals usually have a very high ash content, frequently in the range of 25 to 30 percent, with corresponding large emissions of particulates, unless controlled.[56] With minor exceptions, the sulfur content of the coals is low, making it less necessary to control sulfur oxide emissions. Unfortunately, the electrostatic precipitators normally used for controlling particulates frequently do not perform well when the sulfur content is low.

The pollutants emitted from different types of sources in Calcutta, India's largest city, are shown in Table 6.15.[57] Electricity generation, mainly from coal, is the largest source of particulates, sulfur dioxide, and nitrogen oxides. The industrial sector, including refineries, is the main source of hydrocarbons, and coal-based industries also contribute much of the particulate load. The largest source of carbon monoxide is the transportation sector, which also contributes substantial amounts of hydrocarbons and nitrogen oxides. Emissions from diesel-using buses and trucks are higher than they need to be, because of overloading and improper maintenance, and are likely to be much more difficult to control than emissions from power plants and large industries.

Possibly the most controversial issue emerging from the environmental dimension of energy is the siting of an oil refinery at Mathura, about 75 km from the historic Taj Mahal at Agra. Environmental groups have claimed that the air pollution caused by the refinery, primarily sulfur oxides, would have a harmful effect on the marble of the Taj Mahal. An Expert Committee appointed by the government of India came to the conclusion that the existing levels of sulfur dioxides at Agra were already high due to the use of coal in foundries, railroad yards, and power plants.[58] It concluded that the additional contribution to ambient levels of sulfur dioxide would be insignificant and that the monument would be better protected by the reduction of emissions from existing sources. This conclusion has not been universally accepted, and the controversy continues.[59]

Another controversial energy-environment issue of importance centers around the plan to build a hydroelectric project in Silent Valley, Kerala. Environmental groups have stated that Silent Valley represents one of the last virgin rainforests in India and contains unique species of plants and animals, including the lion-tailed monkey (<u>Maccaca silenus</u>). They have cited studies conducted by the Botanical Survey of India and by a Committee

TABLE 6.15
Pollutants Emitted from Different Types of Sources in Calcutta City

Source	Particulates		Sulfur Dioxide		Carbon Monoxide		Oxides of Nitrogen		Hydrocarbons	
	mt/d[a]	Percent	mt/d	Percent	mt/d	Percent	mt/d	Percent	mt/d	Percent
1. Power generation	150.0	52.8	46.0	71.0	2.3	1.2	17.2	36.4	1.1	1.8
2. Industrial	107.5	37.9	11.5	17.8	2.0	1.1	7.3	15.4	31.3	53.0
3. Domestic	25.3	8.9	5.9	9.2	33.7	19.2	7.2	15.2	8.4	14.2
4 Transportation	1.0	0.4	1.3	2.0	138.7	78.5	15.6	33.0	18.3	31.0
Total	283.8		64.7		176.7		47.3		59.1	

Source: J. M. Dave, N. Ramanathan, and T. A. Siddiqui (eds.), *Proceedings of the Conference on Air Quality Management and Energy Policies*, Baroda, India, February 16-25, 1981 (to be published).

[a] mt/d = metric tons per day.

of the Ministry of Agriculture identifying several rare plant species in the valley.[60] The Kerala State Electricity Board has claimed that only about 845 acres of the forest would be affected and that the environmental effects of using alternate ways of producing electricity, such as coal or nuclear power, would be worse, in addition to the possibility of more frequent disruptions of supply.[61] The issue has become polarized, with the environmentalists refusing to become members of an official panel to safeguard the environment at the planned power site.[62]

Almost by definition, controversial issues are those where accommodation between the groups has not been easy. In most cases, compromises are reached that enable energy developments to go forward in ways that will be less harmful to the long-term sustainability of the environment.

CONCLUSION

In this brief survey, it has only been possible to touch on some of the major environmental aspects of energy developments in Asia and the Pacific. In some cases, such as the construction of a large hydroelectric scheme, the impact on the environment is large and immediate but confined to a relatively small area, and can be handled at a national or provincial level. In other cases, such as the commissioning of a number of coal-fired power plants, the emissions into the atmosphere could affect other nations, and potentially the global climate. Scientists and policy makers from throughout the region need to include such factors in the formation of energy policies and to explore on a collaborative basis the environmental implications of pursuing alternative energy strategies, such as increased use of coal, or nuclear power, or offshore oil exploration or production. These topics cannot be adequately discussed in isolation. In the next chapter, environmental concerns are included in an overview of the range of topics that are faced by energy planners and policy makers in the Asia-Pacific region.

NOTES

1. Paul Ehrlich, Anne Ehrlich, and John P. Holdren, Ecoscience (San Francisco: W. H. Freeman, 1977).

2. Earl Cook, Man, Energy, Society (San Francisco: W. H. Freeman, 1979), p. 478; Carol Steinhart and John Steinhart, Energy: Sources, Use and Role in Human Affairs (North Scituate, Mass.: Duxbury Press, 1974).

3. Kirk R. Smith and Harrison Brown, "From Pakistan to Japan: The Energy Problems of Asia," OPEC Review 4 no. 3 (Autumn, 1980): 19-54; Toufiq A. Siddiqi, "Energy Policy Issues and the Environmental Agenda for the 1980s," in Current Issues: The Yearbook of Environmental Education and Environmental Studies, ed. Arthur Sacks (Madison, Wis.: National Association for Environmental Education, 1980).

4. World Coal Study, 2 vols. (Cambridge, Mass.: Ballinger, 1980).

5. Ibid.

6. Jill Williams, ed., Carbon Dioxide, Climate, and Society (Oxford: Pergamon Press, 1978).

7. Wilfrid Bach, "The Potential Consequences of Increasing CO_2 Levels in the Atmosphere," in Williams, Carbon Dioxide, Climate, and Society, pp. 141-167.

8. J. Stansel and R. E. Huke, "Rice," in Impacts of Climatic Change on the Biosphere, CIAP Monograph 5, Pt. 2.

9. Smith and Brown, "From Pakistan to Japan," pp. 19-54.

10. Irvin L. White et al., North Sea Oil and Gas (Norman, Okla.: University of Oklahoma Press, 1973).

11. World Oil, 35th International Outlook Issue 191, no. 3 (August 15, 1980): 1-296.

12. United Nations, Environment Programme [hereafter, UNEP], The Environmental Impacts of Production and Use of Energy, Part I: Fossil Fuels; Part III: Renewable Sources of Energy (Nairobi, Kenya: United Nations, 1979, 1980).

13. Chua Thia-Eng and Joseph K. Charles, Coastal Resorces of East Coast Peninsular Malaysia: An Assessment in Relation to Potential Oil Spills (Penang: University Sains Malaysia, 1980), p. 507; see also papers presented at the Workshop on Shipping, Energy and Environment, East-West Environment and Policy Institute, Honolulu, December 10-12, 1980.

14. John Gilbert et al., "Draft Environmental Guidelines for Offshore Oil Prospecting, Development Drilling, and Production," working paper (Honolulu: East-West Environment and Policy Institute, 1981).

15. Charles A. S. Hall et al., "Environment Impacts of Industrial Energy Systems in the Coastal Zone," Annual Review of Energy 3 (1978): 395-476.

16. Ibid.

17. See notes 12, 13, and 15.

18. U.S., Department of Energy and U.S., Environmental Protection Agency, Energy/Environmental Fact Book (Cincinnati, Ohio, 1978).

19. See note 12.

20. Alan Gilpin, The Australian Environment: Twelve Controversial Issues (Melbourne, Australia: Macmillan, 1980).

21. Ibid.

22. Ibid.

23. Ibid.

24. W. Paul Weatherly and John H. Arnold, Jr., Environmental Assessment of Stage II of the Mahaweli Ganga Development Project (Washington, D.C.: U.S. Agency for International Development, 1977).

25. David Hughart, "Prospects for Traditional and Non-Conventional Energy Sources in Developing Countries," working paper (Washington, D.C.: World Bank, 1979).

26. Erik Eckholm, Planting for the Future: Forestry for Human Needs (Washington, D.C.: Worldwatch Institute, 1979).

27. J. E. M. Arnold, "Wood Energy and Rural Communities," Natural Resources Forum 3, no. 3 (April 1979): 229-252.

28. World Bank, Forestry: Sector Policy Paper (Washington, D.C.: World Bank, 1978), p. 65.

29. See Norman Myers, Conversion of Tropical Moist Forests (Washington, D.C.: National Academy of Sciences, 1980), p. 205; and Food and Agricultural Organization, Yearbook of Forest Products (Rome: United Nations, 1979).

30. Randy Thaman and Tevita Ba, Energy Needs and Forest Resources of Small Islands (Paper presented at the 49th ANZAAS Congress, University of the South Pacific, Suva, Fiji, 1979).

31. See note 25.

32. Smith and Brown, "From Pakistan to Japan," pp. 19-54.

33. Stephen Budiansky, "Indoor Air Pollution," Environmental Science and Technology 14 (1980): 1023-1027.

34. A. L. Agrawal, "Domestic Source of Air Pollution," in Annual Report of National Institute of Occupational Health (Ahmedabad, India, 1979), pp. 18-22.

35. See note 25.

36. John P. Holdren, "Energy Resources," in Environment, ed. William W. Murdoch (Sunderland, Mass.: Sinamer Associates, 1975).

37. World Bank, Environmental, Health, and Human Ecologic Considerations in Economic Development Projects (Washington, D.C.: World Bank, 1974).

38. Richard Carpenter and William Matthews, "NEPA: Environmental Innocence Abroad," East-West Perspectives 1 (Fall, 1980): 6-11.

39. Ibid.

40. S. Nishioka, Y. Iikura, and T. Siddiqi, "Air Quality Standards and Energy Policies in Japan and the U.S.A.," working paper (Honolulu: East-West Environment and Policy Institute, 1981).

41. The sections on environmental legislation, management, and research institutions in China are based on Qu Geping and Li Jinchan, "Environmental Management in China" (Paper presented at the Workshop on Environmental Management held by the Office of Environmental Protection Leading Group of the State Council, People's Republic of China, and the East-West Environment and Policy Institute, Honolulu, 1980).

42. Ibid.

43. Ellen Winchester, "Letter from China," Sierra (March/April 1981): 7-12.

44. The data presented in this section are based primarily on Bing Lin Wu and Guan Ju-Gen, "Energy Policy and Air Quality Management in the People's Republic of China," in Proceedings of the Conference on Air Quality Management and Energy Policies, Baroda, India, February 16-25, 1981. In press.

45. Ibid.

46. Ibid.

47. India, Department of Science and Technology, Report of the Committee for Recommending Legislative Measures and Administrative Machinery for Ensuring Environmental Protection (New Delhi, 1980).

48. Ibid.

49. India, Planning Commission, "The Long-Term Perspective," *Fourth Five-Year Plan, 1969-74*, 2 vols. (New Delhi, 1969).

50. India, National Committee of Environmental Planning and Co-ordination, *Role of the Government of India in Environmental Protection* (New Delhi, 1981).

51. See note 47.

52. India, Central Board for the Prevention and Control of Water Pollution, *Annual Report, 1978-79* (New Delhi, 1979).

53. See note 50.

54. A. Ramachandran and J. Gururaja, "Perspectives on Energy in India," *Annual Review of Energy* 2 (1977): 365-386.

55. Kirk R. Smith, Jamuna Ramakrishna, and Premlata Menon, "Air Pollution from the Combustion of Traditional Fuels: A Brief Survey," in *Proceedings of the Conference on Air Quality Management and Energy Policies*, Baroda, India, February 16-25, 1981 (to be published).

56. India, Ministry of Steel, Mines, and Coal, Department of Coal, *Report 1979-80* (New Delhi, 1980).

57. J. M. Dave, N. Ramanathan, and T. A. Siddiqi, eds., *Proceedings of the Conference on Air Quality Management and Energy Policies*, Baroda, India, February 16-25, 1981. In press.

58. India, Ministry of Petroleum, Chemicals, and Facilities, *Report of the Expert Committee on the Environmental Impact of Mathura Refinery* (New Delhi, 1977).

59. K. S. Rao, R. S. Svin, and T. Shivaji Rao, "Short-Term Pollution Threat of Mathura Refinery at the Taj and Its Environs," *Chemical Age of India* 30 (1979): 903-907.

60. M. S. Swaminathan, *Development of the Silent Valley Reserve Forest* (New Delhi: Ministry of Agriculture and Irrigation, 1979).

61. R. Balakrishna Pillai, *Silent Valley Hydel Project* (Paper presented at the Workshop on Environmental Management at the Administrative Staff College of India, Hyderabad, 1980).

62. B. Bowonder, "Environmental Risk Management in the Third World," *International Journal of Environmental Studies*. In press.

7
Energy in Asia and the Pacific: The Important Questions

Kirk R. Smith, Harrison Brown, Fereidun Fesharaki, Corazon M. Siddayao, and Toufiq A. Siddiqi

We have now examined a number of the most critical energy problems facing Asia and the Pacific. In Chapter 1, we saw how the relatively recent emergence of the less-developed countries and OPEC as important global political and economic groupings has greatly complicated the processes of achieving international agreements involving development issues in general and energy security in particular. Complications arise both because of the increased number of actors and lines of interaction and because the most important characteristics distinguishing these groups from the traditional centers of power (that is, OECD and CMEA) are the relative importance of these issues for them--their level of development and, in the case of OPEC, their capability to export petroleum. In Chapter 2, the role of OPEC in the rapidly changing petroleum market was described in more detail. Significant structural changes are occurring that will result not only in less total international petroleum trade, but also in greater OPEC control over downstream petroleum operations, including petroleum product trade.

The impacts of these major changes on the countries of the Asia-Pacific region have been substantial, as Chapters 2 and 3 demonstrated. There are potential problems not only in obtaining sufficient supplies for the petroleum-thirsty economies of the region but also in accommodating the growing mismatch between output mixes from existing refineries and trends in petroleum product demand. The financial and economic impacts of the changes in petroleum prices on the net-oil-importing less-developed countries of the region have resulted in shifts in balance of payments that have forced some countries to readjust their development plans and, in a few cases, to consider major structural realignments. Alternative energy sources, particularly those relying on indigenous resources, have become significantly more attractive, although tremendous technical and financial uncertainties are involved with many of these options.

One of the most important of these options is nuclear power, discussed in some detail in Chapter 4. Nuclear power, it seems, will become increasingly important in the region during the century, but by a process of intensification in a few countries (principally in East Asia) rather than by a significant extension into the energy programs of many countries. The technical, infrastructural, and financial barriers in most of the less developed countries will probably remain insurmountable. In addition, continued international concern over proliferation and growing worries about nuclear waste disposal have the potential to slow progress even in those countries with ambitious schedules for nuclear power. Thus, the risks this option entails, both financial and political, will remain fairly high.

China may present the major exception to this generalization about nuclear power and, as discussed in Chapter 5, to many general energy trends in the region. The Chinese give indications of embarking on a major nuclear program in spite of being under most of the same constraints as other less developed countries. The Chinese have been able to accomplish much more than other countries in developing intermediate-scale energy technologies and high-quality rural energy supplies. They have, however, also suffered tremendous difficulties in expanding coal production, their principal commercial fuel. During the next decades, it would seem that the need for energy technology may well be the principal force behind much of the growth in China's international political connections and trade. In expanding and modernizing coal and petroleum exploration, production, utilization, and export capacities, as well as in developing an export market capitalizing on its expertise in intermediate-scale and rural energy technologies, China will require significant amounts of technology and capital from the developed countries. Coal, the principal fuel of the region at present, is likely to increase in importance in the future. This will mean that the difficult environmental problem of coal production and use will become an even more important consideration in energy policy than at present. These and other environmental problems associated with energy systems in the region were set forth in Chapter 6, which also illustrated with case studies of China and India how environmental considerations are beginning to be incorporated into government policy making.

Although we believe these topics to be among the most important in understanding and planning for the energy future of the Asia-Pacific region, we must emphasize that they are not an exhaustive list. A number of other topics of equal importance and urgency demand attention. Thus, in the space remaining, we would like to offer a more comprehensive list of topics, a list that incorporates the topics of previous chapters as well as many that have been only mentioned in passing.

To provide focus and to emphasize how much is left to learn in some cases, the topics have been organized in the form of questions--questions that might be considered the most pressing faced by those who believe that rationalizing the energy system is an appropriate means to achieve immediate needs as well as wider human goals. Thus, some questions are broader and longer-term than those faced by national energy policy makers on a daily basis. It is also evident that in many instances energy planning has been and will be conducted before answers are available. Our questions, therefore, represent the directions in which we believe research will be most effective in improving the planning and policy process.

In addition to listing questions, we have tentatively posed a few answers, where the present state of knowledge makes it possible. In a few cases where significant uncertainty exists, we have tried to indicate, besides describing how decisions are presently being made in spite of the uncertainty, how and when the necessary knowledge might be forthcoming. In other cases, we raise the questions even though the answers are so poorly known that little can be said.

We realize that the approach we have taken is somewhat unorthodox. We also recognize that the state of knowledge regarding the proper questions to ask about energy has been changing nearly as rapidly as the energy situation itself. Nevertheless, we believe that the questions and the simple framework in which they are categorized is valuable for organizing a subject such as energy that tends to become diffuse and incoherent because it seemingly touches every aspect of life. Thus, we intend to define the limitations as well as the potential of energy as a tool in human affairs. Finally, we believe that the priorities implied by our questions will not be completely compatible with the conventional energy wisdom found either in the developing or the industrialized countries. By attempting to combine the two perspectives, we hope to increase understanding of the broader resource realities facing the region.

The questions have been organized into six broad categories, which are ordered here by ease of presentation and not necessarily by priority: economic development, environmental quality, social equity, natural resources, security, and government policy making. Within each category is the list of those questions that most directly address the critical information required to understand how the energy systems of the region can be modified to serve human needs and goals.

ENERGY AND ECONOMIC DEVELOPMENT

1. Are the energy demand structures that can be expected to accompany economic growth targets consistent with the availability of energy supplies?
2. What are the optimum mixes of capital, foreign exchange, labor, and energy intensities of various development projects of different countries?
3. In the short term, what are the financial impacts of rapid energy price rises; in the long term, how are the most cost-effective energy alternatives to be financed?
4. What cost-effective increases in energy efficiency are possible for the existing national infrastructures, and what mix of pricing changes, taxation, subsidy, regulation, institutional reform, and persuasion is appropriate to achieve them.

5. What are the economic implications of policies to shift energy demand to forms (solid, gas, or electricity) that more closely match available supplies?
6. What further price escalations can be expected and what risk premiums are appropriate for imported oil and high-technology energy supplies such as nuclear power?
7. What are the relative socio-economic benefits of investing in energy systems supplying the modern sector compared to investing to improve, or substitute for traditional fuels and power sources?

Globally, the most important single lesson learned about energy since 1973 is probably that no rigid link exists between energy use and economic development. Energy has come to be considered as a basic factor in economic production, to be used in varying proportions depending on the structure of production and the relative prices of all factors. Considerable attention is now focused on achieving an economically optimum energy path in each country, as well as on understanding the energy implications of alternative economic development paths.

Unquestionably, the process of economic development has nearly always been accompanied by increased mechanization in the primary sector (such as agriculture); growth in the secondary sector (manufacturing), which is relatively more capital intensive; and the concomittant growth in the tertiary sector (trade and services), spawned by the increasing specialization accompanying secondary sector growth (see Chapter 3). These changes imply increased use of commercial fuels to complement human and animal power and traditional fuels. The structure of a country's energy demand is, therefore, highly dependent on the combinations of activities in these sectors, on the energy intensity of these activities, and on the form of energy used in each. Thus, the relationship of energy to gross national product will depend largely on the mix of different production factors and the effectiveness with which energy is used in each activity. Energy planners in many countries of the Asia-Pacific region are thus increasingly incorporating considerations of structural change and phased growth patterns in their demand forecasts. Some countries, unfortunately, have not developed the data collection apparatus sufficient to project energy demand by anything other than simple ratios of energy to gross national product. Many observers believe that this approach is so simple as to be of little value, however, and perhaps may even be counterproductive.

Some energy studies in the region tend to concentrate on the energy inputs and outputs of production activities (as exemplified in net energy analysis). This is a narrow approach to the whole problem of resource scarcity and the efficient use of all resources. Overall economically efficient allocation of all factor inputs--capital, labor, raw materials (including energy), and land--requires that the goal be the cost minimization of each factor's contribution to the total output and the maximization of that output. (All this, of course, should be conducted within the context of the maximization of the society's welfare.) Most observers believe that net energy analysis by itself offers little improvement over economic analysis for short-term and interim policy decisions.

The petroleum price increases in the 1970s have improved the competitiveness of many energy resources relative to that of imported Middle Eastern oil, and a continuation of the upward trend in prices has been projected, as discussed in Chapter 2. At the same time, foreign exchange constraints that limit international purchasing power of oil-poor developing countries and concerns about the security of supply to maintain economic activity levels or sustain development targets have caused both developing and industrialized countries to look to alternative sources. Such alternatives include the development of indigenous resources, both renewable and nonrenewable, the adoption of sophisticated imported energy technologies, and research into nonconventional ways of using conventional sources such as coal and natural gas. The choices are not always easy or clear, however. When valued in the usual framework, the economic costs appear to exceed the gains for many of the alternatives. Furthermore, the financial risks attached to the development of alternatives are unknown or unclear. Because social optimality transcends economic optimality, many of these energy systems are still attractive to countries within the region. Most countries, however, recognize the need to understand the direct economic costs involved in these choices. In addition, it will be necessary in the short term for many Asian developing countries to continue to borrow heavily to pay for the shortfalls in foreign exchange that accompany sustained levels of importation of both energy supplies and other development-related goods.

Demand and supply management of energy supplies requires a thorough understanding of both economic and technical factors. Too often, microeconomic, technical, and macroeconomic aspects are viewed separately rather than jointly. System boundaries are sometimes drawn so that the actual interactions among these variables are ignored. The rural sector, for example, should not be ignored in overall development planning, which too often emphasizes growth of the urban sector. In the same way, one should not ignore the implications of using each type of technology or the responses of various economic sectors and industrial decision makers to macroeconomic policies that deal with the demand and supply of energy. Thus, such policies as pricing, taxation, and subsidies, intended to deal with specific impacts arising from a changing structure in the energy area, may have repercussions that work against intended goals. In Asia and the Pacific, this has been observed in such areas as the price subsidization of particular fuels, for example. Although many policy makers recognize that a carefully considered mix of policies is required to reduce the negative distortions that may accompany measures directed to the energy sector alone, political considerations have made such integration difficult in many of the region's nations.

Policy decisions with short-term goals often have long-term impacts, such as distortions in energy production, consumption patterns, and infrastructure resulting from pricing decisions.

The long-run social costs of such decisions can be high, even in comparison with the short-run gains. Over the long term, many countries will have to swallow bitter medicine and adopt structural adjustments--such as the recent decisions by South Korea to choose lower rates of economic growth and growth of exports in the 1980s partly to avoid massive increases in oil import payments. For instance, the real rates of economic growth of 10 percent per year and export growth of more than 30 percent per year in the 1970s are being significantly slowed down to an anticipated real GNP growth of 7 to 8 percent and export growth of 20 percent a year in current terms for the 1980s. Even then, many South Korean officials think that such rates may be subject to downward revisions depending on the economic prosperity of their markets--the industrial world--and changes in the oil market. Many developing countries in the region will be forced to seek ways to increase efficiencies in consumption and production; at the same time, they will need to find ways to increase their output and export earnings, even if these methods call for sacrifices.

Energy conservation efforts in the developing nations can be divided into two groups: price policies and nonprice/administrative policies--the impact of each category being dependent on the geography of the country and the structure of the economy. Price policies have been used together with fuel taxes and subsidies in varying proportions to balance economic needs and political forces. In the 1973-1979 period, many countries resisted passing on rising oil import costs to the consumer with the exception of gasoline prices. In most cases excise taxes were reduced to transfer the price impact of gasoline, and in most countries kerosene prices remained unchanged. After the second oil price shock of 1978-79, most nations were forced to pass through the costs and even impose heavier taxes to reduce demand. Within the region, Philippines, South Korea, Sri Lanka, and Thailand have allowed substantial price increases since 1978 by passing through high crude costs.[1] In this group South Korea is now the most aggressive in imposing large taxes on top of international crude price increases. Again, the distribution of price increases have been uneven, with gasoline prices bearing the brunt of the increase. In 1980, South Korea's premium gasoline prices topped $4.50 a gallon--the second highest gasoline price after Zaire at $6.00 a gallon. The lack of political desire to make across-the-board hikes in the prices of all petroleum products has somewhat reduced the effectiveness of high gasoline prices. Prices of kerosene have been kept low in countries like Sri Lanka; diesel oil has been given preferential price treatment in countries like Pakistan and India because of its critical impact on transportation and irrigation. In some LDCs, mostly in Latin America, however, fuel oil prices are still being kept low to encourage industrialization.

Many oil importing LDCs in the Asia-Pacific region have resorted to nonprice administrative tools as short-term conservation tools. These measures have been found to be useful as supplements to other market-based economic policies if rigorously applied. On the whole, however, nonprice conservation measures of non-OPEC LDCs, like those in the OECD countries, have not been stringent. One example of a reasonably successful program is that of Singapore where the growth of vehicle use has been greatly reduced.[2] Some countries, however, have drawn up tough contingency rationing plans that could be activated if crude prices continue to rise or shortages occur.

On the whole, the oil price shock of 1979 has had a far greater impact on energy intensity through both price and nonprice policies than the events of 1973/1974. Many of these policies are unpopular, however, and there are political pressures to relax them. It seems critical that the momentum of conservation be maintained and even speeded up--particularly as the oil market is currently soft--in order to avoid crash emergency programs that disrupt and distort the development process, in the likely event of another set of oil price increases.

ENERGY AND ENVIRONMENTAL QUALITY

1. What is the connection between the uses of traditional fuels (fuelwood, dung, and crop residues) and damage to ecosystems and human health?
2. What approaches are most suitable for developing countries to take in adapting ambient and emissions standards for air and water quality from the industrial countries?
3. What are the economic impacts of the losses in environmental services, such as renewal of soil fertility, climate moderation, fish nurturing, gene pool maintenance, and flood control, that occur because of energy use?
4. To what extent should the environmental impacts affecting noneconomic and nonhealth values, such as impacts on areas of cultural heritage or aesthetic appeal, be considered in diverting resources away from economic development and into environmental protection?
5. What departures from the experience of environmental impacts in developed countries can developing countries expect for such modern energy technologies as nuclear power and synthetic fuels?
6. What international environmental problems, such as deforestation, acid rain, carbon dioxide, nuclear waste, and oil spills are significant enough to warrant engaging a portion of the limited technical personnel available in most developing countries?
7. To what extent is it possible to direct the large growth in energy-producing and energy-using infrastructures, expected in many developing countries during the next decades, along environmentally less damaging paths through relatively inexpensive adjustments, such as land-use planning?

As discussed in Chapter 6, energy use is a valuable, though imperfect, indicator of environmental impact as well as economic activity. The mining, processing, transport, conversion, and use of fuels are responsible for a large portion of the direct disruption to ecosystems, through deforestation for fuelwood and mining of coal, for example, and impacts on ecosystems and human health from pollution. As with the relationship between energy use and economic activity, the connection between energy and environmental impacts is not necessarily direct. There are substantial differences among the environmental impacts of different fuels as well as possible modifications of technologies to decrease the impacts from any particular fuel cycle. Nevertheless, such modifications have technical and economic limits. In particular, the costs of impact reduction become very high as control approaches 100 percent (i.e., zero impact). Since the absorptive capacity of the environment is essentially constant, a growing energy use implies an increasing amount of pollution control expense, or a shift in the types of energy used, if continuing environmental degradation is to be avoided.

The added benefits of greater energy use decline at high-income levels. There is a level of energy use above which the increased costs of environmental control (or the additional costs of environmental damage from no control) exceed the increased benefits brought about by the increased energy use. There is evidence that this point is about to be approached, or perhaps has even been exceeded, by the most industrialized countries in the region. It is difficult to be sure, partly because so many environmental impacts are displaced in time or space and thus are experienced by groups of people other than those experiencing the economic benefits of the energy use.

One of the most significant developments occurring today in environmental assessment methodology is the increasing perception of the need and some increase in capacity to measure the economic costs of environmental impacts. The loss of such environmental services as soil fertility from ecosystem disruption and pollution can be shown to produce a significant economic loss in many cases. Although such techniques are generally quite crude and for the most part have not yet been applied to the developing countries of Asia, they hold the promise of being powerful tools in determining appropriate investments in impact control and are likely to lead to greater and more effective investments in control than has often been the practice in the past.

The vast expenditures on pollution control in Japan, the United States, and other highly industrialized countries have led to a general impression in the developing countries that efforts to maintain and improve environmental quality are very expensive and will divert scarce capital from projects for economic development. The systematic application of benefit-cost analysis, especially involving the health-related benefits of improved air and water quality, could be of considerable help in identifying areas where a given level of expenditure could have the largest public benefit. Benefit-cost analysis, on the other hand, has not been very useful as a decision tool for protecting areas of historical, religious, aesthetic, or cultural interest.

Health care is generally accepted as a basic human need. It is well known that quality health care is not generally available in the rural parts of Asia, where the majority of the people live, and would be quite expensive to provide. A sizable number of illnesses may have their origin in the frequently contaminated water supply or in the air pollutants from cooking to which women in particular are usually exposed. It would be worthwhile to examine the relative health benefits, costs, and practicality of supplying the rural areas with drinking-water wells, less-polluting stoves, and biogas plants for energy and sanitation--as compared to the remote goal of having a medical clinic in every village.

In urban areas, power plants, petrochemical plants, and transport are usually the largest contributors to air pollution, although domestic sources can also be important. Since the effects of air pollutants on large groups of people appear to be, statistically speaking, similar in all countries, it would be reasonable to expect that the ambient air quality goals would be the same regardless of whether the countries are less or more industrialized. It is likely that the strategies to achieve those goals, however, will be different. For example, a power plant and one or two major industries in a city might be responsible for 90 percent of all the particulates emitted. In a developing country, it might be more practical to achieve air quality goals by requiring controls on these two or three major polluters than by placing across-the-board controls on all emitters. Further, the effects of industrial emissions from new plants can be substantially reduced at low cost by appropriate land use planning--for example, by the siting of such industries downwind from populated areas. Land use planning has been applied with some success in India, for example, in siting small and medium industries in Bombay and protecting major areas of cultural value. It may become critical for siting those many energy facilities with large water requirements.

A large portion of the emissions from the transportation sector of developing countries comes from diesel buses and trucks. Encouraging proper maintenance of these vehicles would not only improve air quality but save a considerable amount of oil as well. This may be more immediately practical and beneficial than to require emission control devices on all automobiles.

Some of the important environmental problems related to energy conversion require international rather than local approaches for their solution. These include acid rain, the possible climatic effects of carbon dioxide buildup in the atmosphere, and radioactive wastes from the nuclear fuel cycle. These issues have been discussed elsewhere in this book, but it is worth emphasizing that knowledge in each of these three areas remains highly inadequate and requires a strong multicountry collaborative research effort. To take only one example, China, India, Japan, and South Korea are among countries with ambitious plans to expand their use of coal. In many cases, they have decided to use very tall stacks to reduce air pollution in the vicinity. This policy could result in acid rain in neighboring countries and could lead to international problems such as those encountered in Europe and between Canada and the United States. In the long run, it may turn out to be less expensive to use emission control technology and short stacks than to disperse pollutants with tall stacks. International collaborative efforts will be required to make this determination.

ENERGY AND SOCIAL EQUITY

1. What are the social and economic impacts of present patterns of cross-subsidization among petroleum products and among categories of electric power customers?
2. Compared to differential pricing, how effective are available alternative mechanisms, such as coupons, in assuring access to energy by society's poorest members?
3. Do efforts to provide energy in greater amounts or in different forms represent an effective strategy to help achieve economic and social development in rural areas--the poorest and most disenfranchised parts of nearly all countries?
4. To what extent, if any, do some rural energy systems, such as household biogas plants and rural electrification schemes, help the relatively rich at the expense of the poor?
5. How can energy systems be modified so that the current rapid pace of urbanization in most developing countries is either serviced properly or, if possible, moderated?
6. Would the considerable economies of scale possible during the growth of most developing-country energy systems be significantly counterbalanced by diseconomies of scale, including possible adverse impacts on social equity?
7. What practical and politically feasible means are available within the international community to mitigate the impact of rapidly rising global energy prices on the poorest oil-importing countries?

There is probably no country that does not use its energy system as a tool to achieve social goals. Most obviously, this is done through differential pricing of fuels and electricity. To some extent in nearly every country, each fuel is taxed or subsidized to a different extent, that is, relative prices are not determined entirely on the basis of the costs of production, the market, or energy content. One of the principal reasons usually given for creating such price differentials is to assist poor or otherwise disadvantaged groups. Thus, to some extent social and economic objectives are pursued simultaneously, sometimes resulting in conflicts.

Along with the pricing of food, the pricing of those fuels purchased by individuals is one of the most politically volatile domestic policy decisions facing governments in developing Asia

Indeed, governments have fallen on the basis of such decisions. Cross-subsidization of petroleum products, in which light products are priced high and middle distillates low, has been the rule for many countries. Even though it is recognized that the economic costs are high (over $2 billion per year in Indonesia, by one estimate), that energy efficiency is impaired, and that middle distillates are not entirely used by the needy, it has been difficult for some countries to break out of this policy. An alternative, being implemented in Sri Lanka as a result of recommendations of the international financial community, is to issue ration coupons. These allow the poor to buy a basic allotment of food and fuel at low prices, the rest being sold at market prices. Some observers believe that the administrative difficulties and costs of such programs, however, may in practice counterbalance the economic benefits. In general, although there is a trend away from large price differentials among fuels, cross-subsidization will remain an important factor in energy pricing.

In recent years, a number of critics have pointed out that energy systems can affect social equity in ways other than through pricing of fuels. They argue that some classes of energy technologies may lead into economic development patterns that consolidate economic and, consequently, political power in the hands of a few people. Other types of energy developments, in contrast, tend to lead societies into less centralized and more participatory modes. Although they assume more influence on society by energy systems than perhaps is warranted, these arguments have played a important role in much of the discussion of the potential of small-scale renewable energy systems in both developed and developing countries.

In the past, rural development has been accompanied by replacement of traditional fuels and power sources with electricity and petroleum fuels. The changing availability of petroleum, however, while not causing an immediate impact on those many rural areas that use little fossil fuel, would seem to mitigate against this conventional energy development path. Less certain are the implications of this uncertain access to petroleum for overall rural development strategy. It is not clear whether the energy constraints on rural development are as significant as others, such as capital, crop production, and education. In general, perhaps even more so than in the modern sector, rural energy problems cannot be adequately addressed in isolation from the entire matrix of rural social and economic life.

A common rural energy concern, for example, is that fuelwood requirements may be causing deforestation. Though this may well be the case in some areas, there is as much evidence in some regions of Asia that deforestation is causing a fuelwood shortage rather than vice versa--that is, that expansion of agricultural lands into forests is the principal cause of deforestation. This implies that considering fuel problems by themselves may not be an effective way of dealing with barriers to the development of the rural poor.

On the other hand, the rural energy supply system is an important part of the agricultural system that provides a livelihood for most of the region's people. Developing a sustainable and high-quality material existence for these people will probably require major changes in the quantity and quality of energy supplies. Without such development in rural areas, the current rapid urbanization of many countries is likely to continue.

Much of the bilateral assistance from industrial countries to Asia-Pacific developing countries for energy projects during the last decades has been directed to rural electrification schemes. Although there are important examples of success, in recent years, there has been an increasing suspicion that such a large effort may have been misdirected. Criticism has focused on two observations of existing programs--that they are often both inefficient and inequitable. In many cases, electrification of villages has preceded both significant demand for electricity and the wherewithal to pay for an energy form requiring so much upfront capital cost in consuming equipment. Consequently, load factors and the fraction of houses connected in electrified villages have often remained low. In addition, the households that do benefit have in many cases not been the poor but the rural wealthy. Wealth, of course, is a relative term and there are arguments that increases in economic activity that will also assist the poor can be stimulated by electrification. Although rural electrification and other capital-intensive rural energy schemes, such as household biogas plants and windmills, remain controversial because of their uncertain impacts on rural income equity, most governments promote them in principle.

One of the easiest means of achieving higher energy and economic efficiencies is to increase the scale of production. There are, however, also diseconomies associated with increasing scale, including increased infrastructure requirements. In addition, such increases in scale tend to have organizational and labor requirements substantially different from those available in many developing countries. Thus, they tend to be less appropriate to present conditions. Many countries, therefore, plan toward achieving an integrated mix of sophistication and scale in energy producing as well as other types of technologies. In Asia, the People's Republic of China has apparently been modestly successful in this effort and, as a result, has employed a wide spectrum of its people in a range of phased technical advances.

Equity is a problem among countries as well as within them. There have been proposals to provide special assistance to those poor countries most hurt by oil price increases; the World Bank Special Fund and the OPEC Energy Aid Program are examples. Little real progress in this effort has been achieved to date, however, and most of what has been achieved has focused on large-scale systems.

ENERGY AND NATURAL RESOURCES

1 What are the prospects for enough additions to oil and gas reserves from enhanced recovery or new [illegible]

within the region to delay the transition to a post-petroleum era?

2. By what arrangements can developing countries encourage exploration by multinational oil companies and yet retain rights to a fair share of the economic return, should discoveries be made?
3. At what petroleum prices will synthetic liquid fuels from coal, oil shale, and gas become competitive?
4. How can the sustainable potential of renewable energy resources such as biomass, wind, and sunlight, and the viability of the technologies of conversion be evaluated reliably and efficiently?
5. Are large-scale biomass plantations for high-quality liquid, solid, and gaseous fuels feasible, economic, and desirable in developing countries?
6. Will nuclear power plants become more competitive with other sources of electricity such as coal power plants in developing countries through evolution, for example, into smaller and more reliable units?
7. To what extent should the region participate in the development of long-term and exotic technologies, such as breeders, fusion, ocean thermal energy conversion, and hot-rock geothermal?

The increases in oil prices in the 1970s, with accompanying forceful reminders of the finiteness of oil resources, have dramatically altered the landscape of energy alternatives. As a result, some conventional alternatives, such as natural gas and coal, are likely to have much broader roles in global trade as well as in domestic energy supplies in the medium term. In addition, by the end of the 1970s, a number of energy technologies that in the 1960s were barely more than curiosities had each become the target for hundreds of millions of dollars in research and development efforts. Also during this period, the development of renewable energy sources received such emphasis that in some countries renewability became a goal in itself, to be pursued separately from conventional economic goals.

New finds of the major commercial fuel, oil, are likely in the Asia-Pacific region, since the amount of exploration in the major sedimentary basins has not been extensive by global standards. Growing attention is also being placed by many developing countries and international and bilateral aid agencies in developing incentives for international oil companies to explore and develop fields that have been in the past considered too small. It is unlikely that these finds would be of sufficient size to alter significantly the basic transition away from oil, but they may be of great help to small developing country economies. The Philippines, for example, has been able to develop some small offshore fields and thereby significantly reduce imports. The extraction of oil is at present a physically inefficient process; it is estimated that many depleted fields retain about two thirds of the oil originally present. Secondary and tertiary recovery techniques that have been or will be employed can extend the lifetime of some oil fields for 10 to 15 years, but only at fairly high costs.

Liquid fuels can also be produced from biomass. Ethanol produced from sugar is now in widespread use in Brazil as an automotive fuel. Possibilities of producing alcohol economically from maize as well as from wood are receiving increasing attention in the United States. In terms of resources, the potential for this approach seems to be quite large, particularly in the tropical countries of Southeast Asia. Indeed, alcohol production facilities are already under construction in Papua New Guinea and New Zealand and are being planned in the Philippines. There are, nevertheless, serious limitations. An oft-mentioned concern is the competition for land and water that may occur between food and energy crops. In addition, the large amount of liquid wastes generated in alcohol production present a serious environmental problem. At present, such fuels are not economically competitive with petroleum-based liquid fuels, and only government subsidies make them acceptable in the marketplace.

A second approach to the production of liquid fuels from nonpetroleum sources is the liquefaction of coal or the extraction of oil from shale. Coal liquefaction is now being attempted on a substantial scale in South Africa, and plans are being developed for a joint Japanese-Australian venture in Victoria, Australia. Several pilot plants for the extraction of oil from shale are under construction in the United States, and a small unit is in use in China. It is possible that liquid fuels from oil shale will be utilized on a substantial scale in the United States during the 1990s, but as in the rest of Asia and the Pacific, it seems that only the most favorable deposits are likely to be developed this century.

The importance of coal will increase in the future along with its trade within the region. The demand for coking coal will increase as steel production increases in the region, and coal of lower grade will be used in increasing quantities for space heating and the production of electrical power. Development of coal is hindered not by resources but by handling infrastructure and constraints on use. Charcoal, a high-quality form of biomass, may also find broader application in tree rich areas in the near future.

The gaseous fuel system will be utilized in those countries that have significant reserves of natural gas or that already have or are able to invest in pipeline networks and are prepared to import liquefied natural gas. Natural gas can also be used to power automobiles and buses either directly or by conversion to liquid fuels. There are indications that substantially more natural gas may be found in the region than present reserves indicate, because so little direct exploration has occurred. In addition, techniques are being developed to convert natural gas to methanol or synthetic gasoline. Thus, natural gas may well turn out to be a significant transition fossil fuel for Australia, Bangladesh, India, Indonesia, Japan, Malaysia, New Zealand, Pakistan, the People's Republic of China, South Korea, and Thailand, as well as the United States.

Increasingly, electrical systems will be fueled by coal (in spite of the difficulties discussed in Chapter 6) and, to a lesser extent, by hydropower, much of which is yet to be developed. Nuclear power will intensify in East Asia but is unlikely to spread rapidly to the poorer countries of the region. Even if smaller units become available, technical, infrastructural, and financial barriers will slow the widespread application of nuclear power. In some areas electricity can be generated economically from earth heat, notably the western United States, Hawaii, New Zealand, the Philippines, and, possibly, Indonesia, but this is not likely to loom very large in the energy picture of the region as a whole. Several biomass-fueled electric power projects are also under development in the region (in Papua New Guinea, the Philippines, and Hawaii, for example). In addition, the rapidly improving economies of large-scale wind electric systems may prompt other Asia-Pacific regions to follow Hawaii and California in the development of "wind farms." At present, the costs of these systems will restrict them to particularly advantageous sites.

The term "energy transition" is usually applied to the period between the petroleum era of the mid-nineteenth century and that era beginning beyond 2000 that will rely on some combination of the only feasible long-term energy alternatives--nuclear fission breeders, renewables, and fusion. This definition, which only refers to technologies of supply, is somewhat myopic and parochial. Today, as during all of human history, most of the energy used by most of the people in the world is in the form of traditional biomass fuels (although it is a small fraction of the total global energy use). Thus, in Asia, where most of the world's rural poor reside, another energy transition is equally as important--a transition to the sustainable high-quality fuels needed for rural development.

For rural areas, a number of alternatives hold promise. Some act to improve the quality of existing traditional fuels by conversion into charcoal, biogas, or liquids. Others, such as modified cookstoves, act to increase the physical efficiency of using existing fuels. Still others, such as fast-growing fuelwood tree species, act to increase the supply of existing fuels. Finally, various solar-powered devices, windmills, and microhydro plants increase the efficiency and flexibility of systems tapping traditional power sources. Except in the People's Republic of China, the contribution of these alternatives has been small in terms of total energy supplies. In terms of numbers of people potentially affected, the benefits to increased efforts seem high.

In view of the diversity of considerations we have discussed, the probable course of postpetroleum energy events will differ considerably from nation to nation. Japan will remain a substantial energy importer, generating its electricity from nuclear energy, liquefied natural gas, and coal. Japan has also made remarkable strides in reducing dependence on oil. Its remaining liquid fuel needs may eventually be met in part by synthetic fuels manufactured from coal in Australia, and, possibly, the People's Republic of China, the United States, and Indonesia. Fuel alcohol may be imported from Southeast Asia. Other than the United States, Japan may be the only country in the region to export significant amounts of high-technology energy systems during this century. These may include not only nuclear power plants but such new technologies as ocean thermal energy conversion systems. The latter, however, are unlikely to gain wide acceptance until well beyond the end of the century, if at all.

New Zealand seems destined to enter the postpetroleum era ahead of most industrial countries, stimulated in part by a difficult balance of payments situation and in part by an abundance of other energy resources. Substantial reserves of natural gas have been found, and there are plans to convert it to synthetic gasoline. The country's relatively large geothermal and hydropower resources are being developed. Its abundant forests may be used directly for power generation as well as for the production of alcohol fuels.

Australia has large reserves of coal, natural gas, and uranium which should give the nation a considerable degree of energy independence for some time in the future. Close to 40 million tons of coal are already being exported each year. Further, Australia is a probable candidate for the export of synthetic liquid fuels made from coal. Electrical generating stations will for the most part be fueled by coal, whereas natural gas is likely to be increasingly used for motor transportation. Australia might possibly become an exporter of LNG, with Japan being a major customer. Although there are no nuclear power plants presently under construction or in use, Western Australia is considering nuclear power as an option for the 1990s. In any case, the country will retain a significant influence in the international uranium market.

Both South Korea and Taiwan are major importers of oil and possess marginal coal deposits. Rapid industrialization is being accompanied, as in Japan, by a major growth in nuclear power capacity.

Singapore, which now imports all of its energy in the form of oil, will continue to import most of its energy, but in other forms. Conceivably, nearby Malaysia will generate some electricity from hydropower and, possibly, natural gas specifically for export to Singapore. The city will continue to expand its use of municipal wastes as a boiler fuel, but the total energy potential of this source is fairly small. It is probable that the city will become an importer of coal (from Indonesia and Australia) as well as an importer of coal-generated liquid fuels. It is doubtful that Singapore will generate electricity from nuclear power.

Although India is a significant importer of oil, it has important reserves of coal as well as some domestic oil. The indigenous technological capabilities make it possible that India can be energy independent in the postpetroleum age, using a combination of coal-generated and nuclear electricity. A commercial Indian thorium breeder reactor seems unlikely to be available for at least a generation. Various of the biomass alternatives seem more appropriate for increasing supplies and for improving energy efficiency in rural areas.

Both Bangladesh and Pakistan now import most of their oil. Both countries have reserves of natural gas that now supply close to one half of their commercial energy needs. Once the natural gas has been exhausted, it is difficult to see how the two countries will avoid massive fuel imports unless they greatly expand their nuclear power capacity and can utilize their biomass potential more effectively. Pakistan has some additional untapped hydropower capacity. Of course, they can develop their modest domestic coal resources and can import more coal as well as, perhaps, synthetic liquid fuels, but are likely to have considerable problems paying for the imports. The prospects of energy self-sufficiency for these countries are not good.

A number of countries in the Asia-Pacific region now obtain virtually all of their commercial energy from imported oil. These include the Philippines, Sri Lanka, Thailand, and most of the independent island states of the Pacific. There are continuing hopes for some oil discoveries, and some potential for coal development. Water power is a substantial resource that has yet to be developed, and the possibility for the development of wind power appears to be reasonably good. The Philippines is already the second largest producer of geothermal power in the world, and prospects for additional development of this source are bright. All of these countries have considerable potential for obtaining energy from biomass and Thailand has been developing offshore natural gas deposits. On the whole, they seem destined to be energy importers for some time in the future. Again, the major problems will be paying for imports and raising the capital to develop indigenous supplies.

Indonesia is a sizable oil exporter. Based on present reserves, both its production and exports are expected to increase moderately in the next few years. Although there are hopeful signs for possible new discoveries, the rapidly rising domestic demand for energy suggests that Indonesia needs to develop alternative energy sources soon. The development of its coal and gas resources, both for export and domestic use, could ease the situation considerably. In addition, there are indications that alcohol from biomass and geothermal energy may also contribute to energy supplies on a modest scale.

The People's Republic of China is now energy self-sufficient and, in view of its sizable coal deposits, is likely to remain so for some time. China is already a coal economy, and as oil reserves are depleted, its dependence upon coal will increase still further. It is possible that China will shift from a modest exporter of crude oil to a major exporter of coal and, quite possibly, synthetic fuels. Exports of any significance, however, would entail both great improvement and expansion in the Chinese coal supply system as well as further shifts in the traditional Chinese policy of self-reliance that have kept net energy trade within a few percent of zero for 30 years. As discussed in Chapter 5, China has perhaps been more successful than other poor countries in the region in improving rural energy systems by using local resources and may initiate significant exports of rural and intermediate-scale energy technologies.

In general, it would seem that most of the countries in the region will benefit most from research and development efforts aimed toward more cost-effective handling and conversion of fossil and traditional fuels rather than devoting funds toward advanced systems utilizing new fuels. An exception may be Japan which has an important, although still small, role in nuclear breeder research.

ENERGY AND SECURITY

1. What are the economic and political costs of the options to increase petroleum import supply reliability, including strategic storage, emergency sharing, joint purchasing arrangement, and closer political and economic ties to producers?
2. What are the implications for supply reliability of the rapid changes in the downstream petroleum market, including shifts in transport and refining patterns?
3. What international political arrangements will be effective in convincingly disassociating nuclear power reactors from their potential to make nuclear weapons?
4. Until these arrangements are in place, how reliable are nuclear power systems in light of their vulnerability to trade disruptions in fuel, fuel-cycle services, and technology?
5. How expensive and successful are efforts to achieve energy self-sufficiency likely to be compared with strategies of supplier and fuel diversification?
6. What are the resource security implications of the fuel and equipment trade patterns that would develop as a result of deploying the newer high-technology energy alternative, including synthetic liquid fuels, high-technology solar power, and fusion?
7. What shifts in traditional political arrangements and what changes in international tensions within the Asia-Pacific region are likely to be caused by trends in energy supply and demand relationships?

Resource security is an integral part of national and international security. Among other considerations, security is determined by the interaction of two principal factors that can be affected directly by interruption in energy supplies--economic security and military security. At present, two fuel cycles have characteristics that make them particularly important in terms of national and international security--petroleum and uranium.

Imported petroleum is vital both to the operation of the economies and to the functions of military defense in many countries. For the remainder of the century, international petroleum trade will remain under the control of groups of countries who have shown a willingness to use this control as a political and economic weapon.

Petroleum security involves two separate issues that are often confused with each other. First, problems of a short-term nature that involve oil supply disruptions and rising prices and result from political upheavals or military confrontations. These problems, which sometimes involve massive disruptions, can be handled by stockpiling (both commercial and governmental), import sharing through coordinated action of consumers (such as the International Energy Agency (IEA) import-sharing program), and, in some cases, use of military force.

The second class of problems comes from the declining availability of oil through physical or political ceilings imposed on production and changes in the control over downstream oil activities (transport, refining, and product trade). This set of problems will unfold gradually over time and does not cause emergencies in the usual sense.

Much of the current literature is concerned with the problems of the first kind--how to deal with crisis.[3] The literature usually recommends larger stockpiling, strengthening IEA, designing emergency pricing schemes to prevent oil producers from capturing huge profits during shortages, and, in the case of the United States, developing rapid deployment military forces. The second set of problems is more serious. Oil security, however, is threatened by structural changes that are likely to happen over the next decade or two without major political upheavals or military confrontations. These structural changes result from the following trends:

1. Increased direct sales by the oil exporters and comparable decrease in control of oil by the international oil companies.
2. Political, social, and economic pressure on the Mideast oil producers to reduce output to the volumes necessary for their own economic needs rather than satisfying world demand.
3. Massive movements by oil producers to refining and increased export of refined products instead of crude.
4. Increasing purchases of product and crude carriers by the oil producers, enhancing their control of destinations.

Clearly, the problems and proposed solutions of short-term supply disruptions are not applicable to the structural changes now under way in the petroleum market. Policy makers in the Asia-Pacific region are growing aware that they must devote more thought to devising means to handle such gradual changes. In 1980, for example, the United Nations Economic and Social Commission for Asia and the Pacific conducted a preliminary study of these problems within the region.[4]

Another important aspect of oil security is the impact of oil prices on economic growth, unemployment, and balance of payments. During short-term disruptions, both spot and official prices rise, but official prices are never increased to the spot level. For instance, during the 1979 Iranian Revolution, spot prices rose to $45 per barrel, while official prices were increased to $26/b.

Thus, short-term disruptions do cause higher prices, although high stocks and good management of distribution can significantly reduce the magnitude of increase in an emergency. On the other hand, structural changes in the petroleum market that cause lower availability of crude are likely to lead to persistent but moderate real increases in the price of oil--increases that cannot be controlled by emergency programs. It is useless to try to stop these increases, since oil is, after all, a finite resource. Real prices will rise sooner or later--if not through political control of production, then through physical limits of the resource. How hard a national economy will be hit by the rise in oil prices is, to a great extent, dependent on the characteristics of the economy: its structure, size, export potential, dependence on imported oil, and management.

Another part of oil security is the distribution of oil in the importing country. In the event of short-term disruptions, much can be done to maintain key sectors of the economy and avoid loss of life. Elaborate models of coupons and rationing exist around the world, but again, much depends on internal management. For longer-term oil security, however, structural changes in demand may be necessary. Once again, this depends greatly on how the leaders of the country will manage the structural shift brought about by declining oil availability.

In general, oil security--both short term and long term--rests on the planning philosophy of the importing nations and the capability of internal management of the economy. Oil per se is not the root of a country's economic problems--there were economic problems before 1970--but rather the management and the characteristics of each economy. The energy sector, as an integral part of the national economy, can be expected to be managed only as well as the rest of the economy. The experience of countries in the Asia-Pacific region bears this out. Well-run economies such as Japan's tend to have vigorous and effective energy policies and programs. It seems certain, however, that market forces are not likely to be able to take care of the problem and increased government planning and intervention may well be inevitable.

Nuclear power, as shown in Figure 4.4 in Chapter 4, interacts with national security not only through a country's economy but also through the changes in military relationships that result when countries develop the capability to produce nuclear weapons as an outgrowth of their peaceful nuclear power programs. As with petroleum trade, disruptions of nuclear fuel supplies and technology have been affected by political considerations. Unlike petroleum, however, nuclear power will remain a relatively modest contributor to total energy supplies in most countries for the rest of the century. Where its contribution will be significant, as in Japan, the relative ease of stockpiling allows countries to be less concerned about potential supply disruptions in the nuclear fuel cycle as compared to the petroleum cycle. The principal source of the disruptions in the trade of nuclear fuel and technology is likely to remain the fear of proliferation. Unless the international nonproliferation agreements and safeguarding systems can be improved, this fear will remain in the supplier countries.

To achieve energy security, countries in the region are increasingly attentive to the need to diversify suppliers and respond to shifts in the ownership and control patterns within the major fuel cycles. Most countries have also placed a heavy emphasis on energy self-sufficiency in government announcements as a means of achieving energy security. In practice, however, these programs, which include efforts to develop indigenous supplies and to increase efficiency, have fallen below even what strict financial analysis would find to be optimum, let alone what might be considered to be optimum with an appropriate risk premium applied to imported energy. Two trends seem to be converging in this area. Programs for indigenous supply and increased efficiency are slowly becoming effective at the same time that exaggerated claims about the possibility and desirability of complete energy self-sufficiency are growing less frequent.

One reason that true energy self-sufficiency is unlikely in the near future for many countries is found in the character of many alternative energy technologies. Some alternatives, of course, merely shift dependence from one fuel to another--petroleum to oil shale, for example. These shifts may not lead to self-sufficiency even though they may greatly increase security by shifting to a different set of suppliers. Shifts to high-technology options such as breeders, fusion, solar power towers, and ocean thermal energy conversion systems might lead to much less import of fuel but would increase the requirements for imported sophisticated technical knowledge and equipment in many countries for the foreseeable future. Although, on the face of it, this would probably result in a net increase in energy security, the long-term and overall impacts are uncertain. This uncertainty is reflected, for example, in the accusation by some developing countries that the United States government's solar energy research and development program is aimed toward achieving industrial dominance of the market for renewable energy systems in the third world. Thus, energy security does not necessarily imply energy self-sufficiency, and fuel self-sufficiency does not necessary lead to energy security.

In Asia, two other energy resources stand out among those whose development has potential for international conflict. The first are the Siberian energy resources being developed by the Soviet Union partly with technical and financial assistance from Japan. This could result in significant changes in the political relationships among China, Japan, and the USSR. In addition, the offshore oil and gas potential along the coastline of Asia from Korea to Iran may lead in some cases to serious disputes over lines of jurisdiction and ownership. Most critical in this respect, as discussed in Chapter 5, are the South China Sea and the areas between Japan, China, Taiwan, and the Koreas.

ENERGY AND GOVERNMENT POLICYMAKING

1. Would achieving full or nearly full integration of national policy making on fuels be worth the price involved in overcoming political and bureaucratic obstacles to doing so?
2. How valuable would it be to extend integration of energy policy making to include such departments as rural development, agriculture, education, and health that are not directly involved in the energy sector?
3. What degree of government intervention in energy systems is justified by lags and other imperfections in the economic marketplace?
4. At what level of decision making and in what form is it appropriate to incorporate noneconomic and longer-term considerations, such as environmental impact, national security, and social equity?
5. How can developing-country governments quickly achieve the assessment capability necessary to independently evaluate and influence the proposals of multinational energy companies and the recommendations of international and bilateral donor agencies and their consultants?
6. How should indigenous energy research, development, and associated information collection be organized in developing countries with limited research personnel and funds?
7. What are the most practical and effective ways, if any, for international organizations, such as the United Nations, World Bank, ASEAN, and OECD, to focus on energy problems?

It has been remarked that the sun never sets on energy conferences. The same could be said with equal accuracy of energy ministries, which have sprung up since 1973 like mushrooms after the rain. In most cases, these ministries are combinations and modifications of older agencies and departments handling separate fuels--ministries of oil and gas, atomic energy, and power, for example. In many cases, integration of these ministries into the government infrastructure is not complete, although coordination is attempted through interagency committees or within the prime minister's or president's office. Several of the major international financial institutions have been pressing their clients to consolidate government energy decision making.

Although there is still much demand for increased coordination and more consolidation of energy-related government functions, the 1980s opened with signs of a possible reversal in this trend in the nation with the largest energy consumption. The Reagan administration seems to believe that de-evolution of energy agency consolidation is desirable. In its view, many of the decisions about energy allocation that had begun to be taken up by government could best be returned to the economic marketplace. Although the reason for this U.S. policy change is partly dissatisfaction with the mixed record of energy policy formulation, coordination, and implementation of previous administrations, it also reflects a dissatisfaction with the very idea of energy policy. Fuel cycles, especially those of petroleum and uranium, clearly require government intervention at many points. The concept of energy policy is different; however, it is more than the sum of the fuel policies' parts. Whether, when properly managed, an integrated energy policy leads to a decided improvement over other means of rationalizing human affairs is still an open question. In varying degrees, most countries in the region are actively pursuing the answer by increasing the government's role in coordinating energy supply and demand. Although the first flush of enthusiasm for energy integration has faded, energy management is usually seen as a necessary but not sufficient condition for economic management. The changes in Washington, D.C., however, may well encourage similar reversals in other countries just as the creation of the United States Department of Energy originally helped strengthen the present trend around the world.

Energy is also becoming the focus of international research, development, and policy making. Regional government-level technology meetings, research agreements, and transfers are becoming common. Regional energy research and development organizations have been developed, including the energy technology program at the Asian Institute of Technology in Bangkok, the United Nations regional center for biogas research being considered for China, a regional center for biomass energy research being proposed by the United States Agency for International Development, and the East-West Resource Systems Institute in Honolulu, where this book was written. Most of the countries in the region have expressed interest in participating in such research, although in many cases the numbers of qualified personnel are small. Training and graduate education in energy technologies and policy remain critical needs.

Energy security pacts are emerging within a few existing international groupings, such as OECD, and energy security provisions are an important part of international nuclear nonproliferation agreements. Furthermore, it seems that energy trends may be leading to new international groupings expressly oriented toward energy security, especially including arrangements to increase the security of the petroleum and uranium fuel cycles. In the region, for example, the Association of South East Asian Nations (ASEAN), has instituted a scheme by which its members can share supplies in an oil shortage. Energy security concerns are also a part of the growing interest within some countries of the region (Japan and Australia, in particular) in creating a Pacific economic community.

As discussed in Chapter 5, trade in fuels and energy technologies seems to be one of the major forces impelling China to closer international ties. Indeed, given its size and population, by the end of the century China may well emerge to take on a global role distinct from but as important as the other four major international groupings discussed in Chapter 1. This will result in a further multiplication of potential international interconnections and arrangements but could also conceivably lead to more diversified and resilient global economic and energy systems.

Although it is true, as we have stressed repeatedly in this book, that energy management should be generally subordinate to economic management, there are reasons why it may be valuable to focus particularly on energy, especially in international cooperative arrangements. Here we speak not of the hypotheses that the concept of energy provides a uniquely useful world view and that manipulation of energy flows can accomplish human goals quicker or more efficiently than other more traditional tools, such as financial and monetary policies. While these energy hypotheses remain unproven, it does seem that energy provides an international organizing principle that is substantially differently politicized and polarized than economics. Thus, some of the areas of development that have become too sensitive to engage nations in productive international economic dialogue and cooperation have the potential to be addressed via energy concepts and organizations.

While the eventual success of this strategy is not certain, it is important that every effort be made to explore this avenue as well as others. After all, the two most critical energy challenges in the world are direct reflections of the dark side of the world situation today as a whole. The first challenge is to find some means of supplying the sustainable high-grade energy needed for development of the rural areas of poor countries, areas where most of humanity lives and relies on traditional fuels such as firewood. The other major international challenge is to accommodate the great shifts in wealth and power accompanying the evolution of the international petroleum market and, to a lesser extent, the internationalization of the nuclear fuel cycle. The first challenge is a major part of humanity's goal to bring its poorest members to a reasonable sustainable material standard of living; the second is a major requirement if global nuclear conflict is to be avoided.

To these ends, international cooperative research and development efforts such as the fledgling Asian Energy for Rural Development Network based in Bangkok should be encouraged and strengthened. They provide a means to share information and experience across a wide range of research, development, and policy organizations concerned with the problems of the rural poor. Significantly, they also serve to call attention to these problems in national and international policy and funding decision-making bodies.

In addition to the potential for regional cooperation to achieve increased petroleum security and lower the political, economic, and safety risks of the nuclear fuel cycle, the seriousness of these energy problems can be an impetus to bring together senior national policy makers for informal nonbinding discussion of mutual problems and common solutions. An informal regional organization designed for this purpose, whose meetings are modeled somewhat after those held at the Royal Institute for International Affairs in London, was formed in Honolulu in 1978 and named the Asia-Pacific Energy Studies Consultative Group (APESC).

It is not clear what forms of national and international organizations in the Asia-Pacific region will eventually emerge to examine and handle the broad range of energy problems and trends outlined here. What is clear, however, is that national and international decision making about resources is now going through a transition as rapid and as radical as the energy supply transition itself, never again to return to their traditional forms. Indeed, in an age when the superpowers of the world have the capability of totally destroying each other, international cooperation leading to equitable and stable access to resources and to their collective global management may turn out to be the glue which binds some 150 diverse nations into a single, peaceful world.

NOTES

1. U.S. Central Intelligence Agency, Some Perspectives on Oil Availability for Non-OPEC LDCs (ER80-10493), September 1980.

2. "Energy Efficiency and Conservation in the Asia-Pacific Region," Asia-Pacific Energy Studies Consultative Group Workshop IV, Energy 7, special issue (1982). In press.

3. See for instance, David A. Deese and Joseph S. Nye, Energy and Security: Report of the Harvard Energy Security Project (Cambridge: Ballinger Press, 1981), and H. Rowen, "Report of the Stanford University's Energy Security Project for the U.S.," (Washington, D.C.: U.S. Department of Energy, 1981).

4. United Nations, Economic and Social Commission for Asia and the Pacific, Short-term Economic Policy Aspects of the Energy Situation in the ESCAP Region (Bangkok: United Nations, 1981).

About the Authors

Harrison Brown, the director of the Resource Systems Institute of the East-West Center, received his Ph.D. in chemistry at the Johns Hopkins University in 1941. After working on problems of plutonium chemistry during World War II, he became professor of chemistry in the Institute for Nuclear Studies at the University of Chicago. In 1951, Dr. Brown became professor of geochemistry at the California Institute of Technology and professor of science and government in 1967, and Foreign Secretary of the National Academy of Sciences from 1962 until 1974. Dr. Brown is the author of several books and articles; the most recent of his books, The Human Future Revisited (1978), was published by W. W. Norton.

Fereidun Fesharaki, a research associate at the Resource Systems Institute, holds a Ph.D. in economics from Surrey University in England, has served as the Energy Advisor to the Prime Minister of Iran, and was a delegate to the OPEC Ministerial Conferences. A consultant to national and international organizations, Dr. Fesharaki is a member of the Graduate Faculty of the Departments of Economics and Geography at the University of Hawaii. Dr. Fesharaki is the author of a number of books and articles, the most recent of which is Revolution and Energy Policy in Iran (1980), published by the Economist Intelligence Unit, Ltd.

Corazon M. Siddayao, a research associate at the Resource Systems Institute, holds a Ph.D. in economics from George Washington University, and has served in academia, government, international organizations, and industry in Asia and the United States. She is also a member of the Graduate Faculty of the Department of Economics at the University of Hawaii, where she teaches energy policy analysis. Dr. Siddayao is the author of several books and articles on energy and other topics, the most recent of which is The Supply of Petroleum Reserves in South-East Asia (1980), published by Oxford University Press.

Toufiq A. Siddiqi is a research associate and assistant director of the Environment and Policy Institute, and is also closely involved with the Energy Project at the Resource Systems

Institute. He holds a Ph.D. in nuclear physics from the University of Frankfurt-an-Main in West Germany, and was formerly an associate professor of public and environmental affairs at Indiana University. Dr. Siddiqi has published several articles on science policy, energy, and the environment; his latest publication, "Coal Use in Asia and the Pacific, Some Environmental Considerations," is forthcoming in Energy: The International Journal.

Kirk R. Smith is a research associate and energy project leader in the Resource Systems Institute. With degrees in astrophysics and in environmental health sciences and energy from the University of California at Berkeley, he has been a member of, or consultant to, a number of national and local energy studies, including two major national studies on nuclear waste. Dr. Smith is president of the Hawaii chapter of the Health Physics Society, an international professional organization concerned with radiation safety, and sits on a national commitee charged with developing risk standards for nuclear power convened by the U.S. Nuclear Regulatory Commission. In addition to publications on risk analysis, nuclear power, and energy problems of rural areas in Asia, he has two other books in press during the coming year: Interaction of Time and Technology: Coal and Nuclear Power (Garland Press); and Introduction to Ecoscience (W. H. Freeman).

Kim Woodard, an adjunct research associate at the Resource Systems Institute and the chief consultant for the China Energy Ventures, received his Ph.D. in political science from Stanford University. He has undertaken research on the politics of advanced technology, international energy problems, and nuclear policy. Dr. Woodard is the author of several articles on China's energy policy; his latest book, The International Energy Relations of China (1980), was published by Stanford University Press.

Index

9780367018979